CRITICAL CURRENT LIMITATIONS IN HIGH TEMPERATURE SUPERCONDUCTORS

Progress in High Temperature Superconductivity – Vol. 30

Proceedings of the International Workshop on

CRITICAL CURRENT LIMITATIONS IN HIGH TEMPERATURE SUPERCONDUCTORS

Zaborów near Warsaw, Poland 10 – 13 September 1991

Editors

M. Baran
W. Gorzkowski
H. Szymczak

Institute of Physics
Polish Academy of Sciences

World Scientific
Singapore • New Jersey • London • Hong Kong

Published by

World Scientific Publishing Co Pte Ltd
5 Toh Tuck Link, Singapore 596224
USA office: 27 Warren Street, Suite 401-402, Hackensack, NJ 07601
UK office: 57 Shelton Street, Covent Garden, London WC2H 9HE

British Library Cataloguing-in-Publication Data
A catalogue record for this book is available from the British Library.

CRITICAL CURRENT LIMITATIONS IN
HIGH TEMPERATURE SUPERCONDUCTORS

ISBN-13 978-981-02-0803-5
ISBN-10 981-02-0803-0

The International Workshop on Critical Current Limitations in HTSC
Zaborów near Warsaw, Poland, September 10-13, 1991
organized by the Institute of Physics of the Polish Academy of Sciences, Warsaw

ADVISORY COMMITTEE:

B.N. Goshchitskii	Sverdlovsk
K.V. Rao	Stockholm
J. Stankowski	Poznań
H. Szymczak (chairman)	Warsaw
H.W. Weber	Vienna

CHAIRMAN OF WORKSHOP:

J. Stankowski

ORGANIZING COMMITTEE:

R. Puźniak	chairman
M. Baran	secretary
R. Żuberek	treasurer
P. Gierłowski	
W. Gorzkowski	
W. Kula	
A. Wiśniewski	

FOREWORD

This volume consists of the papers presented at the *International Workshop on Critical Current Limitations in HTSC* held in Zaborów near Warsaw (Poland), September 10-13, 1991. The Workshop was organized by the Institute of Physics of the Polish Academy of Sciences, Warsaw. It was the second international meeting on HTSC organized by our Institute. The first one was *The European Conference on High T_c Thin Films and Single Crystals* organized two years ago in Ustroń, Poland.

The scope of the Workshop exactly corresponds to the title of the meeting, thus it is not necessary to explain it more precisely. Everybody who works in HTSC knows very well that critical currents is one of the key problems for both theoreticians and experimentalists. However, it is clear that the critical currents cannot be treated as an isolated topic. This explains why some papers have a somewhat wider scope.

The topic of our meeting was not as wide as the topic of the previous conference. Thus we decided to change the size of the meeting and its character. This change allowed us to make the contacts among the participants of the meeting more familiar.

The Workshop gathered 63 participants from all over the world. Thirteen countries have been represented, among them USA, Japan, China, South Korea and USSR (the last one by scientists from seven various centers). The lectures were presented by 13 specialists in the area of HTSC. The contributed works, in the form of short communciations, were also presented orally. In addition, many informal and very fruitful meetings took place.

This volume contains the papers of 10 invited lectures and 26 contributed papers. We are disconsolate because we have not received, up to the final deadline, the text of three interesting lectures:

1. *Structure and properties of irradiation defects in* $YBa_2Cu_3O_{7-x}$
   by M. A. Kirk (Argonne National Lab., Argonne)
2. *Critical current enhancement in oxide superconductors*
   by K.V. Rao (Royal Inst. of Technology, Stockholm)
3. *Thermodynamics of the Josephson-junctions system*
   by J. Stankowski (Inst. of Molecular Physics, Poznań)

We would like to express our deep thanks to all the authors who spent valuable time to prepare their manuscripts in camera-ready form.

It is not easy to predict the future of HTSC. Nevertheless, we plan to organize the next international meeting in 1993. It is difficult to define the scope of the next meeting now, but it seems that the problem of critical currents in HTSC will still be at the center of interest of many researchers working in this field. Probably the scope of the next meeting should also concern the same topic.

The Organizing Committee

CONTENTS

Contributed Papers

INVITED LECTURES

CRITICAL CURRENTS IN NEUTRON IRRADIATED HIGH TEMPERATURE SUPERCONDUCTORS

H.W. WEBER

Atominstitut der Österreichischen Universitäten, A-1020 Wien, Austria

ABSTRACT

Neutron irradiation experiments on single crystals and epitaxial thin films will be reviewed. Since all the measurements on single crystals are made by magnetometry, a thorough reexamination of the measuring process in a SQUID magnetometer is needed to establish the appropriate evaluation conditions for the critical current densities and the irreversibility lines. This will be used to discuss the results of a sequential neutron irradiation program on YBCO-123 and BiSCCO-2212 single crystals. The data show considerable enhancements of $J_c^{a,b;H//c}$ and $J_c^{c;H//a,b}$, shifts of the anisotropic irreversibility lines to higher fields and temperatures as well as a reduction of magnetization relaxation rates. All of these observations indicate the presence of "strong" pinning centers, similar to precipitate pinning in the case of melt textured bulk YBCO or pinning by amorphous tracks in heavy ion irradiated YBCO single crystals. Results on the angular dependence of $J_c^{H//a,b}$ in thin films emphasize the role of intrinsic pinning for H//a,b. The presently available information will be put together to formulate a simple model for flux pinning in irradiated high-T_c superconductors.

1. Introduction

The critical current densities within the grains of high-T_c ceramics or within high-quality (non-granular) single crystals are severely limited by inadequate defect structures for flux pinning and, as a consequence, by thermally assisted flux flow[1], magnetization relaxation[2] and complete depinning[3] of the flux lines ("irreversibility line"). Furthermore, the highly anisotropic nature of the intrinsic superconductive parameters, such as the upper critical field H_{c2}[4], as well as obviously anisotropic flux pinning conditions in the "as-grown" state represent serious obstacles for an application of these materials, especially at the desired operating temperature of 77 K. On the contrary, rather high values of J_c have been reported for epitaxial thin films of $YBa_2Cu_3O_7$ (~ 5 x 10^{10} Am^{-2}, 77 K, zero field), if the field is applied parallel to the basal plane. Such differences in J_c, which may reach one or two orders of magnitude, can be ascribed only to the existence of totally different defect structures for flux pinning, a conjecture which has been substantiated recently by the observation of large numbers of screw dislocations[6] in some epitaxial films.

Modification of the original defect structure in bulk high-T_c materials is therefore of paramount importance. The most remarkable progress, also in view of practical applications, has been achieved by a special melt texturing process for YBCO (melt-powder-melt-growth[6]), in which a substantial amount of normal conducting precipitates of the $Y_2Ba_2CuO_5$-phase is introduced into the 123-matrix. The resulting bulk samples are not only significantly textured, but also show high J_c-values, reduced relaxation rates and considerably enhanced irreversibility lines. The only other technique, which has led to similar beneficial effects, consists of an in-situ modification of the defect structure by particle irradiation[7]. It has turned out that the results of such an irradiation treatment depend sensitively on the type and energy of the irradiating particles, i.e. the distribution of recoil energies produced by a single event, and that all of the above mentioned beneficial effects, which are taken as a signature of strong flux pinning centers, occur only after fast neutron[7] and GeV heavy ion irradiation[8,9].

At present, much more (although certainly not yet enough) information is available on fast neutron irradiation effects in high-T_c superconductors as compared to flux pinning effects by the amorphous tracks produced during GeV ion irradiation. Hence, we will restrict ourselves in the following to a brief description of neutron-induced property changes in high-T_c single crystals and thin films and discuss the results on the basis of recent information obtained by electron microscopy on the nature, size and volume concentration of the radiation-induced defects[10,11]. It will become obvious that the improvements of J_c and of the related flux pinning characteristics are limited due to unfavorable defect configurations and a degradation of the intrinsic superconductive parameters (T_c, H_{c2}) at high fluences.

2. Experimental Considerations

Because of the small size of high-T_c single crystals, essentially all of the information on their irreversible (i.e. $J_c \neq 0$) properties is based on the analysis of magnetic moments measured as a function of applied field in SQUID or vibrating sample magnetometers. The usual evaluation procedure consists of several steps. Following the conversion of magnetic moments into magnetizations, the magnetization data for increasing and decreasing fields at *fixed external field* are used as input parameters in the Bean model[12], in order to calculate J_c. Modifications of the original model have been developed, which account for the finite geometry of the sample[13] as well as for the possibility of anisotropic critical current distributions[14,15]. In view of strong demagnetization effects in very thin single crystals and because of the failure of conventional demagnetization corrections[14] in the case of very large demagnetization factors ($D > 0.7$), many published J_c-values are affected by considerable evaluation errors. As pointed out, e.g., in Ref. 16, more detailed considerations on the field and current distributions within these extreme sample geometries are mandatory. The second major problem with this kind of experiments, which cannot be removed because of its physical nature, is related to distinct magnetization relaxation effects, i.e. a continuously "changing" critical state. Hence, the results of a magnetization experiment are clearly dependent on the timing of the instrument, which may be particularly slow in SQUID magnetometers equipped with superconducting high field magnets.

Most of the above mentioned evaluation problems can be solved by reconsidering the actual measuring process in the SQUID magnetometer[17]. The measured magnetic moment can be considered as being the dipole moment of a network of current loops within the superconductor. Consequently, it can be used to calculate the magnitude of the currents flowing in the plane perpendicular to the field direction in a Bean-model-type way (J_c = const.). This is currently being done for situations, where the fully penetrated state has been reached, but can be done in principle for the entire magnetization cycle. Furthermore, based on this value of J_c, the average *induction B within the sample*, i.e. the quantity needed for all evaluations in terms of flux lattice spacings and bulk pinning forces, can be calculated. Contributions of the reversible magnetization can be detected easily by comparing the J_c's calculated for the two branches of the magnetization loop. Demagnetization corrections are incorporated by replacing the external field H in the usual relation $B = \mu_0 (H + m/V)$ by the effective field $H_{eff} = H + DM$, i.e. $B = \mu_0 [H + (1 - D) m/V]$, and using either the measured demagnetization factor (from the Meissner slope) or the calculated one for the appropriate sample geometry[18]. These corrections for the induction B are generally small, except for low fields where the moments are large. This leads to an unambiguous assessment of the J_c-B dependence, when the sample is magnetized parallel to the c-direction ($J_c^{a,b;H//c}$) under the assumption of a homogeneous current distribution in the basal plane. The most significant alteration of this approach compared to the original Bean model for infinite geometries lies in the fact, that field (i.e. induction) independent J_c's do *not* lead to a *constant slope* of the flux density gradient in these finite sample geometries.

For H//a,b, a similar approach incorporating the anisotropic Bean model with two differently steep flux density gradients can be taken. Since the critical current densities involved in this case are $J_c^{c;H//a,b}$ and $J_c^{a,b;H//a,b}$, which are both unknown a priori, information on these quantities obtainable from the angular dependence of J_c in thin films[19] and some computer modelling have been used. In the thin film work, the critical current densities flowing along the a- or b-directions can be assessed both for H//a,b and for H//c (cf. section 5). The results show that the ratio $J_c^{a,b;H//a,b}/J_c^{a,b;H//c}$ is always larger than 2, sometimes even very much larger. Hence, we assume that *one* of the two flux density gradients in the magnetization experiment with H//a,b is very steep corresponding, e.g., to $J_c^{a,b;H//c}$, 2 or 5 x $J_c^{a,b;H//c}$, or to infinity. It turns out that the results determined under these assumptions are practically identical, i.e. that the measured magnetization hysteresis in entirely determined by the *less steep* flux density gradient, i.e. by $J_c^{c;H//a,b}$. Hence, $J_c^{c;H//a,b}$ can be assessed without further assumptions. The only critical current density which is not accessible to a magnetization experiment on presently available YBCO single crystals is therefore $J_c^{a,b;H//a,b}$, i.e. the critical current density along the a- or b-directions for H//a,b.

With regard to the measurement of the irreversibility line in a SQUID magnetometer, the following procedures can be employed. Firstly, the irreversibility field H_{irr} can be determined from the disappearance of the magnetization hysteresis at constant temperature. Secondly, the irreversibility temperature T_{irr} can be determined from the appearance of hysteresis in fc/zfc measurements at fixed applied field. Thirdly and preferably, in both of these operating conditions of the magnetometer the original response curves of the SQUID can be analyzed. Since the crystal is moved slowly between two pick-up coils in a slightly inhomogeneous magnetic field, which is equivalent to the superposition of a small "ac ripple field" onto the dc field, the response curve of the SQUID is slightly distorted (Fig. 1) by the action of the "ac field" on pinned flux lines, if hysteretic magnetization conditions prevail. The absence of these distortions represents clear evidence for reversible magnetization conditions on a relatively fast time scale ($\delta t \sim 35$ s) and with good experimental resolution ($J_c < 10^4 - 10^6$ Am^{-2}).

The critical current densities in thin films were determined by direct transport measurements on patterned epitaxial $YBa_2Cu_3O_{7-\delta}$ using a voltage criterion of 10 μVcm^{-1}. The angular dependence of J_c was assessed by continuously rotating the film in an 8 T magnet with the current being always perpendicular to the field.

As a final remark the irradiation conditions should be specified. All the irradiations were made in the central irradiation facility of the TRIGA MARK II reactor in Vienna at ambient temperature (sample temperature during irradiation: $\lesssim 50^\circ$ C). In view of the distinct dependence of the damage produced in a material, on the energy distribution of the neutrons, detailed information on this quantity is required for any damage calculations[20] as well as for comparisons with other types of radiation. As an example, the "displacements per atom", which are suitable as an overall characterization

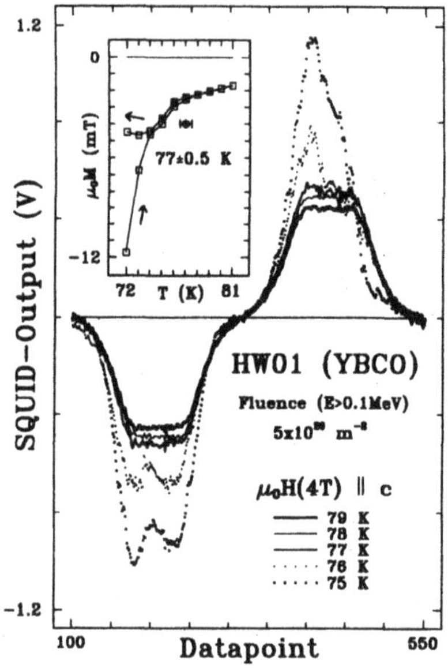

Fig. 1: SQUID output voltage as a function of time. The data refer to full scans at fixed magnetic fields, but different temperatures near the irreversibility temperature.

of damage, have been calculated for the TRIGA reactor[21] and are found to be 1.42, 1.45 and 1.44 x 10^{-3} for $YBa_2Cu_3O_{7-\delta}$, $Bi_2Sr_2CaCu_2O_8$ and $Bi_2Sr_2Ca_2Cu_3O_{10}$, respectively, at a fast neutron fluence of 1 x 10^{22} m^{-2} (E > 0.1 MeV), i.e. practically identical.

3. Critical Current Densities

Neutron irradiation experiments on conventional superconductors[22], even on compounds which are sensitive to disorder, have shown that the transition temperature T_c is only slightly reduced at low fluence levels ($\lesssim 10^{22}m^{-2}$, E > 0.1 MeV), whereas significant T_c-depressions occur later on. This behavior is basically confirmed for the

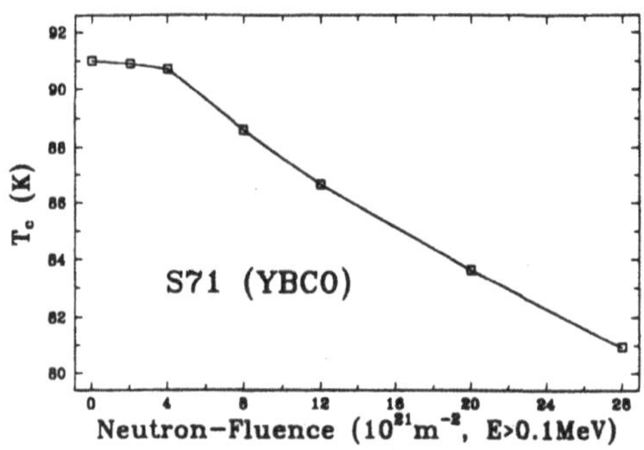

Fig. 2: Transition temperature T_c of single crystal S 71 as a function of fast neutron fluence.

high-T_c compounds investigated so far in a systematic way. As an example, the T_c-dependence on fluence is shown for an YBCO single crystal in Fig. 2. Hence, in the low fluence regime ($\leq 4 \times 10^{21} m^{-2}$) radiation-induced changes in the magnetization curves can be related primarily to *changes in the defect structure* responsible for flux pinning rather than to changes in the intrinsic superconductive parameters.

Most of our recent systematic work on fast neutron damage in YBCO has been done on a different single crystal (HW01, a ~ b ~ 1.2 mm, c = 47 μm) with a very sharp T_c at 91.1 K. At high temperatures (e.g. 77 K, Fig. 3), clear indications of an anomalous pinning feature, which is often referred to as "fishtail" magnetization[23], can be seen.

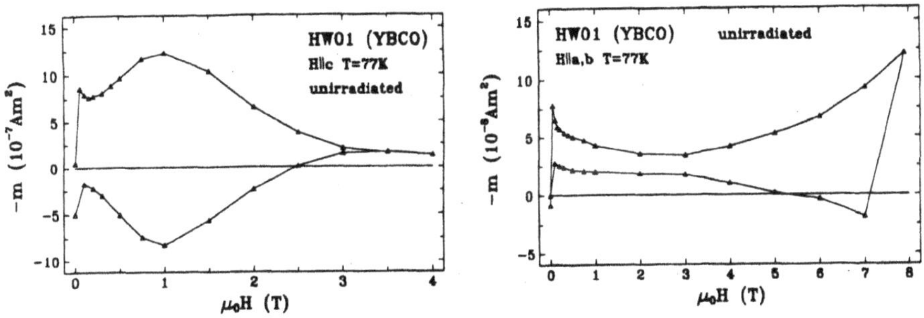

Fig. 3: Magnetic moments versus external field for single crystal HW01 in the unirradiated state (T = 77 K), left: H//c, right: H//a,b

The enhancement of hysteresis at high fields is clearly anisotropic and much less pronounced for H//c. No detailed explanation of this effect is at hand presently, but the appearance of some field-induced normal conducting pinning sites, in particular for currents flowing along the c-direction, cannot be excluded.

The results of magnetization measurements carried out prior to irradiation and at four different fluence levels up to $4 \times 10^{21} m^{-2}$ (E > 0.1 MeV) are summarized in Fig. 4, where the magnetic moments at 5 and 77 K are plotted versus field for both orientations of the crystal (H//c, H//a,b). The observed radiation effects are strikingly different at these two temperatures as well as for the two field orientations. At 5 K and for H//c, the width of the hysteretic magnetization loops increases continuously with neutron fluence. For H//a,b, we observe only a small increase at the lowest fluence and larger enhancements later on. At 77 K, J_c increases by a factor of about 100 at low fields (H//c), but disappears for fields $\geq$ 3 T due to thermal activation of flux lines. On the contrary, large enhancements are observed for H//a,b at high fields. Since the defects produced by neutron irradiation and their distribution are completely isotropic, these observations can only be explained by proposing (a) the existence of two distinctly different pinning mechanisms for H//c and H//a,b, respectively, (b) the occurrence of radiation-induced modifications of the pre-irradiation defect structure, and (c) the appearance of strong additional pinning centers.

The results for the critical current densities obtained according to the procedures outlined in section 2, are shown in Fig. 5. It should be noted that the magnitude of $J_c^{a,b}$, i.e. the critical current density in the basal plane, is very large at 5 K in the irradiated state ($4 \times 10^{21} m^{-2}$) and approaching results reported for thin films. On the other hand, J_c^c is about two orders of magnitude smaller, but enhanced more significantly by the irradiation. At 77 K, the most significant enhancements of $J_c^{a,b}$ occur at inductions below 1 T, followed by a drop towards the irreversibility line. However,

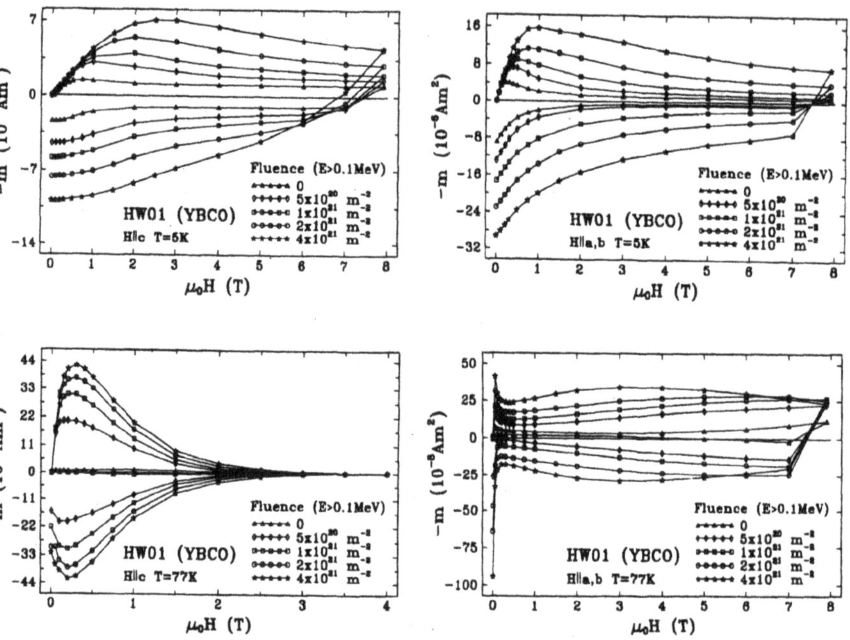

Fig. 4: Magnetic moments versus external field for single crystal HW01 at various neutron fluences (T = 5 K, 77 K). Left: H//c, right: H//a,b

because of the fluence dependence of this line (cf. section 4), significant J_c-values are observed at higher fields in the irradiated state, e.g. ~ 10^8 Am^{-2} at 3 T. The "fishtail" character of the data seems to be completely eliminated as a consequence of the irradiation. The field dependence of J_c is rather complicated at 77 K and seems to be influenced to a larger extent by the pre-irradiation pinning features. It should be noted, however, that the field dependence of J_c is much less pronounced in the irradiated state (except for the anomalous enhancement near B = 0, which is not understood).

High fluence irradiations are not yet available for this single crystal. However, results obtained on other single crystals[24,25] (cf. also Fig. 1) demonstrate that the critical current densities even at 5 K start to decrease at fluence levels of ≥ 1.2 x 10^{22} m^{-2} for H//c and of ≥ 2 x 10^{22} m^{-2} for H//a,b. More drastic effects occur at these fluence levels at high temperatures because of the significant depression of T_c. Hence, both this change of intrinsic superconductive parameters *and* the formation of an unfavorable defect structure observed at this fluence level by electron microscopy[11] are responsible for limitations of further J_c-enhancements by fast neutron irradiation in YBCO.

With regard to BiSCCO-2212 single crystals, magnetization measurements for both field orientations have been made only at low fluence levels (0.5 and 2.5 x 10^{21} m^{-2}) so far[26]. Evaluations of J_c and the bulk pinning forces P_v show that the lowest fluence irradiation (5 x 10^{20} m^{-2}) has led to no remarkable changes (except for some effects at very low fields and on the irreversibility line, cf. below), thus demonstrating that a certain minimal amount of defects per unit volume is needed to effectively interact with the flux line lattice. After the second irradiation step, significant enhancements of P_v

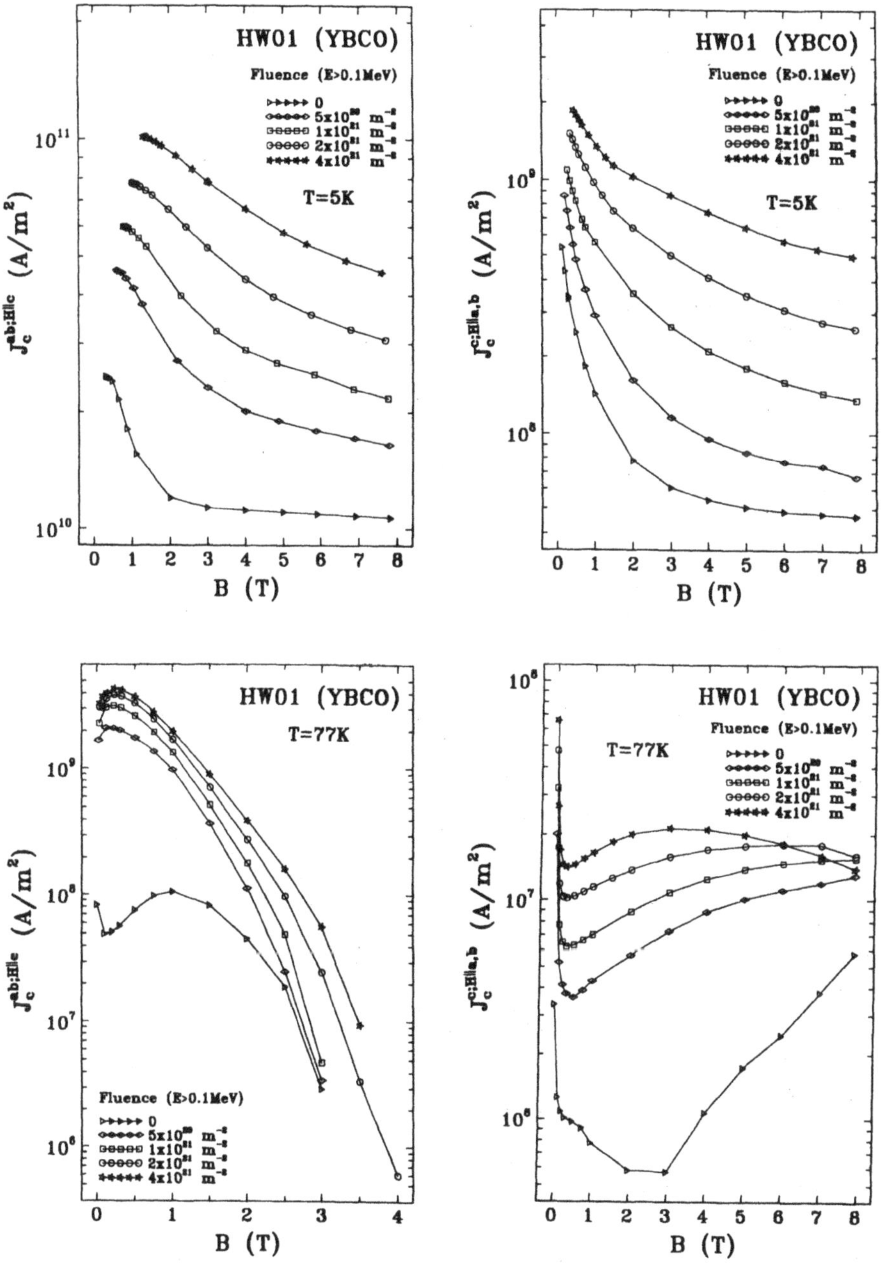

Fig. 5: Critical current densities at 5 K and 77 K calculated from the data of Fig. 4.
Left: H//c ($J_c^{a,b}$), right: H//a,b (J_c^{c}).

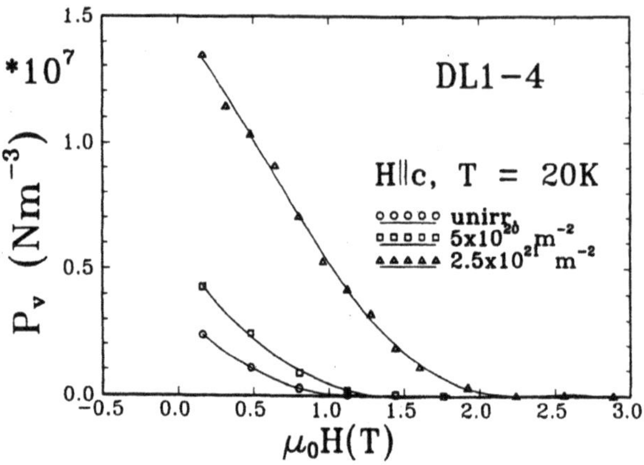

Fig. 6: Volume pinning force for a BiSCCO-2212 single crystal at T = 20 K at various neutron fluences.

have been observed in all cases (Fig. 6), which extend far into the field range of complete magnetization reversibility prior to irradiation. However, the overall magnitude of J_c is still very low (e.g. ~5 x 10^6 Am^{-2} at 1 T and 20 K, H//c).

Irradiation experiments made on two different single crystals by Gerhäuser et al.[27] emphasize the important role of the pre-irradiation defect structures. Furthermore, they demonstrate that the maximum J_c-enhancements achievable in this superconductor (H//c) is reached *below* fluence levels of 1.1 and 1.6 x 10^{22} m^{-2} (E > 0.1 MeV) for these two single crystals, respectively, in rather close agreement with the results reported above for YBCO.

4. Irreversibility Lines and Magnetization Relaxation

The disappearance of flux pinning at certain fields and temperatures by thermal activation of the flux lines, as well as the strong relaxation effects at lower fields and temperatures which are obviously caused by insufficiently deep pinning potentials, are of course subjects of major concern in view of potential applications of high temperature superconductors. Whereas depinning occurs at Tesla-fields and rather close to T_c in YBCO, the depinning field $\mu_0 H_{irr}$ already falls below ~ 1 T at 25 K in BiSCCO single crystals. In both materials the irreversibility lines are clearly anisotropic[25,26,28] and always occur at higher fields and temperatures for H//a,b.

As has been shown previously[29,30], the position of the irreversibility line can be changed significantly by neutron irradiation. However, many interesting features of these shifts, which may shed some light on the physics involved, can only be detected by studying the fluence dependence in a systematic way. Results of this kind are summarized for YBCO in Fig. 7. For H//c, the irreversibility line shifts at first to *lower* fields and temperatures, then back again and significantly beyond the initial position at a fluence of 4 x 10^{21} m^{-2}. On the other hand, a large shift to *higher* fields and temperatures is observed for H//a,b at the lowest fluence and barely changes up to a fluence of 2 x 10^{21} m^{-2}. At the next fluence level a reversal of this trend sets in. This completely different response for H//c and for H//a,b, respectively, points again to the existence of different flux pinning mechanisms as mentioned in section 3. In fact, details of this fluence dependence can be understood qualitatively in terms of a recent

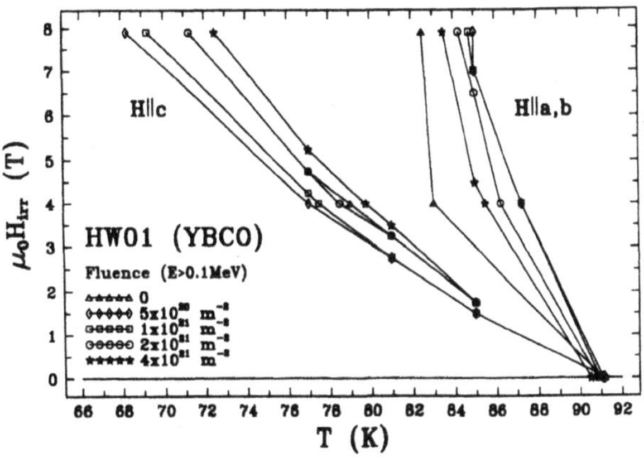

Fig. 7: Irreversibility lines for an YBCO single crystal (HW01) at various neutron fluences and for H//c and H//a,b, respectively.

model[25,31] involving at least three different flux pinning mechanisms in irradiated YBCO single crystals (cf. also below). According to these considerations the initial reduction of the irreversibility line for H//c is attributed to a large decrease in the number of pre-irradiation defects, while the concentration of strong radiation-induced pins is still insufficient. This trend is reversed at higher fluence levels because of the increase in cascade concentration. For H//a,b on the other hand, already a few large defects can change the depinning behavior significantly because of the quasi one-dimensional current flow in the presence of intrinsic pinning. With regard to the temperature dependence of H_{irr} we find reasonable agreement with predictions based on a simple activation model[3], i.e. $H_{irr} \propto (1 - t)^{3/2}$ with $t = T/T_c$, but only for H//c, thus confirming again the differences in the pinning/depinning processes for these two field orientations.

Less complete information is available for BiSCCO at present. The results summarized in Fig. 8 demonstrate firstly, that the temperature dependence of the irreversibility fields differs completely from the $(1 - t)^{3/2}$-law in both cases. Secondly, the data pertaining to the lowest fluence irradiation, which had only led to small effects on J_c, clearly show quite remarkable shifts of the irreversibility lines although only at fields up to 1 T. This is compatible with the conjecture that the density of strong pins is too low to pin the flux line lattice effectively in the entire range of magnetic fields. However, at very low fields where the flux line spacing is very large, a change in the depinning behavior becomes observable and shifts of the depinning temperature occur, e.g. at a field of 100 mT by as much as 14 K. Finally, if the superconductor is subjected to a five times larger neutron fluence, large shifts of the irreversibility lines are observed at all fields, in particular for H//a,b. As shown in Fig. 8, the results are still not satisfactory from an applied point of view. The data on higher fluence neutron irradiation[27] mentioned in section 3 show that the irreversibility line for H//c still shifts to higher temperatures, but the effect becomes very small for $T/T_c > 0.5$. Further results of our sequential irradiation program will have to be awaited before final conclusions can be drawn.

As a final remark, first results on changes in the magnetization relaxation rates observed in the irradiated state by torque magnetometry[32] should be mentioned. The presently available data refer to two experiments made on YBCO single crystals which were

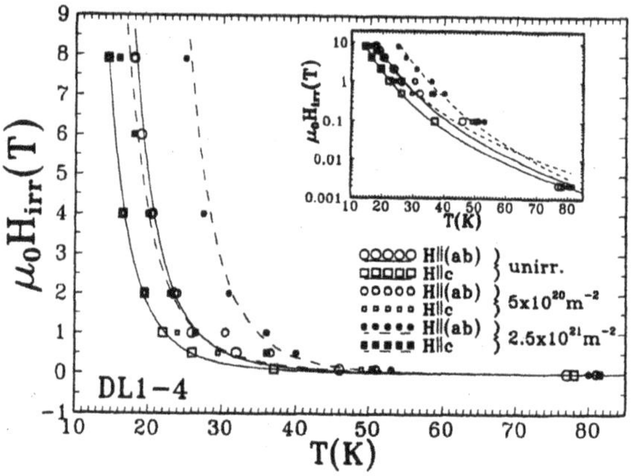

Fig. 8: Irreversibility lines for a BiSCCO-2212 single crystal at various neutron fluences and for H//c and H//a,b, respectively. The insert show the same data on a semi-logarithmic scale.

irradiated to fluences of 0.5 and 2 x $10^{21}m^{-2}$, respectively. The measured normalized relaxation rates following magnetization to a field of 1.5 T are in both cases drastically reduced and nearly temperature independent up to ~ 60 K. At higher temperatures, an increase of the relaxation rate is observed in complete opposition to the behavior in the unirradiated state. This difference in radiation response (low versus high temperature) has already been emphasized in section 3. With regard to the relaxation behavior, it must be borne in mind that the relaxation is of course sensitive to *all* the defects in the crystal and that the removal of a significant amount of weak pinning centers as a consequence of the irradiation may explain the observed effects in a regime where the flux line lattice is soft and where flux lines pinned by defects with deep pinning potentials do not contribute to the measured relaxation at all.

5. Thin Films

Epitaxial thin films show in general critical current densities J_c which exceed those of single crystals by one or two orders of magnitude. Hence, the defect structure of these two forms of the same material ($YBa_2Cu_3O_{7-\delta}$) must be completely different. Furthermore, the anisotropy of J_c in the basal plane, i.e. $J_c^{a,b;H//a,b}$ and $J_c^{a,b;H//c}$, can be correlated to the film quality[19] and may become very significant. As an example, results on a high-quality (magnetron sputtered) film are shown in Fig. 9, where J_c is plotted as a function of rotation angle ϑ ($\vartheta = 0^\circ$: H//a,b; $\vartheta = 90^\circ$: H//c). It should be noted, that the angular dependence of J_c becomes very sharp, especially at high temperatures and fields. This can be interpreted as a signature of intrinsic flux line pinning[33], where the flux lines are confined in between the Cu-O layers for H//a,b due to a significant variation of the superconducting order parameter.

Neutron irradiation experiments on a high-quality (laser ablated) YBCO film[34] have shown only moderate enhancements of J_c at low fluences (4.9 x $10^{21}m^{-2}$) and a decrease of J_c in most cases (except for H//c at low temperatures) at a fluence of 2 x $10^{22}m^{-2}$. In view of the practically optimized defect configuration for flux pinning prior to irradiation, only small improvements, if at all, can be expected. In fact, all the irradiation experiments done by us on magnetron sputtered YBCO films so far have led to negative results, i.e. either unchanged or slightly deteriorated J_c-values in the low fluence regime. All of these films had rather low T_c's ($\leq$ 81 K) but very high critical current densities. An interesting example of this data is shown in Fig. 10 for a film irradiated sequentially to 2 and 7 x 10^{21} m^{-2}. Whereas J_c remains practically unaffected for H//c, a drop of J_c by a factor of ~ 3 is observed for H//a,b, which leads to a nearly complete removal of J_c anisotropy. Because of the high cascade density at this fluence level (cf. below), severe damage of the pre-irradiation crystal structure has occurred, which seems to affect the intrinsic pinning mechanism in a particularly significant way.

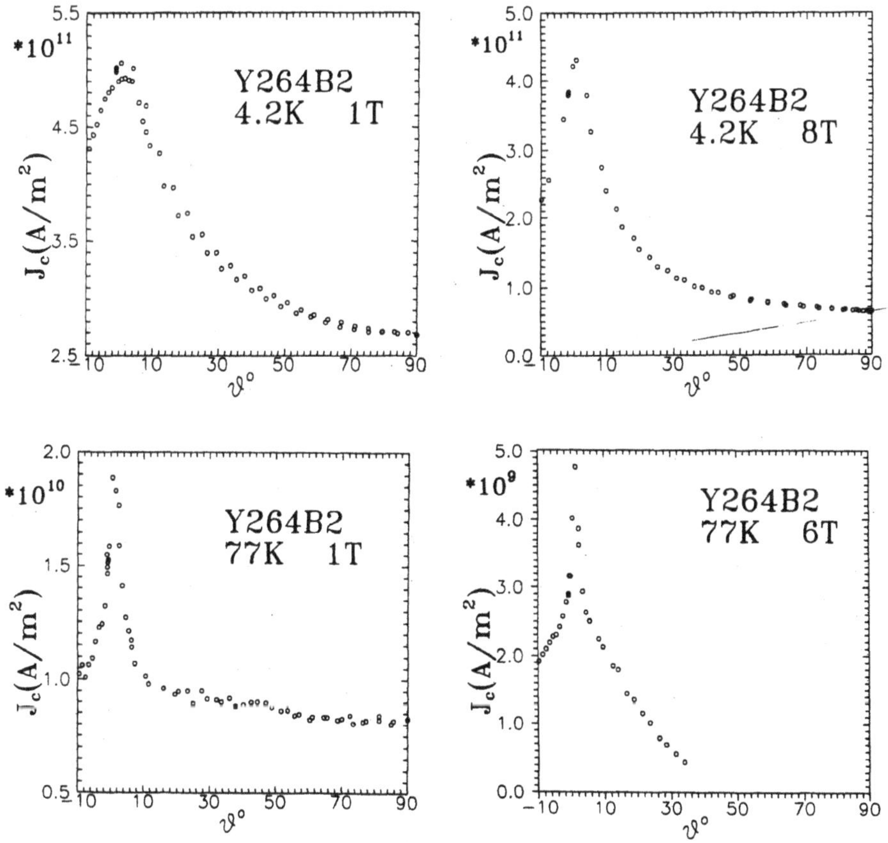

Fig. 9: Angular dependence of J_c in an YBCO thin film at 4.2 K (μ_0H = 1 and 8 T) and at 77 K (μ_0H = 1 and 6 T).

6. Discussion and Conclusions

As outlined in the previous sections, the main features of radiation-induced property changes in single crystals consist of considerable J_c-enhancements, shifts of the irreversibility lines to higher fields and temperatures and reductions of relaxation rates. All of them indicate the presence of *strong* flux pinning centers. The nature of these defects is currently being investigated by electron microscopy[10,11]. Individual

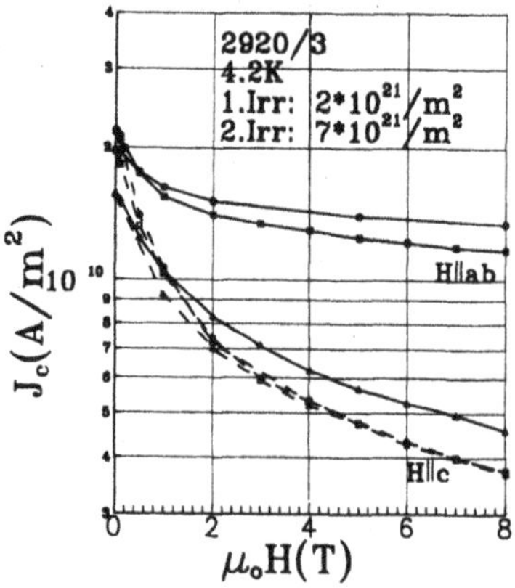

Fig. 10: Field dependence of J_c for H//a,b and H//c in the unirradiated state (OOO) and following neutron irradiation to 2 x 10^{21} m^{-2} (□ □ □) and 7 x 10^{21} m^{-2} (ΔΔΔ).

defect cascades with diameters of 5 - 10 nm have been identified in YBCO single crystals. At a fluence of 2 x 10^{21} m^{-2} the density of cascades was found to be ~ 10^{22} m^{-3} (which corresponds to an average defect spacing of 46 nm), in another crystal irradiated to 8 x 10^{21} m^{-2} it was ~ 4 x 10^{22} m^{-3}, nicely consistent with the previous result. At higher fluences (2 x 10^{22} m^{-2}), significant cascade overlapping and large disordered areas have been observed. Hence, the size and the density of these defects are very well suited for strong flux pinning at low fluences ($\leq 10^{22}$ m^{-2}) but no longer at higher fluences. Damage calculations have shown that the fraction of high-energy recoil atoms leading to cascade formation is very large in the case of fast neutron irradiation. Low energy recoils are of course produced by a certain neutron spectrum as well, which will lead to smaller defects, point defects and some local release of oxygen. Hence, *weak* pinning centers (e.g. oxygen deficient lattice cells), which are characteristic of the pre-irradiation defect structure, may be *removed* through oxygen recovery or direct replacement by a large cascade.

This picture, i.e. the interplay between a gradual build-up of strong pinning centers and a simultaneous (but more rapid) removal or weakening of the pre-irradiation defect structure, is qualitatively adequate to explain the J_c-enhancements, the observed reduction and subsequent increase of the irreversibility line as well as the temperature dependence of the relaxation rates for H//c. For H//a,b, however, flux pinning by the intrinsic pinning mechanism must be taken into account in addition to the above mentioned mechanisms, leading to quasi one-dimensional flux line motion *along* the Cu-O layers. A significant weakening of this mechanism is expected only at high fluence levels where the long range order of Cu-O planes is severely disturbed, such as in the single crystal irradiated to 2 x 10^{22} m^{-2} and investigated in the electron microscope (Fig. 1 of Ref. 11). In fact, the results presented for YBCO in figures 4 and 7 (H//a,b) confirm that the first signs of a trend reversal just appear at the highest fluence level investigated so far, i.e. 4 x 10^{21} m^{-2}. The role of intrinsic pinning and its weakening with increasing disorder is further supported by the thin film results.

In conclusion, it should be pointed out that certain limitations for the enhancement of critical current densities by fast neutron-induced defects are presently emerging from systematic sequential irradiation programs. The limitations are caused primarily by considerable changes of the intrinsic superconductive parameters by radiation-induced disorder at higher fluence levels as well as by the development of a defect structure (cascade overlap), which is unsuitable for flux pinning. However, the evidence for strong flux pinning even at high temperatures provided by these experiments is considered to represent an important step towards the development of suitable defect structures for flux pinning. Further work on the subject as well as on flux pinning by precipitates and amorphous tracks will certainly improve our understanding of flux pinning mechanisms in high-T_c superconductors in the near future.

Acknowledgements

I wish to express my gratitude for the dedicated efforts of my group, in particular Dr. F.M. Sauerzopf, M.C. Frischherz, W. Kritscha, R.M. Schalk, M. Wacenovsky and H.P. Wiesinger. The fruitful cooperation with Drs. G.W. Crabtree and M.A. Kirk (Argonne) as well as with Dr. J.E. Evetts (Cambridge) and Prof. R. Griessen (Amsterdam) is acknowledged. This work has been supported in part by Fonds zur Förderung der Wissenschaftlichen Forschung, Wien, under project no. 7970 and by the EC SCIENCE program under contract no. 89200167.

References

1. P.H. Kes, J. Aarts, J. van den Berg, C.J. van der Beek and J.A. Mydosh, *Supercond. Sci. Technol.* **1** (1989) 242
2. C. W. Hagen and R.P. Griessen, in *Studies of High Temperature Superconductors*, ed. A.V. Narlikar (Nova Science Publishers, New York, 1989), Vol. 3, p. 159
3. A.P. Malozemoff, T.K. Worthington, Y. Yeshurun, F. Holtzberg and P.H. Kes, *Phys. Rev. B* **38** (1988) 7203
4. U. Welp, W.K. Kwok, G.W. Crabtree, K.G. Vandervoort and J.Z. Liu, *Phys. Rev. Lett.* 62 (1989) 1908
5. J. Mannhart, D. Anselmetti, J.G. Bednorz, C. Gerber and D.G. Schlom, *Supercond. Sci. Technol.*, in press
6. M. Murakami, M. Morita, D. Doi and K. Miyamoto, *Jap. J. Appl. Phys.* **28** (1989) 1189
7. H.W. Weber and G.W. Crabtree, in *Studies of High Temperature Superconductors*, ed. A.V. Narlikar (Nova Science Publishers, New York, 1991), Vol. 9, in press
8. L. Civale, A.D. Marwick, T.K. Worthington, M.A. Kirk, J.R. Thompson, L. Krusin-Elbaum, Y. Sun, J.R. Clem and F. Holtzberg, *Phys. Rev. Lett.* 67 (1991) 648
9. W. Gerhäuser, H.W. Neumüller, W. Schmidt, G. Ries, O. Eibl, G. Saemann-Ischenko and S. Klaumünzer, *Physica C*, in press
10. M.C. Frischherz, *Ph.D. Thesis* (TU Wien), 1991
11. M.A. Kirk, M.C. Frischherz, J.Z. Liu, L.R. Greenwood and H.W. Weber, *Phil. Mag. Lett.* **62** (1990) 41
12. C.P. Bean, *Rev. Mod. Phys.* **36** (1964) 31
13. A.M. Campbell and J.E. Evetts, *Critical Currents in Superconductors* (Taylor & Francis, London, 1972)
14. F.M. Sauerzopf, H.P. Wiesinger and H.W. Weber, *Cryogenics* **30** (1990) 650
15. E.M. Gyorgy, R.B. van Dover, K.A. Jackson, L.F. Schneemeyer and J.F. Waszczak, *Appl. Phys. Lett.* **55** (1989) 283
16. M. Daeumling and D.C. Larbalestier, *Phys. Rev. B* **40** (1989) 9350
17. H.P. Wiesinger, *Ph.D. Thesis* (TU Wien), 1991
18. J.A. Osborn, *Phys. Rev.* **67** (1945) 351
19. R.M. Schalk, H.W. Weber, Z.H. Barber, C.E. Davies, J.E. Evetts, R.E. Somekh and D.H. Kim, *Super. Sci. Technol.*, in press
20. H.W. Weber, E. Unfried, H. Böck and L.R. Greenwood, *J. Nucl. Mat.* **137** (1986) 236

21. L.R. Greenwood, *private communication*
22. H.W. Weber and B.S. Brown, in *Concise Encyclopedia of Magnetic and Superconducting Materials*, ed. J.E. Evetts (Pergamon Press, Oxford, 1991), in press
23. M. Daeumling, J.M. Seuntjens and D.C. Larbalestier, *Nature* **346** (1990) 332
24. F.M. Sauerzopf, H.P. Wiesinger, W. Kritscha, H.W. Weber, G.W. Crabtree and J.Z. Liu, *Phys. Rev. B* **43** (1991) 3091
25. F.M. Sauerzopf, H.P. Wiesinger, H.W. Weber and G.W. Crabtree, *Adv. Cryog. Eng.* **38**, in press
26. W. Kritscha, F.M. Sauerzopf, H.W. Weber, G.W. Crabtree, Y.C. Chang and P.Z. Jiang, *Supercond. Sci. Technol.*, in press
27. W. Gerhäuser, H.W. Neumüller, W. Schmidt, G. Ries, G. Saemann-Ischenko, H. Gerstenberg and F.M. Sauerzopf, *Physica C*, in press
28. W. Kritscha, F.M. Sauerzopf, H.W. Weber, G.W. Crabtree, Y.C. Chang and P.Z. Jiang, *Europhys. Lett.* **12** (1990) 179
30. H. Küpfer, C. Keller, R. Meier-Hirmer, K. Salama, V. Selvamanickam and G. Tartaglia, *IEEE Trans. Magn.* **MAG-27** (1991) 1369
31. F.M. Sauerzopf, H.P. Wiesinger, H.W. Weber, G.W. Crabtree, M.C. Frischherz and M.A. Kirk, *Supercond. Sci. Technol.*, in press
32. J.G. Lensink, R. Griessen, H.P. Wiesinger, F.M. Sauerzopf, H.W. Weber and G.W. Crabtree, *Physica C*, in press
33. M. Tachiki and S. Takahashi, *Solid State Comm.* **70** (1989) 291
34. W. Schindler, B. Roas, G. Saemann-Ischenko, L. Schultz and H. Gerstenberg, *Physica C* **169** (1990) 117

RADIATION INDUCED DISORDER AS A UNIQUE METHOD FOR STUDYING ELECTRONIC STATES OF HTSC's AND MODIFYING THEIR PROPERTIES

B.N.GOSHCHITSKII, V.I.VORONIN, S.A.DAVYDOV,
A.E.KARKIN AND A.V. MIRMELSTEIN
Institute for Metal Physics,
620219 Sverdlovsk GSP-170, USSR

ABSTRACT

Investigation results of the influence of fast neutron irradiation at liquid nitrogen temperature on physical properties of HTSC's are discussed. It is shown that investigation of radiation disordering effects favours better understanding of the nature of these compounds and yields the possibilities for modifying their basic properties.

1. INTRODUCTION

The paper by Bednorz and Muller initiated investigations of irradiation-induced effects in HTSC's. Now it is clear that extremely high sensitivity of HTSC to structural and chemical disorder is a distinctive feature of these materials. And understanding of the physical nature of this sensitivity may essentially help to solve the problem of the physical mechanism of HTSC. Apparently, radiation-induced disordering is the most "pure" from the physical viewpoint. Chemical composition of the samples in this case remains unchanged and consequently, the mean number of electrons per atom also does not change. Therefore, investigation of the radiation-induced effects is a fruitful method to study disordered states, which are naturally connected with the fundamental properties of ordered crystals. This conclusion is supported by a great number of radiation effect investigations in A-15 structures, Laves and Chevrel phases (see, e.g. Ref.1). It was shown that in these materials disorder leads mainly to smearing of the fine structure of the electron and phonon spectra. In terms of the available theoretical representations this allows to explain the experimentally observed variations of the superconducting properties and to clear up the role of the electron and phonon subsystems in the formation of a superconducting state. Applied to HTSC's this method gives a unique possibility for investigating their electronic states. Unlike intermetallic superconductors there has not been a theory now which may satisfactory describe the physical properties of HTSC's both in superconducting and normal states. Hence, the electron system of HTSC is "a black box" for researchers. Studying its response to external influences (e.g.,radiation disordering) one may obtain important information about the electron system itself.

Recently there have appeared the papers the authors of which used fast neutron irradiation for increasing critical current j_c in bulk HTSC's thanks to creation of additional centres of pinning in a superconducting grain proper. In this case j_c increases two orders of magnitude. Experimentally observed essential increase of intragrain critical current after irradiation is undoubtedly of special importance since it supports principal possibility to create effective centres of pinning in bulk HTSC's and thus allows to hope that they may used in devices with strong currents. Transport critical current the value of which is determined by weak links decreases under irradiation at B=0. In more detail this problem is discussed below (in 2.5.2).

Main investigation results on the influence of fast neutron irradiation, mainly the YBCO system, will be briefly considered below. Available data on the influence of low-temperature irradiation of HTSC's by ions and electrons coincide in general with the data on neutron irradiation. However, one should bear in mind that there are certain difficulties in adequate comparison of the experimental data. In most cases experimental results obtained with equal quantities of displacements per atom (dpa) are compared. The quantity of dpa describes the level of initial defects while real defect concentration at some given temperature may be different because of annihilation, migration to sinks, i.e. it may to a great extent depend both on the initial state of the samples and initial microstructure of of radiation defects.

2. PROPERTIES OF RADIATION-DISORDERED HTSC

A great number of HTSC properties in normal and superconducting states differ greatly from those of conventional materials with wide energy bands. A distinctive feature of HTSC's is their "boundary position" between such incompatible, at first sight, states of a solid as , for example, a superconductor and insulator, a superconductor and ordered magnet. Charge transfer between the structural blocks of high-temperature superconducting oxides, which plays an important role in their properties, is characteristic of ionic crystals rather than of metals. Bearing in mind the foregoing, it is not a surprise that HTSC's are so sensitive to different kinds of atomic scale defects, even at small concentrations. Moreover, such a sensibility most likely reflects the unique peculiarities of the electronic properties of HTSC's. Disorder (i.e. defects) readily breaks unusual electronic states of HTSC's transforming thus their electronic system into a more stable nonsuperconducting state (e.g., an antiferromagnetic insulator). Just because of this investigation of atomic disorder effects in HTSC's is of great interest.

First studies of radiation-induced effects in LSCO and YBCO revealed a very strong degradation of T_c at irradiation. In well ordered stoichiometric samples with increasing fluence T_c decreases, transition width increases and the volume of the crystal phase with lack of short range order increases. In this case the end of a superconducting transition, determined by weak links, varies essentially stronger than the beginning. Fig.1 shows T_c vs fluence of fast neutrons for different HTSC's. It is seen that the rate of T_c

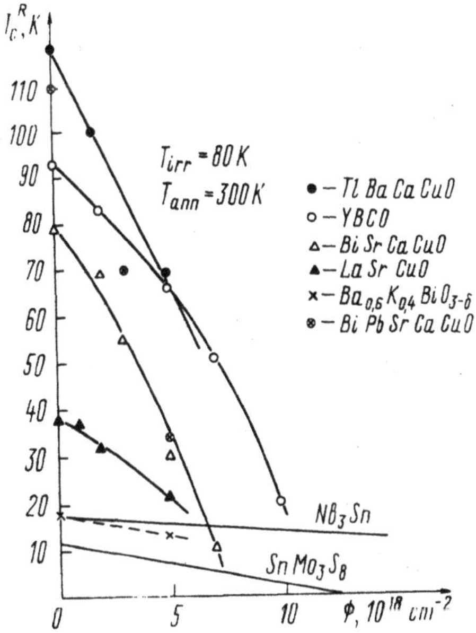

Fig.1.
T_c vs fast neutron fluence.

decreasing with increasing fluence (defect concentration) is approximately the same for YBCO, BSCCO and TBCCO and amounts in the average to ~4 $K/10^{18} cm^{-2}$ that is approximately an order of magnitude larger than for A-15 superconductors. In the "electron" high-temperature superconductor $Nd_{2-x}Ce_xCuO_4$ (not shown in Fig.1) the rate of T_c degradation is the same as in the mentioned "hole" HTSC's. In a cubic $Ba_{0.6}K_{0.4}BiO_3$ it is an order of magnitude less. It is to noted that experimentally it is found that the rate of T_c degradation depends on the initial defectivity of the samples (e.g., oxygen deficiency, impurities, crystalline structure imperfection): in highly defected samples T_c decreases faster than in the samples with optimal chemical and phase composition without defects. Annealing of irradiated samples recovers initial values of T_c provided that the oxygen content is unchanged. Most of authors observed linear (or almost linear) fall of T_c with fluence.

In conclusion it should be noted that strong T_c degradation under radiation disordering is accompanied by dramatic changes of many properties in a normal state. This concerns , first of all, the properties which characterize the electron states. In particular radiation-induced disorder results in the metal-insulator transition in HTSC's. This transition occurs also by decreasing oxygen or strontium content in YBCO and LSCO, respectively, on introduction of substitution impurities to copper or yttrium sublattices. A specific character of the radiation-induced disorder is that irradiated samples keep their full oxygen content and stoichiometric composition. This allows to make more unambiguous conclusions about the possible reasons of variation of the HTSC's properties under atomic disorder and about the peculiarities of their electronic structure. Let us consider the main effects in HTSC's caused by the radiation-induced disorder.

2.1.Crystalline Structure

Investigations of the crystalline structure of HTSC's irradiated by fast neutrons, light and heavy ions and electrons revealed some specific features for each type pf particles. Nevertheless, we may distinguish some general volume changes of the HTSC crystalline structure typical for all of the above mentioned types of irradiation. They are as follows:

- Crystalline structure is preserved within the range of fluences <0.3 dpa; crystal lattice expands isotropically; amplitudes of static mean-square displacements of all atoms increase with fluence and then enter saturation; at damage level of > 0.3 dpa translational symmetry breaks and the material becomes fully amorphous.
- In the temperature range 150–350 K radiation-induced defects are to a great extent mobile; the defected structures of polycrystals and single crystals do not differ appreciably.
- In YBCO for 0.02–0.08 dpa, the crystalline phase exhibits the orthorhombic-tetragonal transition caused by statistic redistribution of oxygen atoms in the Cu(I)–O planes over the crystallographic O(4) and O(5) sites,superconductivity disappears in an orthorhombic phase at $(b-a)\cong 0.05$Å;
- For strong T_c degradation in radiation disordered YBCO the bonding length L of the Cu(I)–O(1) atoms does not practically change as compared to other defected states[2].

Analysis of the X-ray and neutron diffraction data evidences for the reservation of the chemical content of the samples irradiated at low temperatures, particularly, it concerns the oxygen content in a crystal lattice. Crystalline electric field (CEF) studies of ErBCO with inelastic scattering of thermal neutrons[24] also showed that low temperature fast neutron irradiation does not change oxygen content in a unit cell. Since under irradiation chemical composition remains unchanged, variation of physical properties of HTSC's is connected with disorder itself, i.e. with a random potential in a crystal lattice. An isotropic expansion of a crystal lattice evidences about the absence of charge redistribution between the planes in radiation disordered YBCO. Analysis of the CEF spectra of the radiation disordered 1-2-3 exhibits the absence of charge redistribution between the Cu-O chains and CuO_2 planes. Due to high mobility of oxygen atoms in HTSC's radiation defects are mainly formed in oxygen sublattice. They may be different indifferent systems (e.g., oxygen redistribution over the O(4) and O(5)sites in YBCO, location of oxygen atoms in interstices in LSCO and etc.). However, qualitative changes of the electronic subsystem of HTSC's are, most likely, determined by crystal lattice distortions (static atomic displacements) which occur on introduction of these defects.

To understand the mechanism of pinning in irradiated HTSC's it is important to know a concrete type of specific radiation defects, especially, at small damages when j_c increases. Since real defected structure of the irradiated samples strongly depends on their initial state, let us confine ourselves to two recent papers[3,4], where the defected state of high quality YBCO single crystals irradiated by fast neutrons was investigated by employing the technique of transmission electron microscopy in imaging and diffraction modes.

The authors of Ref.3 studied the damaged structure of the YBCO single crystals (T_c=91K, $\Delta T^{\chi} \sim$ 1K) irradiated by fast neutrons at 330K (Φ = 2 x 10^{18}cm^{-2}, E $\cong$ 1 MeV). In the electron-microscope image the structure consisted of strongly diffracted microcrystals or cells with the effective size 50 to 100 Å, surrounded by highly damaged or amorphous (i.e., nondiffracted) layers. At $\sim$ 0.003 dpa when this structure occurs degradation of T_c is small (<10 K). Qualitatively the same picture was observed for the YBCO single crystals irradiated by the Kr^+ (1.7 MeV) and Ne^+ (1.5 MeV) ions. In this case, however, the cellular structure appears at much more higher level of damage (~0.07dpa), when there is no superconductivity.

The authors of Ref.4 also studied the damaged structure of the YBCO single crystals (T_c = 91 K, T_c(R=0) = 89 K) at lower damage level (Φ = 10^{16} cm^{-2}, E =0.5MeV, T_{irr} = 370 K). The irradiated single crystals have two types of extended defects: linear defect stripe regions with a characteristic width of $\sim$ 14 Å and arranged along stripe direction agglomerates with the size from 100 to 300 Å.

The authors attribute the first type of defects to a high-energy stage (thermal spike) of a collision cascade, while the second to a low-energy (a displacement spike) one. The defected regions of both types are parallel to the (001) planes and stretched along the (010) direction, i.e. along the b-axis of the ortho-phase. Similar defects were never observed in unirradiated crystals. The authors believe that the defects of the first type represent microregions of local remove of the Cu-O layers, while those of the second type represent microregions of local displacement and rearrangement of oxygen atoms in the layers (nucleation of oxygen vacancies and interstitials).

It is not excluded that with increasing fluence (i.e.,concentration of radiation defects) such a defected structure transforms into a cellular one that was observed in Ref.3. It should be noted that in both the cases [3,4] the samples were irradiated at relatively high temperatures (i.e., 330 -370 K), when the mobility of radiation defects is high (at least, it concerns oxygen atoms).Hence, at low-temperature irradiation (T<80 K) the real defected structure may be different.

2.2. Electrical Resistivity

In all the HTSC's investigated (i.e., LSCO, BSCCO, TBCCO, NdCeCuO) the electrical resistivity changes under radiation disordering in a different way as compared with simple metals or conventional superconductors [5]. Introduction of defects,including radiation defects, into the crystal lattice of simple metals increases residual electrical resistivity ρ_0, but the temperature dependent contribution ρ_T being practically unchanged (Matthissen rule). ρ_0 increases linearly with defect concentration C_d .

In low-temperature superconductors, LTSC,(e.g.,Nb, intermetallic compounds with A-15 and C-15 structures, Chevrel phases) disordering also leads first to a strong linear increase of ρ_0, reaching saturation for (150-200) $\mu\Omega$ cm at high C_d . However, at very strong disorder ($\Phi > 10^{20}$ cm^{-2}) in Mo_6Se_8 we observe an Anderson type metal-insulator transition [6] that manifests itself in a large increase of ρ_0 after an approximately constant value within the "saturation" range. The temperature dependence of electrical resistivity in LTSC's weakens with increasing C_d , and at large defect concentration the temperature factor becomes negative [1].

In HTSC's electrical resistivity exhibits different behaviour from that mentioned above. At fixed temperature (e.g., at 100 K) it grows exponentionally with increasing fluence (at small fluences $C_d \sim \Phi$), starting from the smallest C_d .The linear dependence $\rho(T)$ remains at $T>T_c$,while the value of $d\rho/dT$ increases.The $\rho(T)$ curves simultaneously contain parts with $d\rho/dT>0$ (high temperatures) and $d\rho/dT<0$ (low temperatures). At high C_d the linear dependence $\rho(T)$

transforms to an exponential curve that is typical of hopping conductivity over localized states. In HTSC's variation of ρ as a function of T and Φ may be described by the empirical formula

$$\rho(T, \Phi) = f(T)\exp(k\Phi/T^{1/4}) \qquad (1)$$

where f(T) is a function that varies smoothly from linear in an ordered state to temperature independent in a strongly disordered one.

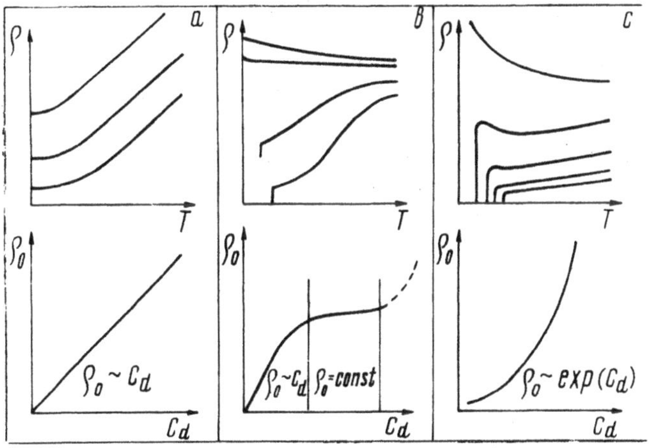

Fig.2. ρ vs T and C_d (scheme): a - simple metals; b - LTSC (---Mo_6Se_8); c - HTSC.

For illustration, Fig.2 shows the variation of electrical resistivity for simple metals, LTSC's and HTSC's under disorder. As can be seen, the HTSC electron system response to atomic disorder differs dramatically from that of the systems with metallic conductivity, in which at small defect concentrations electrical resistivity always grows proportionally to C_d. It is reasonable to associate this behaviour of $\rho(T, \Phi)$ with charge carrier localization effects in HTSC's which we observed earlier in Mo_6Se_8. An analysis of the HTSC dependences $\rho(T, \Phi)$ shows that there is no sharp metal-insulator transition under disorder (exponentional growth of ρ begins from the smallest C_d), i.e. localized states exist also in ordered high-T_c superconductors. At first sight this is in contradiction with the "metallic" character of conductivity ($\rho \sim T$) of ordered HTSC's. This contradiction has not been removed as yet. Its existence, however, points to unusual electronic properties of HTSC's.

Exponential increase of electrical resistivity ρ_{ab} in the basal Cu-O planes as a function of C_d exists also in HTSC single crystals of the YBCO, BSCCO and NdCeCuO systems and is an intrinsic property of

HTSC's themselves[7,15]. Fig.3 shows the dependence $\rho_{ab}(\Phi)$ for single crystals of YBCO , NdCCO irradiated by fast neutrons and electrons. The question arises whether this exponential response to atomic disorder, found first in HTSC's, is typical only for these materials or also for other nonsuperconducting oxides with a similar structure.To answer this question we studied the dependences ρ (T,Φ) of $BaNb_4O_6$[8] and $K_{0.3}WO_3$[15] with metallic conductivity. The former is a cellular niobium oxide with a single perovskite bronze layer while the latter has a quasi-planar structure. The electrical resistivity variation due to introduction of defects is determined in these compounds by a carrier free path decrease. The response of BNbO and KWO to disorder is typical of ordered compounds with metallic conductivity and relatively weak electron-phonon interaction and differs qualitatively from the response of HTSC's. On the basis of available experimental data we can assume that the closer the electronic system to the metal-insulator transition the higher T_c may realize.

Fig.3.
log ρ_{ab}(80K) vs fluence:

Fast neutrons (T_{irr}=80K):
YBCO (△,▲), NdCCO (□).

Electrons (30 MeV,T_{irr} = 100K):
YBCO (*,■) after Petrusenko et al.

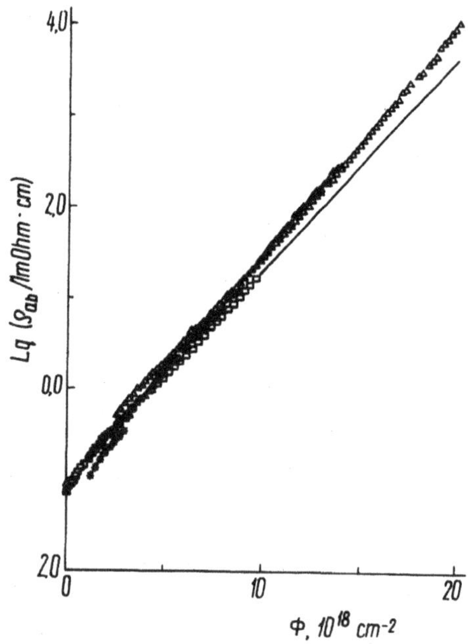

Increase of ρ , linear dependence ρ(T), decrease of the slope dρ/dT at relatively small fluences at $T_c(\Phi)/T_c(0)$=0.6, the metal-insulator transition at high fluences were observed practically in all the experiments on HTSC's irradiated by neutrons, ions and electrons. Of principal importance is the question whether the dependence of type (1) is universal for all kinds of irradiation or it is typical only for neutron irradiation and concrete samples.

Most of authors observed an exponential $\rho(\Phi)$ dependence or a dependence stronger than linear. The linear dependence $\rho(\Phi)$ was observed in YBCO irradiated by electrons (E=3 MeV)[9] and oxygen ions (E=25 MeV)[10,12] and LSCO irradiated by electrons (E=3 MeV)[11]. Note, that at slow rates of fluence and high enough irradiation temperatures the $\rho(\Phi)$ curve may shift from the exponential dependence in samples with different initial defective states because of high mobility of radiation defects in HTSC's. Experiments from Ref.25 support this assumption. Besides, for the initial small range of fluences it is rather difficult to distinguish between the exponential and quasi-linear dependences. Perhaps, it is because of this that different researchers observe the linear dependence $\rho(\Phi)$ under electron irradiation (small fluences). It is to be noted that recently Yu.Petrusenko et al. has observed dependence of the type $\rho \exp(\Phi)$ in good single crystals of YBCO (Fig.3) irradiated by electrons of high fluences (E_e=30 MeV, T_{irr}=100 K) (the paper is in print). At the same time the linear dependence $\rho(\Phi)$ observed in YBCO irradiated by oxygen ions at low temperatures within a wide (up to the metal-insulator transition [12]) range of ρ(100K) requires special explanation.

Radiation-induced defects which occur in HTSC's irradiated by fast neutrons at T=80 K are relatively unstable: isochronous annealing at 200K and 300K for 20 min leads to a partial recover of T_c and ρ. A two-week annealing at room temperature results in further recovering of T_c and other properties.

2.3. Anisotropy of Electrical Resistivity and the Upper Critical Field. Hall Effect

Anisotropy of kinetic properties found experimentally in single crystals is a specific character of HTSC's which varies within a wide range for different HTSC systems. Thus, in YBCO anisotropy of electrical resistivity accounts for $\sim 10^2$, while in BSCCO it is approximately 10^5. Hall coefficient, thermoelectric coefficient, the upper critical field H_{c2} and a number of other properties are also, to a great extent, anisotropic. With radiation disorder one may trace the interplay between T_c and anisotropy to find out the role of two-dimensionality in forming of the superconductivity of HTSC's.

In more detail the effect of disordering on anisotropy of electrical resistivity and that of the upper critical field is studied on single crystals of YBCO[7]. As mentioned above, ρ_{ab} grows exponentially starting with the smallest C_d as in polycrystalline samples. At the same time ρ_c at small C_d grows slower than ρ_{ab} and only at comparatively large C_d do they grow with the same rate. Accordingly the anisotropy ρ_c/ρ_{ab} at 80K drops rapidly to some "residual" value (= 30 at $\Phi=10^{19}$ cm^{-2}) which remains unchanged with further increase of C_d (Fig.4). The larger the disorder the weaker the temperature dependence of the anisotropy in disordered samples. The absolute value of ρ_c/ρ_{ab} in this case decreases strongly (several times) at T<100 K while at room temperature it decreases some tens percents. Such a behaviour of anisotropy as a function of C_d and T is typical also of other investigated single crystals (i.e., BSCCO, NdCCO), and does not depend on the initial defected state of single crystals and the initial anisotropy (from 1.5 x 10^2 to 3 x 10^4). The experimental data on the influence of disorder on T_c and anisotropy ρ_c/ρ_{ab} of HTSC's under study point to universal correlations between these quantities.

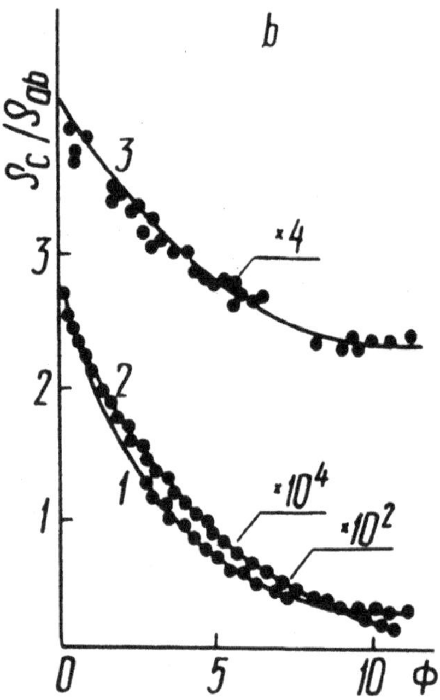

Fig.4.
ρ_c/ρ_{ab} vs fast neutron fluence
(T_{irr} = T_{meas} = 80K):
YBCO (1),
NdCCO (2),
KWO (3).

We observed decrease of anisotropy ρ_c/ρ_{ab} under radiation disorder not only in HTSC's but also in other strongly anisotropic crystals with metallic conductivity. According to a simple relation $\rho \sim 1/\ell S_F$ (where S_F is the Fermi surface, ℓ is the free path), anisotropy of ρ may be connected either with the anisotropy of S_F (i.e.,mass anisotropy of charge carriers m_e) or with the anisotropy of ℓ (e.g., for the case of flat defects which are perpendicular to the c-axis). In KWO anisotropy of ρ is mainly determined by the anisotropy of S_F while in graphite by that of ℓ. In the former decrease of ρ_c/ρ_{ab} under irradiation is relatively small (Fig.4) while in the latter it is very high. Up to now the nature of the anisotropy of ρ in HTSC's has not been elucidated in details. But it is likely that it is not singly the anisotropy of m_e and ℓ since there are data pointing to a possible charge transfer by different quasi-particles within the basal plane and along the c-axis.

Anisotropy of the upper critical field in the HTSC single crystals also decreases with increasing disorder. In single crystals of YBCO the derivative $dH^{\parallel}_{c2}/dT$ (the field is parallel to the ab-plane) decreases, while $dH^{\perp}_{c2}/dT$ (the field is perpendicular to the ab-plane) increases essentially. $H^{\parallel}_{c2}(T)$ and $H^{\perp}_{c2}(T)$, linear in the initial (unirradiated) state, excluding the range of $(T_c-T)<0.5K$ become essentially nonlinear after irradiation. Hence, correct determination of $dH^{\parallel,\perp}_{c2}/dT$ is impossible. Nevertheless, the anisotropy of the upper critical field $H^{\parallel}_{c2}/H^{\perp}_{c2}$, represented as the ratio of T_c differences in external fields H=0 and H=5T, applied parallel and perpendicular to the ab-plane, rapidly decreases with increasing defect content disappearing practically in the samples with T_c < 7K, while the anisotropy of electrical resistivity ρ_c/ρ_{ab} at 100 K is still high (0.5 as compared to the initial value). This result is difficult to understand within the frames of the conventional theoretical models since it should be $(H^{\parallel}_{c2})'/(H^{\perp}_{c2})' = (\rho_c/\rho_{ab})^{1/2}$ in the "dirty" limits. Besides, within the same limit $(H^{\parallel}_{c2})' \sim (\rho_{ab}\cdot\rho_c)^{1/2}$, but from the experiment it follows that an essential decrease of $(H^{\parallel}_{c2})'$ is accompanied by a strong increase of $(\rho_c\cdot\rho_{ab})^{1/2}$. Thus, the theory for traditional superconductors even qualitatively does not describe experiment in the case of HTSC.

In terms of the conventional interpretation, charge carrier density per unit of a cell for simple single-band metal $n_H = 1/R_H e$ as well as its value do not practically depend on temperature. In a general case n_H may not coincide with real concentrations if simultaneously electrons and holes with approximately the same concentration and mobility serve as charge carriers. In HTSC's $1/R_H$ either increases linearly with temperature (YBCO, BSCCO, TBCCO) or slightly depends on T (LSCO).

First investigations of the $1/R_H(T)$ curves in radiation-disordered high-T_c superconductors were made on ceramics(LSCO, YBCO) irradiated by fast neutrons [13] and on epitaxial films of YBCO irradiated by the ions Ne^+ (E=1 MeV)[14]. Disorder weakens the temperature dependence of $1/R_H$, which remains linear just as in the initial ordered samples and almost disappears at high C_d. At T<100K $1/R_H$ does not practically vary while superconductivity exists and decreases for all temperatures in strongly disordered samples (superconductivity is absent). Nevertheless, even in the latter case $1/R_H$ (100K) does not differ greatly from the initial value with exponential increase of ρ(100 K). Remind, that at comparable T_c in the oxygen deficient YBCO samples $1/R_H$ decreases several times both at low and high temperatures. It may be assumed that $1/R_H e$(100 K) at high enough degree of disorder, when its temperature dependence is weak, corresponds to a true concentration of charge carriers in YBCO. In the YBCO single crystal[15] the dependences of $n_H^{\perp}$ (the field is perpendicular to the ab-plane) on temperature and defect concentration are in good agreement with those obtained for ceramic samples. $n_H^{\parallel}$ (the field is parallel to the ab-plane) is negative in the initial state (electron conductivity). After irradiation by fluence $\Phi = 1 \times 10^{19} cm^{-2}$ no superconductivity) it becomes positive (hole conductivity) and comparable with n_H.

A comparison of the dependences $1/R_H(T)$ and $\rho(T)$ for samples with different defect content shows that in HTSC's the metal-insulator transition is due to a decrease in charge carrier mobility rather than to their "freezing out". The low-temperature constancy of $1/R_H$ in samples with different defect content agrees with the assumption concerning the charge carrier localization in high-T_c superconductors[5]. On the basis of the experimental data we may assert that the behaviour of $1/R_H$ correlates with the value of T_c; strong temperature dependence is characteristic only for a superconducting state and becomes practically "normal" when superconductivity disappears. Further increase of disorder practically does not affect the dependence $1/R_H(T)$ and only slowly decrease its absolute value. Note that within the same region (no superconductivity) $n_H^{\parallel} \simeq n_H^{\perp} > 0$, i.e., unusual states disappear and the "usual" three-dimensional conductivity by localized states is realized in a crystal.

2.4. *Magnetic Susceptibility and Low-Temperature Heat Capacity*

Magnetic susceptibility of high-T_c superconductors either decreases slightly with decreasing temperature (LSCO) or is temperature independent at all (YBCO). In conventional superconducting intermetallic compounds with high enough T_c = 16-20 K (e.g., A-15) χ increases greatly with decreasing temperature. This is connected with a narrow peak (or sharp peculiarities within the peak) of $\sim 10^{-2}$eV width in the d-electron state density at Fermi level. Note, that in isostructural intermetallic compounds with low $T_c \sim$ 2K the temperature dependence of χ is absent.

Under radiation disorder the character of behaviour of χ and the corresponding changes of T_c differ greatly in A-15[1] and high-T_c superconductors (Fig.5). In A-15 a decrease in T_c (V_3Si, V_3Ge, initial T_c = 17 - 18 K) under disorder is accompanied by a decrease in χ and weakening of its temperature dependence, while the T_c increase (Mo_3Si, Mo_3Ge, initial $T_c \sim 2K$) is accompanied by increasing χ and strengthening of its temperature dependence. The main variations of magnetic susceptibility of high-temperature superconductors under radiation disordering are as follows[5]:

- In the temperature range T_c - 300 K the temperature dependence of the Curie-Weiss type appears $\chi(T) = \chi_0(\Phi) + C_k(\Phi)/(T - \theta(\Phi))$, which increases with increasing irradiation fluence; C_k is proportional to Φ (or C_d); in YBCO it grows three times faster than in LSCO (or is the same when recalculating per Cu atom).

Fig.5. χ vs T (scheme) for: V_3Si (a), YBCO (b). Insert: $C_k(\Phi)$ and $\chi_0(\Phi)$ for YBCO and LSCO.

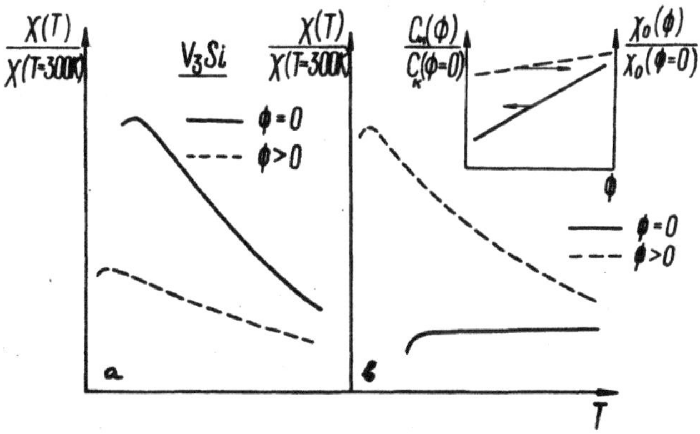

Similar variations were also observed in YBCO irradiated by 2.7MeV electrons at 40K[26]. The appearance of the Curie-Weiss temperature dependences is connected with variations in the nearest surroundings of copper atoms Cu(2), giving rise to localized magnetic moments (LMM) on these atoms due to short-range character of antiferromagnetic spin fluctuations. In a highly disordered YBCO the average distance between LMM is of the order of the distance between the nearest Cu-Cu neighbours in the Cu(2)-O planes. High LMM concentration results in the interaction between LMM and formation of the short-range magnetic order at low temperatures which is supported by the NQR data[17]. Appearance of the short-range magnetic order leads to the experimental deviation of the $\chi(T)$ curve from the Curie law at T<10 K in strongly disordered samples.

The change of the HTSC magnetic state under radiation disorder should manifest itself in the behaviour of the heat capacity vs temperature curve at T<10 K. In Ref.5 the anomalous behaviour of the C(T) curve is observed in the irradiated YBCO samples: at T<7 K the value of C/T increases sharply with decreasing temperature. This low-temperature increase of C/T also exists in ordered LSCO (in oxygen-deficient $YBCuO_{7-\delta}$ as well as in stoichiometric ordered YBCO $C/T \sim T^2$ below 10 K). In Fig.6, the temperature dependence of the additional contribution to heat capacity $C_{add} = C_{exp} - \beta T^3$ is shown for

Fig.6.
C_{add} vs T and

$S_{add}/R\ln 2$ vs C_K:

Δ - Φ = 0.5×10^{19} cm^{-2},

0 - Φ = 5×10^{19} cm^{-2},

● - Φ = 5×10^{19} cm^{-2}
plus short annealing

□ - Φ = 5×10^{19} cm^{-2}
plus prolonged annealing

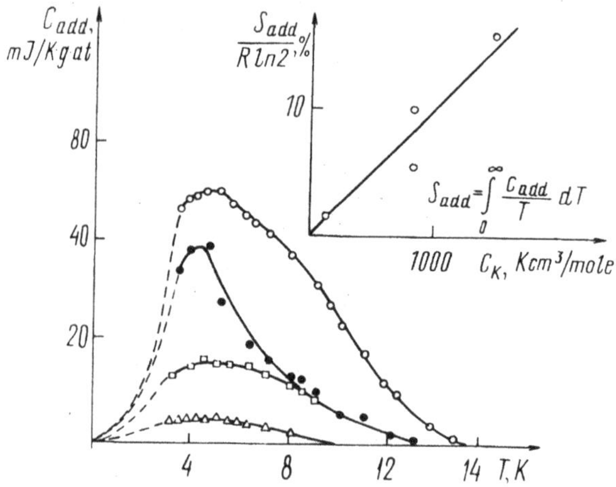

the YBCO samples with different defectivity. It is seen that the "extra" heat capacity separated from the total heat capacity is similar to that of a two-level system (Shottky anomaly). Increase of disorder increases the extra contribution "intensity", but it does not change its temperature scale peak position which is approximately equal to splitting energy.

Increase of the "extra" contribution to heat capacity of LSCO and YBCO with disorder may be interpreted as an increase in the number of two-level systems of a magnetic nature with the spin s=1/2 (the number of such systems in YBCO is three times larger that that in LSCO)). If a unit cell has one two-level system, the corresponding entropy will be S=Rln2. The insert in Fig.6 shows the normalized "extra" entropy vs Curie constant as determined from magnetic susceptibility measurements for YBCO samples with different defect content. The linear correlation between these values implies that the contribution to the low-temperature heat capacity of atomic disordered HTSC's is of a magnetic nature. In this case in high-quality LSCO and YBCO samples the low-temperature heat capacity singularities are not connected with the presence of other phases and are, therefore, the inherent properties of a superconductor itself. The extra contribution is nonlinear in temperature , has magnetic nature and comes, most probably, from the atomic scale defects, which exist even in samples with maximum T_c. The contribution of these defects to magnetic susceptibility is negligible and cannot be detected at low temperatures because of the diamagnetism.

It is found that concentrations of the Cu atoms with LMM determined from the data on low-temperature heat capacity and magnetic susceptibility coincide numerically: the number of the two-level systems coincides with the number of the Cu atoms with LMM = 1 μ_B. Full suppression of superconductivity in HTSC's occurs at the concentration of the "defected" Cu atoms of the order of several percents, i.e. occurrence of a copper atom with LMM per approximately 10 unit cells is enough to change drastically all the HTSC properties that characterize their electronic states.

2.5.Critical Currents

Investigations performed on conventional irradiated superconductors have shown that introduction of radiation defects may appreciably vary the behaviour of the flux pinning, thus, changing the critical current density vs temperature and field (in an applied field), $j_c(T,H)$. At small fluences, when T_c does not practically decrease under disorder, j_c increases thanks to appearance of extra centers of pinning. With further increase of fluence j_c begins decreasing both as a result of T_c decreasing and, seemingly, of the defective structure homogenizing. This phenomenon is known as "peak-effect". It should be noted that in optimal current-carrying superconducting elements with maximum j_c "peak effect" is absent, and under irradiation j_c decreases because of "dissolving" of pinning microprecipitations of admixture phase, which is obtained under certain thermal treatment, and of T_c decreasing.

In high-temperature superconductors the dependences $j_c(T,H)$ a redetermined by the peculiarity of a vortex structure and the flux motion. High transition temperatures cause relatively high vortex energy under thermal motion allowing to overcome potential barriers comparatively easy. Small correlation lengths appreciably weaken the pinning forces (i.e., lower the potential barrier height) which hold vertex in a fixed position. High anisotropy (quasi-two-dimensionality) results in anisotropic behaviour of vortex structures and, hence, of critical currents depending on mutual magnetic field orientation and evolved crystallographic direction (e.g., the c-axis). High initial defectivity of high-T_c superconductors is also a specific character. In this case one should differentiate between the values of j_c calculated from the experimental magnetization in Bean's[18] model ("physical" critical current j_c^M in a superconducting material proper) and those obtaine from the resistivity measurements (transport critical current j_c^R determined in particular by the weak links).

2.5.1. Critical Magnetization Current j_c^M

Inspection of the magnetization curves (hysteresis loop) of the HTSC poly- and single crystals of Y and Tl systems irradiated by fast neutrons[19,22], electrons and heavy ions has revealed that with increasing fluence magnetization hysteresis grows that is typical of critical states of hard superconductors. In a superconducting material proper radiation defects are either the extra centers of pinning or enhance pinning of already available centers. Both the facts increase j_c^M. Apparently, these factors as well as T_c variation determine j_c^M vs H and Φ curves.

In spite of the differences in the initial states of the samples (i.e. poly- and single crystals of different quality) and irradiation conditions, we may note a number of general trends in the behaviour of j_c^M under radiation disorder (Fig.7):

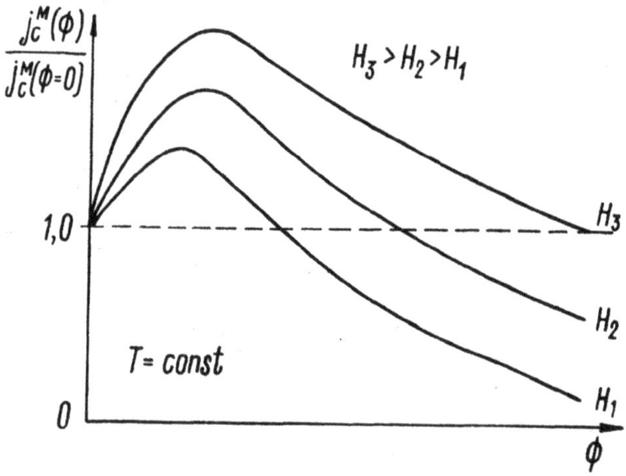

Fig.7. $j_c^M(\Phi)/j_c^M(\Phi=0)$ vs Φ for different external magnetic fields (scheme).

- At first j_c^M increases with fluence for all the measurement temperatures and all the fields studied (at least up to 6–9 T). The larger the field and the measurement temperature the higher increase in j_c^M reaching one to two orders; according to the data of Ref.21, in the YBCO melt-textured ceramics irradiated by fast neutrons of fluence $6 \times 10^{17} cm^{-2}$ $j_c^M(5K;\ 5T) = 8 \times 10^6 a/cm^2$.
- For all the irradiated temperatures and fields under study $j_c^M(\Phi)$ exhibits "peak effect" (i.e.,drop of j_c^M after growth at small Φ) caused either by a T_c fall or destruction of available pinning centers; temperature and field dependences of j_c^M weaken with increasing fluence;
- In single crystals j_c^M increases for all the measurement temperatures and fields under study, with the field parallel and perpendicular to

the c-axis; anisotropy of $j_c^M(H\|c)/j_c^M(H\perp c)$ decreases. According to the data from Ref.26, in the YBCO single crystal irradiated by fast neutrons $j_c^M(77K;\ 0.9\ T) = 6 \times 10^5 a/cm^2$.

2.5.2. Transport Critical Current j_c^R

From point of view of practical use of high-T_c superconductors, in most cases the most important are the parameters of the transport critical current under different external conditions. The value of j_c^R is limited by weak links both in poly- and single crystalline bulk HTSC's as well as in thin (textured or single crystalline) films.

Variation of j_c^R at different temperatures and external fields after irradiation by fast neutrons and ions was investigated in detail on the YBCO poly- and single crystalline films. According to the data of most authors, in HTSC films irradiated by H^+, He^+, O^+, Ar^+, Xe^+ at different T_{irr} the j_c^R (along the film) decreases in a zero external field in the range 4.2–77 K. Drop in j_c^R in this case starts from the fluences T_c is practically unchanged. Some data evidence that with increasing fluence the dependence $j_c^R(T)$ weakens. This behaviour is characteristic of the YBCO films with the high initial value j_c^R. According to available data, in BSCCO and TBCCO at equivalent damage level variations in $j_c^R(B=0,\ T)$ are more noticeable.

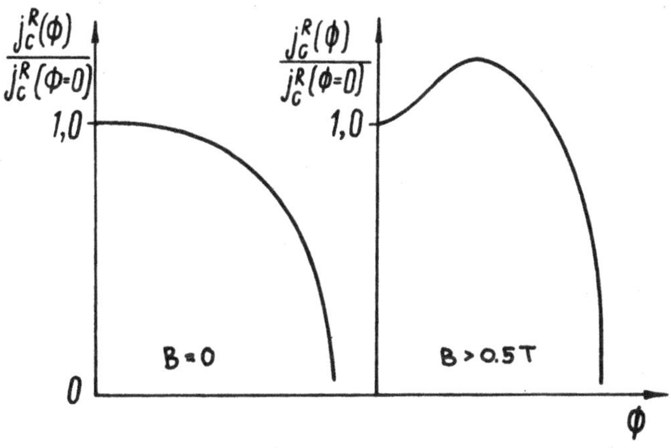

Fig.8. $j_c^R(\Phi)/j_c^R(\Phi=0)$ vs Φ for different external magnetic fields (after [10,23] data).

An alternative effect was observed in epitaxial YBCO films irradiated by fast neutrons[22]. The film with initially high $j_c^R=2 \times 10^7 a/cm^2$ (4.2 K; H=0–8 T) was irradiated by fluences of $5 \times 10^{17} cm^{-2}$ and $2 \times 10^{18} cm^{-2}$ at 35°C. The decrease in T_c was 1.3 K and 5.6 K, respectively. At smaller fluence j_c^R increases one and a half at all temperatures (4.2, 60, 77 K) without external field. With increasing fluence j_c^R grows slower depending on temperature, or falls ("peak effect" by fluence).

A characteristic feature of high-T_c superconductors (at least, of YBCO) is an increase of j_c^R in the range T=4.2–77 K for the fields larger than 0.5–1 T after irradiation by small fluences of ions and neutrons (in the latter case increase of j_c^R in the field turns out to be more higher). Within this range of fluences j_c^R increases both in the field parallel and perpendicular to the ab-plane. Though, in the case of ion irradiation[27] of very high energies (Xe^+, 173 MeV, T_{irr} =80 K) resulting in well-formed tracks in an irradiated film j_c^R increases in the fields higher than 0.5 T when B is parallel to tracks (i.e., perpendicular to the ab-plane) and decreases when B is perpendicular to tracks (i.e. parallel to the ab-plane). In the latter case the decrease is smaller than at B=0.

At high fluences increase in j_c^R becomes slower. Its value essentially depends on temperature, direction and the value of external magnetic field. And at last at high damages the density of critical current falls below the initial value for all temperatures and fields. Fig.8 schematically shows variation of the relative value of the critical current density at 60 K as a function of fluence of the ^{16}O ion irradiation (25 MeV, T_{irr} =77 K) with and without magnetic field (according to the data of Refs.10, 23). Note once more, that a number of authors mention not only growth of j_c^R in magnetic field after irradiation but also weakening of the dependence $j_c^R(B)$.

3. CONCLUSION

Analysis of the experimental data on radiation-induced variation of electronic and lattice properties of high-T_c superconductors revealed a well defined two-dimensional character of charge carriers. Disordering causes the variation of the electronic states which may be called three dimensionalization.

In ordered HTSC's the motion of charge carriers perpendicular and parallel to the ab-plane has different character. This is manifested in qualitatively different temperature dependences of electrical resistivity, Hall coefficient, thermoelectric coefficient and etc. for these directions. It is possible that the charge along and across the ab-plane is transferred by different quasiparticles (i.e.spinons-holons, bosons-fermions etc.). At $T_c \Rightarrow 0$ (radiation disordering) these discrepancies disappear and we deal with three-dimensionality of electronic states.

Such a disorder-induced "2D-3D" transition is likely to be a general property of anisotropic crystals. However, in high-T_c superconductors localization effects of carriers are also of importance. Disordering of HTSC leads to a transition from weak to strong carrier localization , quasi-two-dimensionality, however, being destroyed. The normal state properties of a disordered system are not anomalous any more concerning the temperature dependences, though the disordered state itself is characterized by a pronounced localized (dielectric) type of conductivity and short-range magnetic order. Hence, it may be assumed that it is the two-dimensional character of the carrier motion and their weak localization that are responsible for the main properties of the normal state of superconducting oxides with high T_c and investigation of these peculiarities will favour further elucidation of the nature of HTSC's.

Apparently, theoretical analysis of the behaviour of critical fields and currents in radiation-disordered HTSC's based on modern understanding of pinning and flux creep has become an acute problem nowadays. Besides, investigation of the dependences $j_c(B, T, \Phi)$ in irradiated HTSC's (probably mainly j_c^M) should provide a unique information about the mechanism of pinning and flux creep in the proper HTSC materials as well as about the microstructure of defects occurring under radiation disorder.

Investigations available by the present time have revealed that typical radiation-induced defects in HTSC's have much more higher pinning potential as compared to ordinary defects resulting thus in suppression of the flux creep in irradiated samples. Since pinning is strong when the centers are of the order of the coherence length (10–80 Å in YBCO), clusters of defects or strongly damaged zones are likely to serve as the centers in the irradiated crystals[3,4]. Interaction between the radiation defects and pinning centers in the initially ordered crystals also plays an important role in the formation of the current parameters of HTSC's.

REFERENCES

1. B.N.Goshchitskii et al., *Sov.Sci.Rev.Sec.A.,Phys.Rev.* 8 (1987) 520
2. V.I.Voronin, I.F.Berger, A.E.Karkin et al.. *SFCT* 3 (1990) 75.
3. M.A.Kirk et al., *Phil.Mag.Lett.*62 (1990) N1, p.41.
4. R.Janan et al., *Sol.St.Comm.* 77 (1991) N4, p. 259.
5. B.A.Aleksashin, V.I.Voronin et al., *JETP* 95 (1989) 678.
6. V.I.Voronin, V.A.Arkhipov et al., *FMM* 67 (1989) 115.
7. S.A.Davydov, B.N.Goshchitskii et al., *Physica* C161 (1989) 549.
8. V.I.Voronin et al., *Rev.Mod.Phys.* B4 (1990) 1531.
9. Kensuke Shiraiski, Hiroshi Iton, Osamu Yoda et al., *J.Appl.Phys.Jap.*27 (1988) L2339.
10. B.Roas, B.Hensel, G.Saemann-Ischenko et al., *Appl.Phys.Lett.*54 (1989) 1051.
11. C.H.Chen et al., *Appl.Phys.Lett.*54 (1989) 1178.
12. P.Schmidt, H.Adrian et al., *IEEE Trans.Magn.*25 (1989) 2241.
13. B.N.Goshchitskii, S.A.Davydov, A.E.Karkin et al., *Physica* C162-164 (1989) 997.

14. J.M.Vallers, A.E.White et al., *Phys.Rev.*B39 (1989) 11599.
15. S.A.Davydov, A.E.Karkin, A.V.Mirmelstein et al.,in *Effects of Strong Disordering in HTSC* (Sverdlovsk, USSR, 1990).
16. A.Hoffman, P.Schule et al.,*Physica* C153-155 (1988) 341.
17. Yu.N.Zhdanov et al., *Physica* C165 (1990) 475.
18. C.P.Bean, *Phys.Rev.Lett.* 8(1962) 250.
19. A.Wisniewski,M.Baran, Z.Koziol et al.,*Physica* C170 (1990) 333.
20. W.Schindler, *Appl.Phys.Lett.* (1991), in print.
21. P.H.Hoz, Z.J.Huang, L.Gao et al., *Mod.Phys.Lett.B* 4 (1990) 703.
22. W.Schindler,B.Roas,G.Saemann-Ischenko, *Physica* C169 (1990) 117.
23. B.Roas, B.Hensel, G.Endres et al., *Thin Solid Films* 174 (1989) 179.
24. A.A.Podlesnyak et al., *Intern. Conf. M^2S-HTSC III*, *Rep. N* 4A-10 (22-26 July, 1991, Kanazawa, Japan).
25. A.D.Marwick, G.J.Clark et al., *Phys.Rev.*B39 (1989) 9016.
26. R.V.Van Dover, E.N.Gyorgy et al., *Nature* 342 (1989) 55.
27. B.Roas, B.Hencel et al., *Europhys.Lett.*11 (1990) 669.

INFLUENCE OF FAST NEUTRON IRRADIATION ON VARIOUS TYPES OF HTSC MATERIALS: COMPREHENSIVE STUDY

A. WIŚNIEWSKI, R. PUŹNIAK, M. BARAN and H. SZYMCZAK
Institute of Physics, Polish Academy of Sciences
Al. Lotników 32/46, PL 02-668 Warsaw, Poland

ABSTRACT

The influence of fast neutron irradiation on the inter- and the intragrain critical currents, the critical temperature, the first and second critical field, the normal state resistance is discussed in details on the base of experimental results obtained on $YBa_2Cu_3O_{7-\delta}$ and $Bi_{0.7}Pb_{0.3}SrCaCu_{1.8}O_x$ ceramics. The response of the 90 K- and the 60 K-phase of $YBa_2Cu_3O_{7-\delta}$ ceramics, irradiated with the same fluences, especially with regard to the increase of the intragrain critical current and the influence on the irreversibility line is also compared. The damage of the weak links during irradiation is the reason of a decrease of the intergrain (transport) critical current in ceramic materials. The problem of the weak links could be overcome in melt textured materials. The results of neutron irradiation of the high critical current density $YBa_2Cu_3O_{7-\delta}$ prepared by powder melting process are discussed. The critical current anisotropy of $Bi_2Sr_2CaCu_2O_x$ single crystal before and after irradiation is studied.

1. Introduction

The values of critical temperature, T_c, and upper critical field, H_{c2}, of high-T_c superconductors (HTSC) are significantly higher in comparison with the conventional superconductors. On the other hand, the third parameter - very crucial from the point of view of applications - critical current density, j_c, is still too low at higher temperatures. More strictly, values of j_c as high as 10^5 A/cm^2 or more in 1 T at 77 K could be obtained now in specially prepared and/or treated single crystals (in both bulk materials and epitaxial thin films). However, one should notice that the form of single crystals limits strongly range of applications, e.g. for thin films the critical current density is quite high but the total critical current is small. The presented paper concerns the influence of one of special treatment methods of materials - fast neutron irradiation of HTSC - on their physical properties. The disadvantage of HTSC materials, low critical current density, arises from the low pinning energy, the large relaxation and appearance of the reversible magnetic behavior in the mixed state region, well below the diamagnetic upper critical field H_{c2}. These phenomena are caused by a short coherence length and a large anisotropy of HTSC compounds. One can also noticed that, the temperature of 77 K, the most suitable temperature from the practical point of view for applications of HTSC compounds, is relatively high working temperature. Namely, 77 K refers to $0.84T_c$ for YBaCuO compounds (T_c = 92 K) and to $0.62T_c$ for Bi or Tl based compounds (highest T_c = 125 K). In comparison, 4.2 K - the standard operating temperature for conventional superconductors refers to $0.47T_c$ for Nb-Ti alloy (T_c = 9 K) and $0.23T_c$ for Nb_3Sn (T_c = 18 K). Another problem is connected with the fact that various materials (ceramics, single crystals, thin films) exhibit significantly different values of transport critical currents. So, from the point of view of applications at least two problems exist. First is connected with the weak links between the superconducting grains, which are responsible for mentioned above differences of transport critical currents. Various approaches, including melt textured processes, doping of ceramic by silver and producing of epitaxial thin films, were used to overcome this difficulty. Second problem concerns the best way of introducing new effective pinning centers.

The fast neutron irradiation is one of the best methods of increasing the defects concentration, in a controlled way, over the whole sample volume. Determination of the sensitivity of HTSC parameters to the lattice imperfections could result in a better understanding of the mechanisms which limit the critical currents and therefore could enable obtaining samples with j_c's suitable for applications.

Studies of the influence of neutron irradiation on different HTSC materials have been carried out by several groups. A summary of our present knowledge in this field, mainly with regard to single crystals, can be found in a review paper by Weber and Crabtree [1], a survey of the relevant literature is compiled by Ozawa [2]. The present work has been aimed at a comparison of neutron irradiation effects in the different HTSC materials, mainly ceramic materials, melt textured growth materials and single crystals. All the measurements discussed below concern samples irradiated at room temperature, the values of fluence were estimated with an accuracy of 10 % and they correspond to the neutron energies above 1 MeV. For a comparison with the experiments using $E > 0.1$ MeV the fluences for $E > 1$ MeV should be multiplied by a factor of about 2.20 and for a comparison with the experiments using $E > 0.5$ MeV the fluences for $E > 1$ MeV should be multiplied by a factor of about 1.44.

2. Ceramics

In the case of ceramics we can distinguish two critical currents - the intragrain critical current which corresponds to the critical current of single crystals (bulk or thin film), and the intergrain critical current limited by the grains connections. It has to be borne in mind that an increase of the intragrain critical currents (as deduced from magnetic measurements) after neutron irradiation is in general not accompanied by an increase of the intergrain currents, determined from the transport j_c measurements. This is caused by the fact that neutron irradiation enhances flux pinning inside the grains, but damages the intergrain junctions at the same time. However, apart from the weak links problem, magnetic measurements on ceramics can tell us to what extend the intragrain j_c's can be enhanced by neutron irradiation and what changes of temperature and field dependence of j_c will occur under these conditions. Other points of interest, which can be covered by this type of experiments, are the effects of neutron irradiation on T_c, the normal state resistance as well as on the irreversibility line. Such studies enable establishing the general features of the response of HTSC compounds to neutron irradiation. However, of course the anisotropy of the measured properties could not be inferred from these experiments.

Studies of neutron irradiation effects on ceramics have been carried out, by several groups, on different HTSC compounds, namely LaSrCuO, e.g. ref. 3, YBaCuO, e.g. ref. 4-9, Bi(Pb)SrCaCuO, e.g. ref. 10-11 and also BaKBiO, e.g. ref. 12. In the present paper, on the base of the earlier works, the influence of fast neutron irradiation on the magnetic and transport properties of $YBa_2Cu_3O_{7-x}$ and $Bi_{0.7}Pb_{0.3}SrCaCu_{1.8}O_x$ is summarized. Also the comparison of irradiation effects on the 60 K- and the 90 K-phase of YBaCuO reported in ref. 13, is discussed. In this case, because of the essential role of the oxygen in YBaCuO compounds, the comparative study of radiation effects in both phases are worthwhile.

The influence of irradiation on T_c for the 90 K-phase of YBaCuO and BiPbSrCaCuO samples was studied in a rather wide range of fluences, namely from $4.0 \cdot 10^{16}$ up to $3.3 \cdot 10^{18}$ n/cm^2 (YBaCuO) and from $3.3 \cdot 10^{16}$ up to $3.0 \cdot 10^{18}$ n/cm^2 (BiPbSrCaCuO). Critical temperature drops more abruptly for BiPbSrCaCuO system (for unirradiated sample $T_c = 108 \pm 1$ K) than for the 90 K-phase of YBaCuO ($T_c = 92.3 \pm 0.5$ K), the decrease of T_c is about 10.3 K per 10^{18} n/cm^2 and about 2.4 K per 10^{18} n/cm^2, respectively. For the 60 K-phase of YBaCuO a range of investigated fluences was narrower, from $5.0 \cdot 10^{16}$ to $1.5 \cdot 10^{17}$ n/cm^2 and it seems that T_c drops quicker than for the 90 K-phase. The lower critical field determined from a linear part of $M(H)$ at 4.2 K decreases with fluence both for YBaCuO and BiPbSrCaCuO samples. In the first case H_{c1}, 380 ± 50 Oe for unirradiated sample, drops down to 210 ± 50 Oe after irradiation with fluence of $8.7 \cdot 10^{17}$ n/cm^2. For BiPbSrCaCuO H_{c1}, 600 ± 90 Oe before irradiation, decreases to 300 ± 50 Oe after irradiation with fluence of $3.0 \cdot 10^{18}$ n/cm^2. The magnetic and the magnetooptic measurements carried out on $Bi_2Sr_2CaCu_2O_x$ single crystals, ref. 14, confirm also the decrease of H_{c1} with fluence. Since the H_{c1}, the penetration depth, λ, and the correlation length, ξ, are connected through the formula:

$$H_{c1} = \Phi_0 \cdot \ln\varkappa / 4\pi\lambda, \tag{1}$$

where $\varkappa = \lambda/\xi$, one could not exclude a priori the changes in the parameters λ and ξ after neutron irradiation. Studies of irradiation effects on the penetration depth λ are currently under way in our group. However, it is also possible, that in fact the surface barrier (Bean-Livingston barrier), see ref. 15, is affected by irradiation instead of H_{c1}. This problem was discussed by Konczykowski during his presentation at IW CCL in Zaborów [16].

The influence of irradiation on the intragrain critical current density could be easily inferred from $M(H)$ hysteresis loops. However, an accurate determination of the absolute values of j_c is difficult for ceramic samples. The main reason is that the exact information on the sizes and shapes of grains (diameters of supercurrents loops) would be needed. Since the sizes of the grains were not controlled in our processing technology, one can use only approximate mean diameters. In the case of ceramics, data could be only analyzed in terms of the isotropic Bean model which introduces additional uncertainties because of the anisotropic nature of these compounds. With respect to irradiation effects the most reasonable way is to compare the widths of the hysteresis loops before and after irradiation, i.e. $\Delta M^{irr}/\Delta M^{unirr}$, which are proportional to the corresponding fractional change of j_c, i.e. j_c^{irr}/j_c^{unirr}. Comparison of the influence of irradiation on the 90 K-phase of YBaCuO and on BiPbSrCaCuO ceramics at different temperatures is shown in Table 1.

As it was discussed above one can make only a rough comparison of j_c's of these two compounds by assuming that the mean diameter of the supercurrent loops is the same in each case. For instance, if we take into account an average grain size of 10 μm for the 90 K-phase of YBaCuO samples irradiated with fluence of $6.1 \cdot 10^{17}$ n/cm^2, j_c is $4.2 \cdot 10^6$, $1.3 \cdot 10^6$ and $8.4 \cdot 10^4$ A/cm^2 at 4.2, 42 and 77 K, respectively. For BiPbSrCaCuO samples irradiated with the fluence of $4 \cdot 10^{17}$ n/cm^2, j_c is $6.3 \cdot 10^6$ A/cm^2

at 4.2 K and $3\cdot10^5$ A/cm^2 at 42 K. It should be pointed out that on the contrary to the 90 K-phase of YBaCuO, in the case of BiPbSrCaCuO samples, at 77 K, even after irradiation a reversible behavior is observed in the hysteresis loops in higher fields. The similar situation is observed at 40 K for the 60 K-phase of YBaCuO. In addition, j_c's decrease more quickly with temperature and field, what is in agreement with the fact that the irreversibility line is shifted to the region of lower fields and temperatures in comparison with the 90 K-phase. This problem and the influence of irradiation on the position of irreversibility line will be discussed below.

Table 1.
The influence of irradiation on T_c and the intragrain critical currents measured at 4.2, 42 and 77 K in magnetic field of 1 T for polycrystalline YBaCuO and BiPbSrCaCuO systems.

				$YBa_2Cu_3O_{7-\delta}$		
	T (K)	fluence (10^{18} n/cm^2)				
		0.18	0.37	0.61	2.7	3.3
j_c^{irr}/j_c^{unirr}	4.2	1.8	1.8	2.0	1.6	1.7
	42	7.0	6.4	11.7	5.1	5.7
	77	9.9	5.9	9.7	0.4	0.2
ΔT_c (K)		-0.9	-1.4	-2.2	-6.6	-8.4
				$Bi_{0.7}Pb_{0.3}SrCaCu_{1.8}O_x$		
	T (K)	fluence (10^{18} n/cm^2)				
		0.033	0.16	0.4	0.66	3.0
j_c^{irr}/j_c^{unirr}	4.2	1.2	1.8	2.0	1.9	0.9
	42	2.0	7.6	20.5	13.0	0.5
ΔT_c (K)		0	-3.0	-4.1	-6.5	-31.0

One could noticed that the influence of irradiation is rather similar in the case the 90 K-phase of YBaCuO and BiPbSrCaCuO, namely increase of j_c's is rather significant (for fluences up to $6.6\cdot10^{17}$ n/cm^2) and particularly large at higher temperatures and lower fields. In another experiment, ref. 13, the 60 K- and 90 K-phase of YBaCuO ceramics have been subjected to fast neutron irradiation up to the fluence of 0.35, 0.76 and $1.05\cdot10^{17}$ n/cm^2. The effects of irradiation were found to be significantly different in the two phases. Whereas the enhancement of j_c is very pronounced in the 90 K-phase (even up to the factors of 40 at 77 K and 1 T) and particularly large in higher temperatures and lower fields, the increase of j_c is rather moderate in the 60 K-phase (factors of 1.5 to 5) and only weakly temperature and field dependent. The sensitivity of the irreversibility line to the nature and density of radiation-induced defects is of considerable interest at present - compare for example ref. 17 and 18. The influence of irradiation on irreversibility line in the 60 K- and the 90 K-phase of YBaCuO was reported in ref. 13. Irreversibility lines were studied by means of ac- susceptibility measurements (0.1 mT, ν = 21 Hz). Although the irreversibility temperature at constant field is usually defined through the peak in the imaginary component of susceptibility χ'', in the case of discussed measurements the tangents of real component χ' were chosen due to the following reasons. In ceramics the χ'' component usually shows two peaks which are related to the intra- and the intergrain currents, respectively. After irradiation, due to destruction of weak links between grains, this problem will be discussed below, the intragrain peak is masked by the continuously broadening intergrain peak. On the other hand the χ' signal remains reasonably sharp and could be used for determination of a position of the irreversibility line with sufficient accuracy. The results for both the 60 K- and the 90 K-phase of YBaCuO are shown in figure 1. Following neutron irradiation a large shift of irreversibility line to higher fields and temperatures in the 60 K-phase is observed, whereas this effect is very small in the 90 K-phase. Comparison of the influence of irradiation on increase of j_c and position of the irreversibility line in the 90 K- and the 60 K-phase shows that the shift of the irreversibility line is obviously not correlated with the enhancement of the critical currents. Presumably it means that, created defects have the pinning potentials similar to the potentials of defects existing before irradiation in the case of the 90 K-phase, and greater than defect potentials before irradiation in the case of the 60 K-phase. However, further measurements of magnetization relaxation, currently underway, are needed for a better understanding of this problem. The different response of the both phases to the same fluences means that same defects introduced into the 60 K-phase are not efficient pining centers, whereas they are efficient in the case of the 90 K-phase. This fact can indicates that coherence length, ξ, in the 60 K-phase is bigger in comparison with the 90 K-phase.

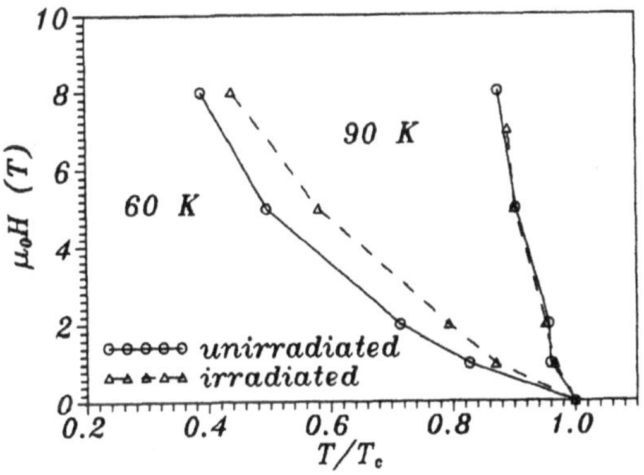

Figure 1. Irreversibility lines for the 60 K and the 90 K-phase of polycrystalline $YBa_2Cu_3O_{7-\delta}$ before and after fast neutron irradiation with the dose of $0.76 \cdot 10^{17}$ n/cm^2.

A characteristic feature of ceramic samples is the existence of low field hysteresis loops, far below H_{c1} of grains. This low field irreversibility can be explained as a result of flux trapping in intergrain connections and of decoupling of grains in magnetic fields in which shielding current flowing through the weak links exceeds its critical value. Figure 2 presents the field dependence of the susceptibility σ/H_{int} of sintered YBaCuO sample with typical grains size of 30 μm and the differential susceptibility of pulverized sample when every grain was mechanically separated forming a powder with 30 μm grains. In the first case we measure the shielding current of the whole sample while in the second case the sum of the shielding currents of separated grains. This provides the difference of mass susceptibility measured after zero field cooling for the bulk sample and the powder sample. The difference of mass susceptibility is a consequence of a difference of a volume of the bulk material and the grains forming such material. The lack of difference of susceptibility at fields higher than few tens of Oersteds indicates that the weak links of bulk material are disconnected by magnetic field.

We will use the name 'small' hysteresis loops for the hysteresis loops measured at magnetic fields with maximal field sufficient to break connections between grains but still smaller than H_{c1}. Such loops give us quite good qualitative and also under some assumptions quantitative information about the weak links existing in ceramics.

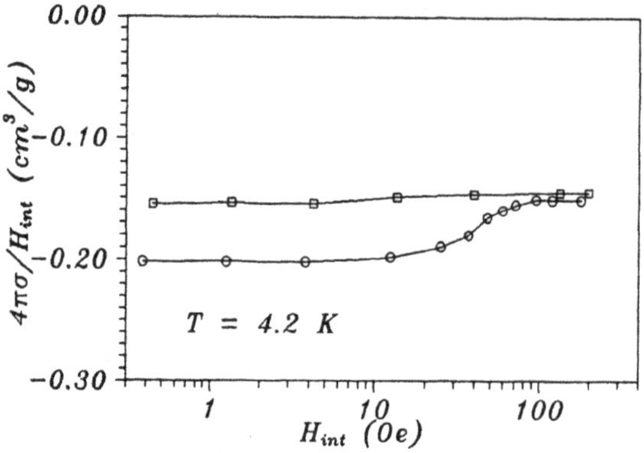

Figure 2.
The field dependence of the susceptibility σ/H_{int} of sintered YBaCuO sample with typical grains size of 30 μm and the susceptibility σ/H_{int} of pulverized sample when every grain was mechanically separated forming a powder with 30 μm grains.

The interpretation of small hysteresis loops in terms of critical state model allows to determine the intergrain critical current density, as was shown in ref. 19, and was developed by Senoussi [20, 21]. The intergrain critical current density can be easy determined from the loops, when maximal applied field is sufficient to break every or at least most of grains connections but is still smaller than intrinsic H_{c1}. The intergrain j_c determined from such magnetic experiment is proportional to the difference of magnetization for decreasing M^- and increasing magnetic field M^+. For determination of j_c it is necessary to separate the grain boundary contribution from measured *M-H* dependence. This can be done by extracting the superconducting grains contribution (equal χH at fields below intrinsic H_{c1}) from *M-H* loop. *M'-H* loop, where *M'* is the magnetization of grains boundary, defined as:

$$M' \cdot V' = (M - \chi H) \cdot V \qquad (2)$$

and *V'* and *V* are volume of grains boundary and total volume of sample, respectively, corresponds to the intergrain current only. The grain boundary contribution to the magnetization is equal to the difference of measured magnetization and the magnetization of superconducting grains (χH) divided by a ratio of the volume of grains boundary to the volume of the sample. This ratio is equal to the difference between ideal diamagnetic susceptibility and measured volume susceptibility χ of ceramic sample at field sufficient to break connections between grains but below intrinsic H_{c1}. Thus,

$$M' = (M - \chi H)/(1 + 4\pi\chi). \tag{3}$$

From such M'-H loops we can calculate the critical current density using the Fietz formula:

$$j_c = 30\cdot(M'^{-} - M'^{+})/d, \tag{4}$$

where, in this case, d is the diameter of the sample. The critical current determined in such a way corresponds quite well to values obtained from transport measurements. The absolute value and field dependence of intergrain critical current are quite sensitive to the quality of grains connections. For example, figure 3 presents field dependence of magnetically determinate intergrain critical current for two ceramic $YBa_2Cu_3O_\delta$ samples with different density (higher density corresponds to better grains connections and in consequence to higher critical current).

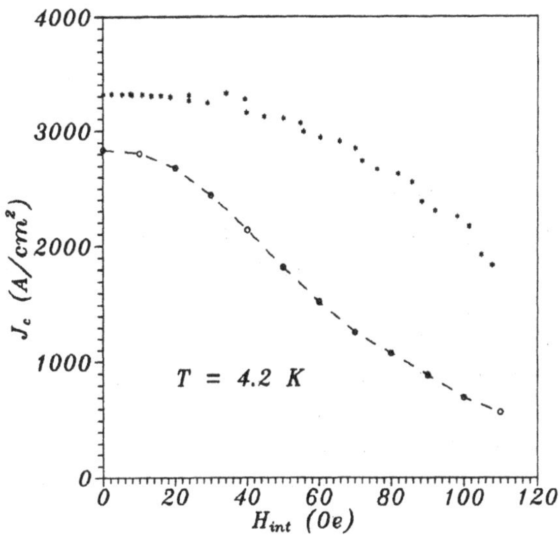

Figure 3.
The field dependence of the intergrain critical current determined from the magnetic measurements for two ceramic $YBa_2Cu_3O_\delta$ samples with different density (higher density corresponds to better grains connections and in consequence to higher critical current).

The apparent decrease of 'small' hysteresis loops with raising fluence shows that the decoupling of weak links increase following irradiation. The values of j_c for YBaCuO ceramics [19] are for example following: at 4.2 K in magnetic field of 60 Oe j_c is equal 1220 A/cm^2 and 210 A/cm^2 for unirradiated sample and sample irradiated with fluence of $2 \cdot 10^{16}$ n/cm^2, respectively. These values agree well with transport measurements and the decrease of the intergrain critical current following irradiation is clearly seen. The same decrease of intergrain critical current is observed in Bi-based ceramic materials. The vanishing of 'small' hysteresis loops of ceramic BiPbSrCaCuO is presented at figure 4.

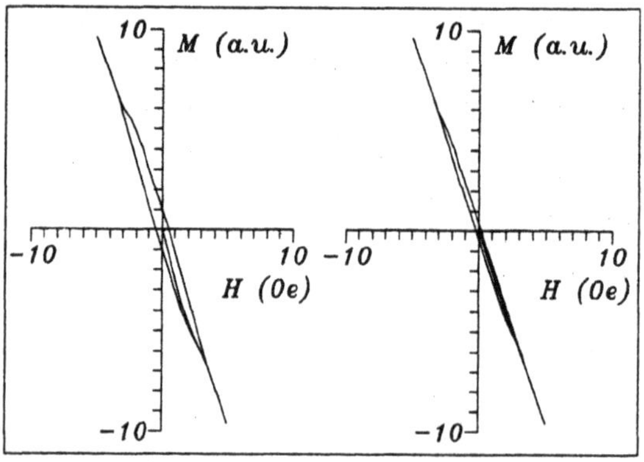

Figure 4.
Low field hysteresis loops for BiPbSrCaCuO ceramic measured at 77 K before (left) and after (right) fast neutron irradiation with the fluence of $4 \cdot 10^{17}$ n/cm^2.

The transport properties are strongly affected by irradiation. In this paper only main conclusions will be discussed, more detailed studies one can find in the previous papers [9, 22]. For YBaCuO and BiPbSrCaCuO ceramics the following features are common. In normal state resistance, R_n, systematically increases following irradiation. For example in the case of YBaCuO samples room temperature resistivity increases from 2.1 Ω·cm (the unirradiated sample) up to 6.6 Ω·cm for sample irradiated with fluence of $8.7 \cdot 10^{17}$ n/cm^2. Simultaneously, temperature dependence of the normal state resistance is more flat after irradiation. A ratio of the resistance at room temperature and at 100 K R_{rt}/R_{100K} is 2.5 for the unirradiated sample and 1.5 after irradiation of $8.7 \cdot 10^{17}$ n/cm^2. In the case of BiPbSrCaCuO system R_n increases also systematically with raising fluence, e.g. at 180 K from 3.5 Ω·cm for the unirradiated sample to 7.5 Ω·cm for fluence of $3.0 \cdot 10^{18}$ n/cm^2. Since the Josephson coupling energy E_j is inversely proportional to the normal state resistivity the increase of R_n is of course connected with decrease of the critical temperature as determined from transport measurements. The zero resistance of sample occurs when Josephson coupling energy is comparable or greater then the thermal energy. Thus, for lower values of E_j the condition $E_j \geq k_B T$

is fulfilled for lower temperatures.

The most important result of the transport measurements is the direct evidence that the intergrain critical current decrease with increasing fluence as the consequence of a damage of the weak links by irradiation. The field dependence of j_c shows a remarkable decrease even at low magnetic fields. For example, in figure 5 for BiPbSrCaCuO samples a decrease of j_c of about one order of magnitude between zero field and 5 Oe at 77 K is shown. It may be noticed, when one compare figures 4 and 5 that the disappearance of the field irreversibility at about 5 Oe and 77 K corresponds well to the results of transport measurements.

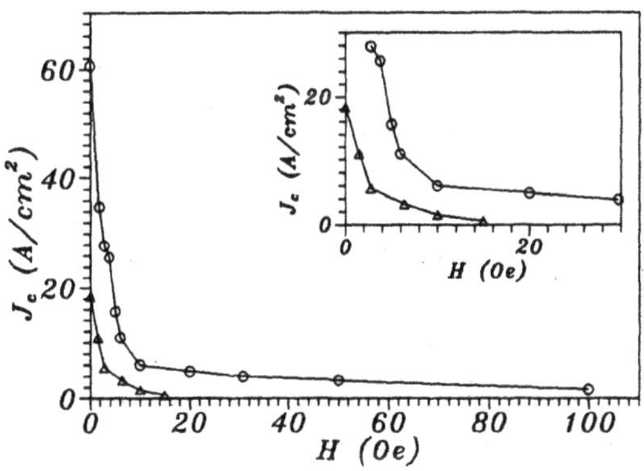

Figure 5. The field dependence of the transport critical current of polycrystalline BiPbSrCaCuO at 77 K for unirradiated sample (circles) and for sample irradiated with neutron fluence of $4\cdot10^{17}$ n/cm^2 (triangles).

An attempt was made to check the influence of fast neutrons on structural properties of YBaCuO ceramics irradiated with fluence between $4.0\cdot10^{16}$ and $8.7\cdot10^{17}$ n/cm^2 (ref. 23). Investigations of the average crystal structure by X-ray and neutron powder diffraction show as an essential change a linear increase of the lattice constants *a*, *b* and *c* for small values of fluences and a decrease at higher fluences. A nearly isotropic variation of all three lattice parameters was observed for fluences below $1.2\cdot10^{17}$ n/cm^2. Any other changes of the microscopic crystallographic parameters (occupancies and positional parameters) were noticed to be smaller than the statistical errors or the systematic differences between the X-ray and the neutron results.

As it was discussed above irradiation caused opposite effects on the intra- and the intergrain properties of HTSC ceramics. It seems that one could find some region of neutron fluences in which it is possible to enhance the intragrain critical currents up to about one order of magnitude or more (in fields of order of 1 T) without significant drop of the critical temperature. On the other hand radiation damages the weak links, what is evidenced by a reduction of the transport critical current and an increase of the normal state resistivity. So from the point of view of applications the only perspective materials are those in which weak links do not exist or are not dominated at least.

It should to be mentioned that $Ba_{1-x}K_xBiO_3$ (ref. 12) compounds behave differently from the other HTSC superconductors since the intragrain critical currents are not affected by irradiation up to fluence of $1.4 \cdot 10^{18}$ n/cm^2.

3. Melt textured materials

Recently, a significant progress has been achieved with regard to the enhancement of the transport critical current density in bulk YBaCuO and BiSrCaCuO materials. The transport critical current of sintered superconductors is limited by intergrain weak links of these materials. The intergrain critical currents of polycrystalline materials can be enhanced by improving the quality of grains connections. The weak links limitations existing in HTSC ceramics produced by standard methods are in fact caused by the following reasons: a random orientation of grains, porosity (density of typical ceramics is significantly lower than theoretical one), local and global deficiency of oxygen, existence of secondary phases. These limitations have been overcome in different melt textured processes. Samples produced in these processes are very dense, grains of needles like shapes are oriented (*a-b* plane along the longest sample dimension).

One can distinguish three types of melt approaches:

- melt textured growth (MTG) process, proposed by Jin et al. [24]. In this process precursor $YBa_2Cu_3O_{7-x}$ sample is heated up to 1200 °C (decomposition to Y_2BaCuO_5 and liquid phase). The melted sample is next cooled down in a temperature gradient of at least 50 °C/cm. Values of j_c at 77 K and 1 T reported in the first papers [24, 25] are about 4000 A/cm^2 (measured by pulse technique, using 2-20 μV/cm voltage criterion). Some modification of this process was proposed by Salama et al. [26]. Namely, proper microstructure of the sample is controlled by very slow cooling (about 1 °C/hour). The values of j_c are about 37 000 A/cm^2 at 77 K and 0.6 T (pulse technique, 1 ms pulse width, 50 μV noise level).
- quench and melt growth (QMG) process, proposed by Murakami et al. [27]. This process consists of the three stages. In the first stage $YBa_2Cu_3O_{7-x}$ precursor sample is rapidly heated up to about 1400 °C (decomposition to Y_2O_3 and liquid

phase) and then quenched. In the second stage sample is heated once more up to 1100 °C (decomposition to Y_2BaCuO_5 and liquid phase). In the last stage sample is slowly cooled in a temperature gradient (about 10 to 20 °C/cm). Samples produced by this methods exhibit j_c exceeding 10 000 A/cm^2 at 77 K and 1 T.

- powder melting process (PMP) proposed by Zhou Lian et al. [28, 29]. In this approach instead of $YBa_2Cu_3O_{7-x}$ powders of Y_2BaCuO_5 and $BaCuO_2$ are used as precursors. After a careful mixing of these powders (homogeneous distribution of Y_2BaCuO_5 particles turns out to be very crucial for obtaining textured samples of high quality) they are heated below 1040 °C and then cooled down. Best samples obtained using this method exhibit j_c of about 70 000 A/cm^2 at 77 K and 1 T and 35 000 A/cm^2 at 4 T (pulse technique, pulse duration 5 ms, voltage criterion 10 μV/cm).

Significant increase of the critical current density observed in melt textured materials in comparison with the standard ceramic materials results from the improvement of the grains connections. However, it is necessary to distinguish two possibilities:

- grains in melt textured materials are better connected than in ceramics materials, but they still can be disconnected by high magnetic field (e.g. 1 T)
- grains are connected really very well and they can not be disconnected by magnetic field.

The second possibility only give us a real improvement of the quality of the material and only in this case the significant progress in applications can be made. Thus, it is very important to know whether really the weak links are absent in melt textured samples. The simplest way to check it is to find if the linear dependence between the magnetization irreversibility and sample size is observed in magnetic field. If this is a case, j_c determined from the magnetization measurements corresponds to the critical current density throughout the whole sample.

We have investigated melt textured samples prepared by powder melting process (PMP) [28, 29]. For magnetic measurements, in three different perpendicular orientations of the sample with respect to magnetic field we have chosen perpendicular parallelepiped sample with following dimensions A = 4.55 mm, B = 1.85 mm and C = 0.95 mm. The crystallographic planes of PMP YBaCuO lay in principle in A-B, A-C and B-C planes of the sample, however due to some scattering of directions of crystallographic axis c (inferred from X-ray measurements) none of the sample axes A, B or C does not correspond exactly to the crystallographic c-axis. For this reason calculated below j_c does not correspond to the currents which flow exactly in the crystallographic planes. After the measurements with magnetic field perpendicular to each sample plane at temperatures of 60, 77 and 85 K the sample was polished and the sizes were diminished down to $A \cdot B \cdot C$ equal $3.8 \cdot 1.6 \cdot 0.75$ mm^3 and the measurements were repeated. Then the dimensions were decreased once more to $2.05 \cdot 1.35 \cdot 0.48$ mm^3 and the same measurements were done. On the base of the obtained results it was found that the decrease of magnetization hysteresis loops is proportional to the decrease of the sample sizes for all three orientations. Apparent linear dependence of magnetization measured at 1 T and 60, 77 and 85 K is shown in figure 6 for currents circulating in *A-B*, *A-C* and *B-C* planes.

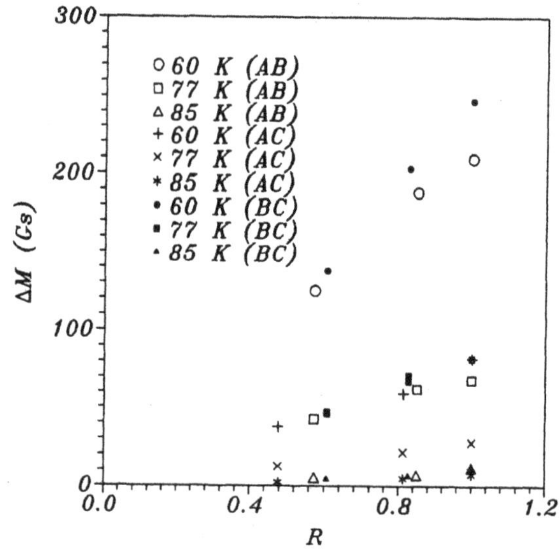

Figure 6. The irreversibility ΔM for decreasing and increasing magnetic field measured in 1 T as a function of R - ratio of geometrical average dimension to the initial geometrical average dimension of the sample wall perpendicular to applied magnetic field. *A*, *B* or *C* (mentioned in brackets) describe actual sample wall.

For comparison of magnetization j_{cm} and transport j_{ct} critical currents some longer samples were used. It was not possible to make magnetic measurements with field along the longest axis. Thus, we have estimated only average critical current j_{cm1} and j_{cm2} in two sample planes parallel to the transport current direction. The agreement between j_{ct} measured along longest axis and j_{cm} was quite good e.g. for sample with $j_{ct} = 2 \cdot 10^4$ A/cm^2, $j_{cm1} = 1.2 \cdot 10^4$ A/cm^2 and $j_{cm2} = 1.1 \cdot 10^4$ A/cm^2 and with $j_{ct} = 1 \cdot 10^4$ A/cm^2, $j_{cm1} = 8.5 \cdot 10^3$ A/cm^2 and $j_{cm2} = 7.3 \cdot 10^3$ A/cm^2.

The samples have been subjected to the fast neutron irradiation with fluence of

$1.32 \cdot 10^{17}$ n/cm^2. It should to be pointed out that lack of weak links before irradiation does not assure of their absence after irradiation. Firstly, following irradiation regions on the borders of grains could be degraded. Secondly, after increasing of the intragrain current due to additional pinning centers inside the grains regions on the border of grains could not carry this higher j_c. Therefore, it should be check out once more whether weak links are absent in the irradiated samples. For this reason measurements on the irradiated sample which dimensions have been decreased in similar way as was discussed above were performed. First measurements were performed for the sample dimensions $A \cdot B \cdot C = 5.3 \cdot 1.6 \cdot 0.9$ mm^3, second after diminishing the dimensions down to $4.0 \cdot 1.2 \cdot 0.6$ mm^3, third for the dimensions $1.6 \cdot 1.0 \cdot 0.3$ mm^3. It was found again that a decrease of magnetization hysteresis loops is proportional to a decrease of sample sizes for all three orientations. We have noticed increase of a width of hysteresis loops by a factor of about 3 after neutron irradiation (see figure 7). Because of the lack of the weak links also after irradiation one could believe that the transport currents of at least $5 \cdot 10^4$ A/cm^2 would flow. Irradiations with other fluences are currently underway and we believe that for optimal fluence j_c of 10^5 A/cm^2 could be achieved.

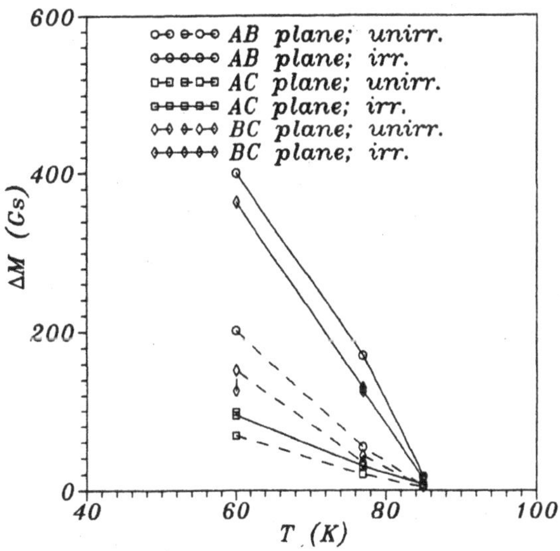

Figure 7. Comparison of the hysteresis width in the magnetic field of 1 T for melt textured YBaCuO unirradiated and irradiated with the fluence of $1.32 \cdot 10^{17}$ n/cm^2.

Until now only two papers concerning effect of fast neutron irradiation of melt-textured samples, were published. Küpfer et al. [30] have investigated YBaCuO samples prepared by Salama et al. [26]. They found that the intragrain j_c increases after irradiation with a fluence of 10^{17} n/cm^2 ($E > 1$ MeV) up to a factor of 4 reaching value $2.7 \cdot 10^4$ A/cm^2 when magnetic field is perpendicular to *a-b* plane and $5.3 \cdot 10^3$ A/cm^2 for field parallel to this plane at 77 K and 2 T. In contrast to the results obtained on the single crystals this enhancement was almost independent on a field orientation. The decrease of current with field was noticed to be much more pronounced after irradiation. The increase of j_c was accompanied by a decrease of the activation energy U_o in the field and temperature region at which U_o passes through a maximum. Following irradiation the irreversibility lines, for both orientations, shift to the region of higher temperatures and fields. The measurements have indicated that the weak links were not introduced during the irradiation. This last conclusion was inhered from the absence of a deviation from uniform current flow in irradiated samples.

Hor et al. [31] have investigated YBaCuO samples, obtained using some modification of the technique proposed by Salama et al. [26], irradiated up to fluence of $6.0 \cdot 10^{17}$ n/cm^2. They have observed an increase of the intragrain j_c up to e.g. $2 \cdot 10^5$ A/cm^2 at 77 K and 1 T and also very pronounced decrease of critical current anisotropy. It was even noticed that ratio of the critical current in *a-b* plane to critical current along *c*-axis $j_c^{ab}/j_c^{c} = 1$ at 77 K and 4.8 T. It was found that the enhancement of j_c^{ab}, after irradiation, is more effective at lower field (< 2 T). In spite of j_c increase the activation energy remains almost unchanged. It was shown that annealing up to 600 ^{o}C removes neutron induced pinning centers and recovers pre-irradiated state.

It is also worthwhile to notice very important results obtained by Neumüller et al. [32] on melt processed $Bi_2Sr_2CaCu_2O_8$ samples irradiated with 400 MeV oxygen ions. They have found that transport critical current increases after irradiation and this effect is more pronounced at higher temperatures. Magnetization irreversibility has shown in contrast a marked increase at lower temperatures. They have also found that irradiation induced defects have activation energies close to that of the intrinsic defects.

4. Thin films

According to our best knowledge the only one experiment has been carried out on neutron irradiated thin films by Schindler et al. [33]. The measurements were performed on the epitaxial, *c*-axis oriented, thin films of $YBa_2Cu_3O_{7-x}$ prepared by laser ablation on $SrTiO_3$ substrates. The samples were irradiated with fast neutrons ($E > 0.1$ MeV) up to fluence of $2 \cdot 10^{18}$ n/cm^2 (for comparison with experiments using E

> 1 MeV Schindler et al. suggest that the fluences for $E > 0.1$ MeV should be divided by a factor of about 3). The most important result of their experiments was that transport critical current (measured resistively) increased after neutron irradiation. For a moderate irradiation fluence of $4.9 \cdot 10^{17}$ n/cm^2 the enhancement of j_c in all magnetic fields of up to 8 T and at temperature of 4.2, 60 and 77 K was noticed. For example at 4.2 K initial $j_c = 2 \cdot 10^7$ A/cm^2, with very weak dependence of only a few percent on magnetic fields up to 8 T increased after this fluence in vanishing magnetic field by 50 % and in magnetic fields up to 8 T by 300 %. The apparent anisotropy of j_c enhancement was noticed. The measurements of j_c at angles of 0, 30, 60 and 90 degrees between the *c*-axis and the *B* vector showed that the greatest enhancement was observed for *B* parallel to *c* axis.

It should be pointed out that the ion irradiation using 25 MeV ^{16}O-ions performed on the same epitaxial thin films [34] showed an increase of j_c only in magnetic fields.

5. Single crystals

The results of fast neutron irradiation on single crystals were discussed in details by H.W. Weber during his presentation at IW CCL in Zaborów [35]. In our presentation we discuss the influence of fast neutron irradiation on field and temperature dependence of anisotropic critical current.

A crystal of a nominal composition $Bi_2Sr_2Ca_1Cu_2O_x$ used for this study was grown by Dr. A. Balbashov at the Moscow Power Institute by floating zone technique using an arc image furnace with Xenon lamp [36]. Resistivity measured by a four-terminal method reached zero value at 85 K. The sample used for magnetic measurements contained two pieces of crystal joined together to increase the magnetic moment. Both pieces had the form of parallelepiped of approximate dimensions $2.74 \cdot 2.74 \cdot 0.51$ mm^3. The *c*-axis was perpendicular to sample plane ($2.74 \cdot 2.74$ mm^2). The sample shows no low-temperature anomalies which might indicate extensive second-phase regions.

The *dc* magnetic measurements were carried out using vibrating sample magnetometers (VSM) PAR-155. Magnetization of zero field cooled sample as a function of applied magnetic field was measured at temperatures range 4.2 - 55 K in field up to 40 kOe at 4.2 K and up to 16 kOe at higher temperatures. The demagnetizing field was taken into account for internal magnetic field, H_{int}, and magnetic induction, *B*. The demagnetizing factor *D* of our sample calculated assuming an equivalent cylinder was equal $4\pi \cdot 0.546$ for measurements in magnetic field along *c*-axis and $4\pi \cdot 0.227$ for measurements in magnetic field in *(a-b)* plane. From magnetization data the principal-axis components of critical current were calculated. The measurements described above were performed for unirradiated BiSrCaCuO crystal and after irradiation with fluence of $1.5 \cdot 10^{17}$ n/cm^2.

For magnetic field H applied along *c*-axis we assume that the components j_c^a and j_c^b are equal and will be further denoted as j_c^{ab}. In this case the formula (4) could be used for calculations of j_c^{ab}. For field perpendicular to *c*-axis two different components j_c^{ab} and j_c^c exist. The critical current j_c^c along *c*-axis usually is much lower than critical current j_c^{ab} along *a,b*-axes of crystal [37, 38, 39]. Therefore, when the sample is not too flat, for magnetic field H applied in *(a-b)* plane of a sample the measured ΔM is limited by j_c^c. It is possible to determine the critical current j_c^c along *c*-axis of a sample from formula:

$$j_c = 20 \cdot \Delta M / d, \tag{5}$$

where d is dimension of a sample in *a-b* plane.

It was found that the field dependence of both principal-axis components of critical current, $j_c^{ab}(B)$ and $j_c^c(B)$, at any temperature is described the best by the exponential relation:

$$j_c(T, B) = j_c(T, B{=}0) \cdot \exp(-B/B_0(T)) \tag{6}$$

The temperature dependence of B_0 is quite strong. We found that $B_0(T)$ is described by the exponential function:

$$B_0(T) = B_0(T{=}0) \cdot \exp(-T/T_0). \tag{7}$$

The $j_c(B{=}0)$ is a function of temperature and we found that is well described by the power-law dependence proposed by Agostinelli et al. [40]:

$$j_c(T, B{=}0) = j_c(T{=}0, B{=}0) \cdot (1 - T/T_c)^A. \tag{8}$$

Similar analysis of field and temperature dependencies of critical currents in melt processed BiSrCaCuO was done by Neumüller [41].

Finally we fitted four parameter function

$$j_c(T, B) = j_c(T{=}0, B{=}0)\cdot(1 - T/T_c)^A\cdot\exp(-B/(B_0(T{=}0)\cdot\exp(-T/T_0))) \qquad (9)$$

to the full set of our experimental data (approximately 600 points at 7 or more temperatures at each case) for both principal-axis components of the critical current for unirradiated and irradiated BiSrCaCuO crystal. The best fitted parameters of the function (9) A, B_0 and T_0, describing temperature and field dependence of the critical current, are presented in table 2. The quality of the fit R-squared was better than 0.994 at any case. The T_c = 86 K, used in calculations, was determined from temperature dependence of magnetization of field cooled sample at external field of 10 Oe.

Table 2.
The best fitted parameters of the function (9), describing both principal-axis components of temperature and field dependence of the critical current $j_c^{ab}(T, B)$ and $j_c^{c}(T, B)$ of $Bi_2Sr_2Ca_1Cu_2O_x$ crystal.

	A	$B_0(T{=}0)$ (Gs)	T_0 (K)
unirradiated sample j_c^{ab} in *(a-b)* plane	7.6	118000	5.3
irradiated sample j_c^{ab} in *(a-b)* plane	5.6	64000	7.3
unirradiated sample j_c^{c} along *c*-axis	5.5	140000	8.2
irradiated sample j_c^{c} along *c*-axis	5.5	70000	18.4

One can see that the temperature and the field dependence of the critical current for both j_c^{ab} and j_c^{c} principal-axis components is well described by the same function. Also the same function could be applied after fast neutron irradiation even when significant changes of critical current density are observed. The j_c anisotropy is reduced after fast neutron irradiation. We observe the enhancement of the field dependence of critical current of the irradiated samples at higher temperatures. It can be easy demonstrated at the plot of critical current density j_c as a function of magnetic induction for unirradiated and irradiated sample. Figure 8 presents field dependence of the critical current j_c^c along c-axis of unirradiated and irradiated BiSrCaCuO crystal at 28 K.

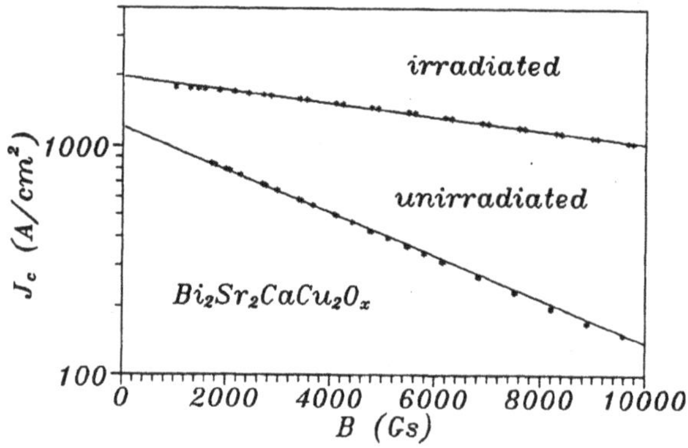

Figure 8. The critical current, j_c^c, along c-axis of BiSrCaCuO crystal unirradiated and irradiated with the fluence of $1.5 \cdot 10^{17}$ neutrons/cm^2 as a function of the magnetic induction, *B*, at temperature of 28 K.

In fact the field dependence of the critical current density is controlled by the function $B_0(T)$. The higher value of this function at given temperature corresponds to weaker field dependence of the critical current density at this temperature. Thus, the influence of irradiation on the field dependence of the critical current density can be calculated at different temperatures and presented at the plot of $B_0^{irr}/B_0^{unirr}(T)$. The values of B_0^{irr}/B_0^{unirr} higher then 1 correspond to the improvement of the field dependence of critical current density as a result of irradiation. Figures 9 and 10 present the temperature dependence of the ratio of B_0^{irr}/B_0^{unirr} for both principal-axis components j_c^{ab} and j_c^c of $Bi_2Sr_2Ca_1Cu_2O_x$ crystal. It is important to notice, that below some temperature for both critical currents j_c^{ab} and j_c^c the value of function B_0^{irr}/B_0^{unirr} was lower than 1 so, also the increase of field dependence of j_c after fast neutron irradiation was observed. In certain range of lower temperatures the influence of irradiation on the critical current in high magnetic fields is rather destructive. This could be compared with results of the reference 30 concerning the influence of fast neutron irradiation on critical current in YBaCuO, where the improvement of field dependence of critical current was not observed. Probably, in this case, the characteristic temperature above which the field dependence of critical current is improved (critical current is less field dependent) is quit high. For BiSrCaCuO this characteristic temperature is much lower than T_c and, as it is presented at figures 10 and 11, is equal to 11.8 K for critical current circulating in *(a-b)* plane and to 10.3 K for critical current along c-axis.

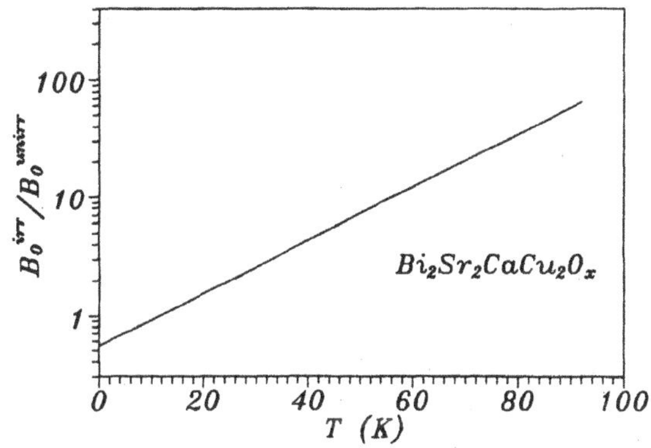

Figure 9. The temperature dependence of B_0^{irr}/B_0^{unirr} for the critical current $j_c^{ab}(B)$ in $(a\text{-}b)$ plane of superconducting $Bi_2Sr_2Ca_1Cu_2O_x$ crystal.

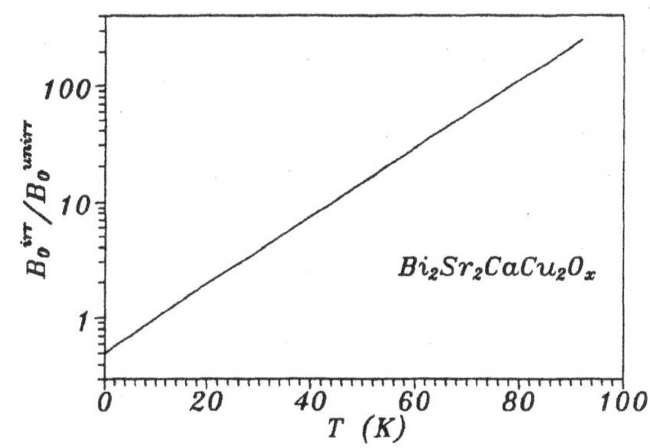

Figure 10. The temperature dependence of B_0^{irr}/B_0^{unirr} for the critical current $j_c^{c}(B)$ along c-axis of superconducting $Bi_2Sr_2Ca_1Cu_2O_x$ crystal.

In conclusions, both principal-axis components of critical current $j_c^{ab}(B)$ and $j_c^{c}(B)$ of superconducting $Bi_2Sr_2Ca_1Cu_2O_x$ crystal in $(a\text{-}b)$ plane and along c-axis are described by exponential-law model at full investigated range of temperatures. The temperature dependence of coefficients $j_c(T, B{=}0)$ and $B_0(T)$ are described by relations $j_c(T, B{=}0) = j_c(T{=}0, B{=}0)\cdot(1 - T/T_c)^A$ and $B_0(T) = B_0(T{=}0)\cdot\exp(-T/T_0)$. The exponential-law model for both principal-axis components of critical current is still true even when significant changes of critical current versus temperature and magnetic field are observed for a crystal irradiated by fast neutrons.

6. Conclusions

One can noticed looking on a 'history' of HTSC that, during first 2 years after discovery of Müller and Bednorz, majority of the activity was concentrated on attempts to achieve new compounds with higher T_c. This is of course still very important. But another problem: namely how to produce HTSC materials with critical currents density of order of 10^5 A/cm^2 or more at liquid nitrogen temperature begin to be in a center of interest of many groups. The problems connected with all mechanisms which control critical currents in superconducting materials have not been solved even in respect to conventional type II superconductors. In the case of HTSC compounds situation seems to be much more complicated. Any defects with sizes comparable to coherence length ξ could, in principle, play a role of pinning centers. In HTSC materials ξ is of order od angstroms or tens of angstroms so there is really a variety of structural defects which could contribute to pinning mechanism. Neutron irradiation is a very useful method which enables one to introduce defects into material in a controlled way. This helps us to improve our understanding of the nature of pinning centers and all related phenomena. With regard to applications of HTSC materials it was shown that it is possible to enhance j_c, after neutron irradiation in materials without weak links. Especially, melt-textured materials seem to be promising from the practical point of view. It is also possible to improve $j_c(T,B)$ characteristic what is very important for applications.

Acknowledgements

The authors would like to thank Dr. M. Wójcik for critical reading of the manuscript.

References

1. H.W. Weber and G.W. Crabtree, *Neutron Irradiation Effects in High-T_c Single Crystals*, in *Studies of HTSC*, vol.9, ed. A.V. Narlikar (Nova Science Publishers, New York 1991).
2. K. Ozawa, *A Bibliography of Irradiation Effect on High-T_c Superconductors* (1990), Energy Research Laboratory, Hitachi, Ibaraki, Japan.
3. S.T. Sekula, D.K. Christen, H.R. Kerchner, J.R. Thompson, L.A. Boatner and B.C. Sales, *Jap. J. Appl. Phys.* **Suppl. 26-3** (1987) 1185.

4. H. Küpfer, I. Apfelstedt, W. Schauer, R. Flükiger, R. Meier-Hirmer, H. Wühl and H. Scheurer, *Z. Phys.* **B 69** (1987) 167.
5. K. Atobe and H. Yoshida, *Phys. Rev.* **B 36** (1987) 7194.
6. P. Müller, H. Gerstenberg, M. Fischer, W. Schindler, J. Ströbel, G. Saemann-Ischenko and H. Kammermeier, *Solid State Comm.* **65** (1988) 223.
7. A. Wiśniewski, M. Baran, P. Przysłupski, H. Szymczak, A. Pajączkowska, B. Pytel and K. Pytel, *Solid State Comm.* **65** (1988) 577.
8. B.A. Aleksashin, I.F. Berger, S.V. Verkhovskii, V.I. Voronin, B.N. Goshchitskii, S.A. Davydov, A.E. Karkin, V.L. Kozhevnikov, A.V. Mirmelshtein, K.N. Mikhaylov, V.D. Parhomenko and S.M. Cheshniskii, *Physica* **C 153-155** (1988) 339.
9. P. Przysłupski, A. Wiśniewski, S. Koleśnik, W. Dobrowolski, A. Pajączkowska, K. Pytel and B. Pytel, *ibid*, 345.
10. A. Wiśniewski, M. Baran, Z. Kozioł, P. Przysłupski, J. Piechota, R. Puźniak, A. Pajączkowska, M. Pękala, B. Pytel and K. Pytel, *Physica* **C 170** (1990) 333.
11. Y.H. Herr, K.H. Lee, C.J. Kim, H.G. Lee, C.T. Kim, G.W. Hong and D.Y. Won, *Jap. J. Appl. Phys.* **28** (1989) L 1561.
12. J.D. Thompson, J.R. Cost, G.H. Kwei and K.C. Ott, *Physica* **C 169** (1988) 265.
13. A. Wiśniewski, R.M. Schalk, H.W. Weber, M. Reissner, W. Steiner, J. Górecka, to be published in *Physica* **C** (*Proc. HTSC-M^2S*, Kanazawa, 1991).
14. R. Szymczak, M. Baran, K. Piotrowski and H. Szymczak, in *this volume*.
15. M.W. McElfresh, Y. Yeshurun, A.P. Malozemoff and F. Holtzberg, *Physica* **A 168** (1990) 308.
16. M. Konczykowski, presented at *IW CCL*, in *this volume*.
17. W. Kritscha, F.M. Sauerzopf, H.W. Weber, G.W. Crabtree, Y.C. Chang and P.Z. Jiang, *Europhys. Lett.* **12** (1990) 179.
18. L. Givale, A.O. Marwick, M.W. McElfresh, T.K. Worthington, A.P. Malozemoff, F.H. Holtzberg, J.R. Thompson and M.A. Kirk, *Phys. Rev. Lett.* **65** (1990) 1164.
19. K.V. Rao, R. Puźniak, D.-X. Chen, N. Karpe, M. Baran, A. Wiśniewski, K. Pytel, H. Szymczak, K. Dyrbye and J. Bøttinger, *Physica* **C 153-155** (1988) 347.
20. S. Senoussi, C. Aguillon and P. Manuel, *Physica* **C 175** (1991) 202.
21. S. Senoussi, C. Aguillon and S. Hadjoudj, *ibid*, 215.

22. P. Przysłupski, A. Wiśniewski, W. Dobrowolski, A. Pajączkowska, K. Pytel and B. Pytel, in *Physics and Materials Science of HTSC*, ed. R. Kossowsky et al. (Kluwer Academic Press, Dordrecht, 1990) p. 641.
23. J. Waliszewski, N. H. Anderson, L. Dobrzyński, J. Ihringer, B. Lebech, W. Prandl and A. Wiśniewski, *Physica* **C 160** (1989) 189.
24. S. Jin, J.H. Tiefel, R.C. Sherwood, R.B. van Dover, M.E. Davies, G.W. Kammlott and R.A. Fastnacht, Phys. Rev. B 37 (1988) 7850.
25. S. Jin, J.H. Tiefel, R.C. Sherwood, M.E. Davies, R.B. van Dover, G.W. Kammlott, Fastnacht and H.D. Keiith, *Appl. Phys. Lett.* **52** (1988) 2074.
26. K. Salama, V. Selvamanickam, L. Gao and K. Sun, *Appl. Phys. Lett.* **54** (1989) 2352.
27. M. Murakami, M. Morita, K. Doi and K. Miyamoto, *Jap. J. Appl. Phys.* **28** (1989) 1189.
28. Zhou Lian, Zhang Pingxiang, Ji Ping, Wang Kequang, Wang Jingrong and Wu Xiozu, *Supercond. Sci. Technol.* **3** (1990) 490.
29. Zhou Lian, *The Melt Process for Fabrication of High J_c YBCO Superconductors*, presented in *TMS Ann. Mtg.: 3rd Int. Symp. on High-Temp. Supercond., New Orleans, U.S.A., Feb. 17-21, 1991.*
30. H. Kupfer, C. Keller, R. Meier-Hirmer, K. Salama, V.Selvamanickam, G.P. Tartaglia, *IEEE Trans. Magn.* **MAG-27** (1991) 1369.
31. P.H. Hor, Z.J. Huang, L. Gao, R.L. Meng, Y.Y. Xue, C.W. Chu, Y.C. Jean, J. Farmer, *Mod. Phys. Lett.* **B 4** (1990) 703.
32. H.W. Neumüller, G. Ries, W. Schmidt, W. Gerhauser, S. S. Klaumunzer, *Journal of Less-Common Metals* **164 & 165** (1990) 1351.
33. W. Schindler, B. Roas, G. Saemann-Ischenko, L. Schultz and H. Gerstenberg, *Physica* **C 169** (1990) 117.
34. B. Roas, B. Hensel, G. Saemann-Ischenko and L. Schultz, *Appl. Phys. Lett.* **54** (1989) 1051.
35. H.W. Weber, in *this volume*.
36. A.M. Balbashov, E.A. Antonova, A.S. Nigmatulin, S.G. Karabashev, K.V. Kiseleva, O.V. Aleksandrov, *Superconductivity: physics, chemistry, engineering* **2** (1989) no.1, p.57.
37. T.R. Dinger, T.K. Worthington, W.J. Gallagher and R.J. Sandstrom, *Phys. Rev. Lett.* **58** (1987) 2687.
38. E.M. Gyorgy, R.B. van Dover, K.A. Jackson, L.F. Schneemeyer and J.V. Waszczak, *Appl. Phys. Lett.* **55** (1989) 283.

39. D.C. Cronemeyer, T.R. McGuire, A.P. Malozemoff, F. Holtzberg, R.J. Gambino, L.W. Conner and M.W. McElfresh, *Proc. Intl. Conf. on Transport Properties of Superconductors*, ed. R. Nikolsky (World Scientific, Singapore, 1990).
40. E. Agostinelli, G. Balestrino, D. Fiorani, P. Paroli, J. Tejada, A.M. Testa, *Proceedings of MRS Fall Meeting, Boston, U.S.A., Nov. 27- Dec. 2, 1989.*
41. H.W. Neumüller, G. Ries, J. Bock and E. Preisler, *Cryogenics* **30** (1990) 639.

CRITICAL CURRENTS AND ANGULAR DEPENDENCE IN MAGNETIC a.c. RESPONSE: TEST FOR INTRINSIC PINNING

L. Krusin-Elbaum, L. Civale, T.K. Worthington and F. Holtzberg

IBM Research, Yorktown Heights NY 10598-0218

ABSTRACT

a.c. magnetic measurement is a convenient non-contact method to probe the irreversible behavior of the high temperature superconductors. We present a simple geometrical construct allowing to extract the temperature dependence of critical currents J_c, which is independent of sample shape. The angular dependence of the a.c. response allows us to probe the *intrinsic pinning* - the pinning by the structure itself, and we show that in single crystals of YBaCuO there is a critical boundary for the onset of the intrinsic pinning. At low fields, in good quality single crystals of YBaCuO, there is a considerable reduction of pinning below a *thermal softening boundary*, at which vortex cores are "smeared" on the scale of the coherence length. This observation supports the idea that the predominant pinning centers in YBaCuO crystals are of the atomic origin.

1. INTRODUCTION

The behavior of critical currents in type II superconductors is most directly probed in transport measurements[1]. Such measurements, however, are an exceedingly difficult task in high temperature superconductors, because of the many serious problems in electrically contacting these materials through surface oxide layers, especially in small and delicate single crystals. Magnetic measurements, since they are non-contact, are convenient. But to extract the information about the critical currents J_c one pays the price of a more complex analysis and a reliance on a model relating magnetization to J_c. This is commonly done using Bean critical state model[2], which explains many features of the high-field magnetic hysteresis loops. The idea here is that in a *homogenuous* superconductor with defects, the hysteretic magnetization is related to J_c via an Ampere's law $dB/dx = (4\pi/c)J_c$, where B is the locally averaged magnetic induction inside the superconductor. In such description, the width of the magnetic hysteresis ΔM can be used to derive J_c values and in a slab geometry of thickness d and field applied along the plane of the slab, $J_c \simeq 40\Delta M/d$[Ref.1,2]. The d.c. M(H) measurements also allow determination of the onset of reversible magnetization, and when performed at various temperatures give an estimate of the "irreversibility line", $H_{irr}(T)$. Empirically, this boundary, which is considerably below the upper critical field[3] H_{c2}, separates two regimes in the Abrikosov mixed state; one in which magnetization is reversible, i.e. the magnetic vortices are capable of viscous flow and another, irreversible regime in which vortices are essentially immobile. The arrest of the vortex motion, responsible for the magnetic irreversibility, is associated with the development of critical currents and nonlinearities in the I-V curves[4,5]. $H_{irr}(T)$ was first reported in ceramic LaBaCuO[6] and subsequently confirmed in single crystals of YBaCuO[7-10] and BiSrCaCuO[10,11]. This boundary is probed with much better accuracy by the a.c. susceptibility[8,9], where the current sensitivity is $\sim$ 20 A/cm², in contrast to $\sim$ 1000 A/cm² in d.c. method[6,7,10]. Moreover, by varying other available parameters in the a.c. technique, such as frequency and the amplitude of the a.c. excitation one can examine the regime not only near and at the irreversibility line, but also of the fully developed critical state well below the irreversibility line.

In this paper we will demonstrate how to extract $J_c(T)$ in the critical state from such simple a.c. measurements and show the consequences of the anisotropic layered structure on the irreversible boundary. The a.c. technique allows us to probe the *intrinsic pinning* - the pinning by the structure itself, which enhances critical currents for the magnetic fields in ab-plane; we show that there is a critical boundary for the onset of the intrinsic pinning. We also show that in good quality single crystals of YBaCuO, there is a considerable reduction of pinning at low fields below a ***thermal softening boundary***, at which vortex cores are "smeared" on the scale of the coherence length. This observation supports the idea that the predominant pinning centers in YBaCuO crystals are of atomic origin.

2. a.c. RESPONSE - DETERMINATION OF $J_c(T)$

The technique consists of placing a sample inside a coil supplying a uniform a.c. field $h_{ac} = h_0 e^{i\omega t}$, which may be superimposed with the d.c. field H_0. In a normal metal or in a superconductor without pinning characterized by a linear resistivity ρ, the a.c. field will induce eddy currents which will decay in a characteristic length known as the skin-depth $\delta_s = (c^2\rho/2\pi\omega)^{1/2}$. The real or in-phase component of the a.c. susceptibility χ' measures the amount of shielding, and the imaginary or out-of-phase component χ'' measures the energy loss due to the induced currents. χ'' exhibits a maximum when the skin-depth δ_s equals the thickness[12] of the sample. This resistive absorption is a *linear* effect and so δ_s does not depend on the amplitude of the ac field h_0.[13] In a superconductor with pinning, I-V curves are nonlinear in some range of fields and temperatures. The condition "penetration depth $\simeq$ thickness" may now occur[13,14] in the linear or nonlinear portion of the I-V curves depending on J, ω and sample dimensions. In the fully developed critical state, Ampere's law implies that the magnetic induction B will decay linearly from the surface, not exponentially as in the ohmic case. The a.c. field will penetrate to a depth $D_c = \frac{c}{4\pi} h_0/J_c$, beyond which the a.c. field is totally screened. Thus in the nonlinear regime, the penetration depth D_c (and χ) will depend on h_0.

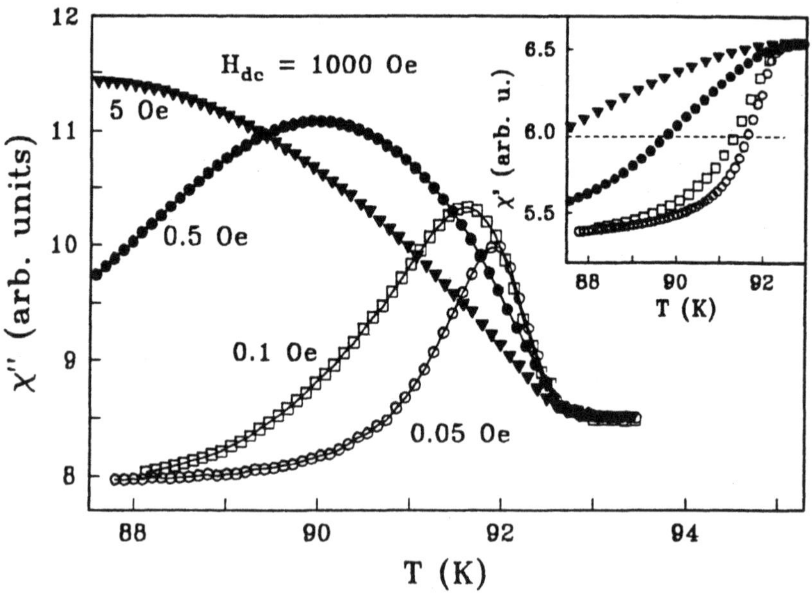

Fig. 1. Out-of-phase component of the ac susceptibility χ'' measured in a single crystal of $YBa_2Cu_3O_7$ at 1MHz in 1000 Oe dc field for several amplitudes of H_{ac}. Inset: Corresponding χ' curves. Half-screening is indicated by the dashed line.

Fig. 1 shows χ'' for several values of h_0. For the 30μm thick crystal, at ω = 1MHz and small excitation amplitudes h_0 in the range of 0.05 - 0.1 Oe, we *observe* that the position of the maximum roughly separates regions of linear and nonlinear amplitude response. For a crystal of different dimensions, the frequency must be adjusted in order to have the transition into a nonlinear response to be at the maximum of χ''. We *define* the irreversibility line at the *onset of nonlinearity* which, of course, is a result of *pinning*. Fig. 1 shows that the maximum in χ'' roughly coincides with half-screening as determined from χ', shown in the inset. For this 30 μm crystal, this condition corresponds to a current density of $\simeq$ 20 A/cm^2. At ω = 1 MHz, the peak in χ'' occurs at a resistivity of $3.6 \times 10^{-7}\Omega$cm. Transport measurements on a crystal[15] similar to those used in this experiment, confirm that under these conditions and for this crystal, the maximum in χ'' will occur very near the temperature where nonlinear behavior is first noted in I-V curves. The complicated I-V behavior in this region leads to a rather complex behavior[13] of χ'', where the high temperature portion of the peak is a measure of the resistivity in the linear, flux flow regime, and the lower temperature portion a measure of the development of *nonlinear* I-V character.

When the frequency of the a.c. excitation is lowered, the nonlinear I-V behavior should be reflected in the a.c. response. For ω = 100 kHz, this crystal exhibits a maximum well into the nonlinear regime. This is shown in Figures 2 (a) and (b), where χ'(T) and χ''(T) respectively are plotted for a d.c. field of 0.5 Tesla and 6 values of h_0, which follow the sequence 1:2:4:8:16:32, which corresponds to h_0 from 0.1 to 3.2 Oe. There is a clear amplitude dependence which is a consequence of the nonlinearity in the electrodynamic response.

The observation of the nonlinear behavior is not sufficient to prove that the critical state model applies. We present now a procedure that allows us to test whether the χ' data can be accurately described by the critical state model[13], and to derive the temperature dependence of J_c *independent* of the particular field distribution or sample geometry[13]. In the critical state for a given sample and coil geometry χ' (and χ'') is only a function of D_c. Therefore, in Fig. 2(a) the horizontal lines, marked α and β are lines of constant J_c. But J_c is only a function of temperature. This last observation implies that $J_c(P) = J_c(Q)$, where P and Q are points indicated in the Figure. The point P belongs to the curve h_0 = 16. The point R, with the same L_p as P, belongs to the curve h_0 = 32, so $J_c(R) = 2\, J_c(P)$.

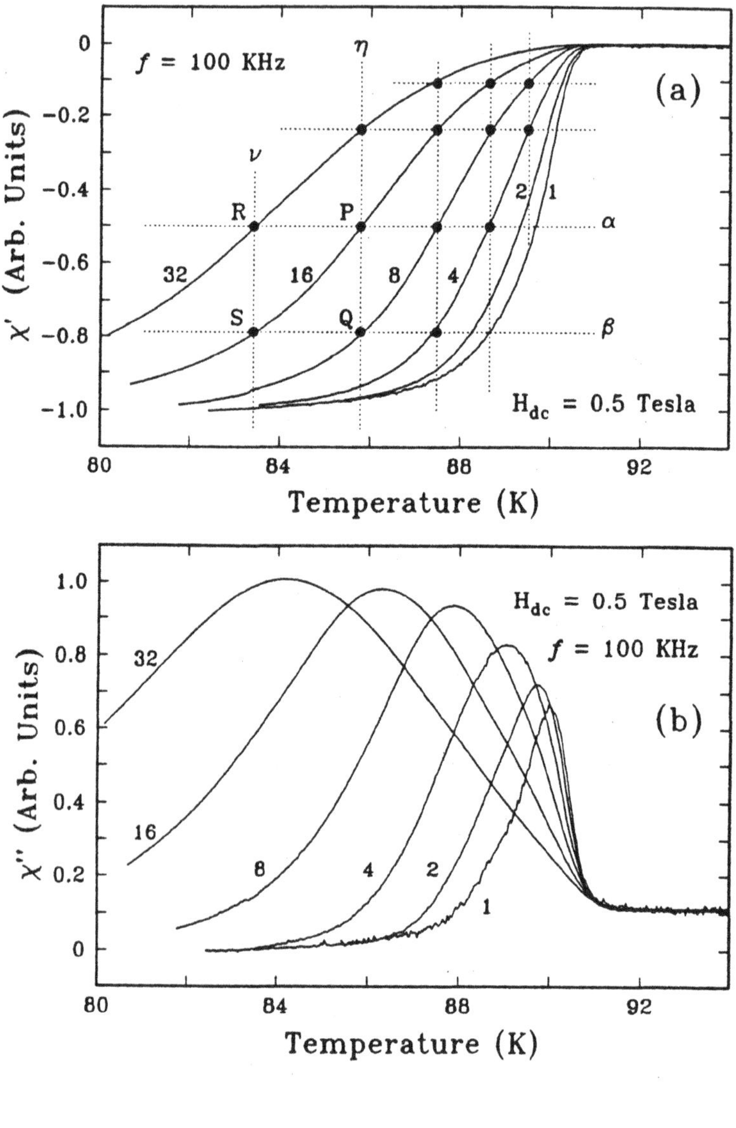

Fig. 2. The a.c susceptibility (a) in-phase and (b) out-of-phase response of YBaCuO crystal measured with $H_{dc}^{\parallel}$ to the c-axis of 0.5 Tesla using a 100 kHz a.c. field at six different a.c. amplitudes. The geometrical construction in (a) is used to extract $J_c(T)$.

Similarly, the point S of the curve of $h_0 = 16$ has the same D_c as the point the point Q of the curve of $h_0 = 8$, hence $J_c(S) = 2\ J_c(Q)$. The consequence, of course, is that $J_c(R) = J_c(S)$, correctly implying that R and S are at the same temperature. That the four points P,Q,R and S form a rectangle stems from the fact that χ' is completely determined by D_c and thus a proof that the critical state model applies. The other points indicated in Fig. 2(a) clearly show that this rectangular construction does not work everywhere; indeed it does not work *at all* at lower a.c. amplitudes as the critical state model *must* fail if h_0 is small enough.

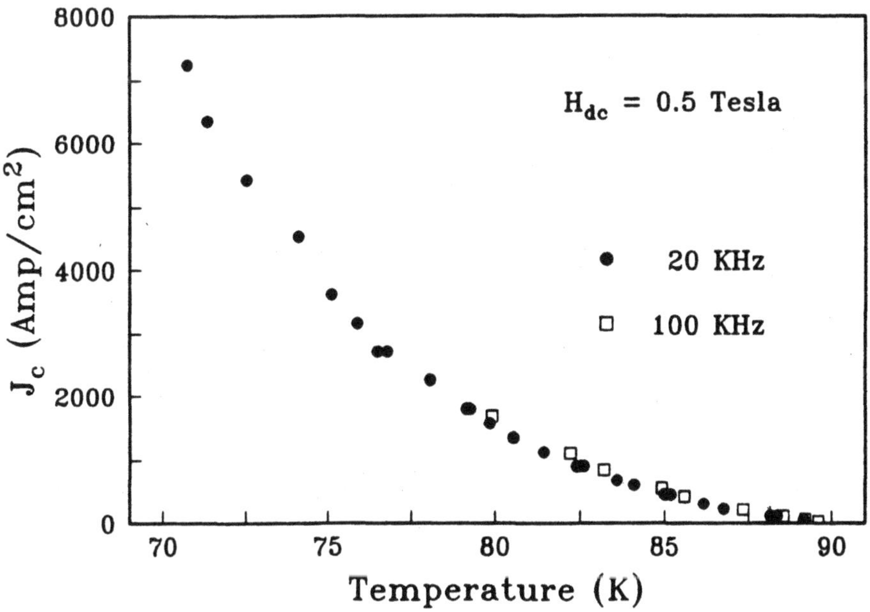

Fig. 3. The critical current density as a function of temperature derived from the a.c. susceptibility data of Fig. 2 and similar data taken at 20 kHz.

The geometrical construction described above gives us a way to determine the temperature dependence of J_c. Suppose in arbitrary units we define that $J_c(P) = 1$. Then following the line α we obtain $J_c(R) = 2$. This process can be followed to extract the temperature where $J_c = 0.5, 0.25, 0.125$, etc. Along the line β we find $J_c(Q) = 1$, $J_c(S) = 2$ and so on. The J_c vs temperature so obtained is shown in Fig. 3. This Figure also shows the J_c obtained by the same procedure at a frequency of 20 KHz. With the geometrical construction described here, *the shape of* $J_c(T)$ *can be obtained regardless of the geometry of the sample or the coil.* The absolute values of J_c can be estimated by assuming that for our geometry, $D_c = d$, the sample thickness, at half-screening. This gives us, for our crystal for example, $J_c \simeq 25$ A/cm^2 at T = 90 K. Again we note here, that for a crystal this size, the sensitivity limit of a commercial SQUID magnetometer is ~ 1000 A/cm^2.

3. ANGULAR DEPENDENCE OF a.c. RESPONSE - INTRINSIC PINNING

The strikingly anisotropic nature of high temperature oxide superconductors has received much attention since many interpretations of observed phenomena, such as Kosterlitz-Thouless[16] and depinning or flux-lattice melting[17,18] transitions depend on possible two-dimensional behavior. Indeed, apart from the underlying microscopic mechanism for the superconductivity in these materials, there is an intriguing possibility of the novel fluctuation effects and vortex structures due to the distinct layered crystal structure and extremely short superconducting coherence length[19]. Remarkable entropic effects have been proposed, due to flux line wandering and entanglement[20,21], which should be considerably more pronounced for the systems with large effective mass anisotropies and which may significantly alter the vortex behavior near H_{c1}. It has been questioned whether a 3D anisotropic Ginzburg-Landau description is valid[22] at low enough temperatures and in the case of weak coupling between Cu-O layers, i.e. when the coherence length in the c-direction, ξ_c, is smaller than the spacing between the layers. A Lawrence-Doniach model[23], incorporating the discrete arrangement of Cu-O layers, with coupling between the planes provided by Josephson tunneling or proximity effect, has been considered[23 – 25]. In this scenario one expects an unusual behavior of the vortex core and indeed, two-dimensional pancake vortices have been discussed[24], residing only in Cu-O layers. The vortex cores

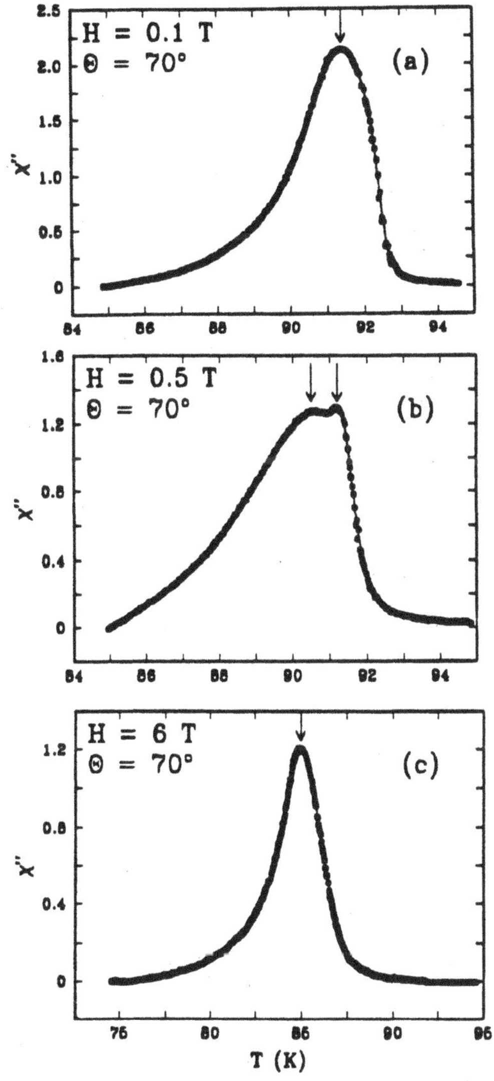

Fig. 4. Out-of-phase component of the ac susceptibility χ'' measured in a single crystal of YBaCuO at 1MHz for (a) H_{dc} = 0.1 T, (b) 0.5 T, (c) 1 T and (d) 6 T fields applied at 70° angle away from the c-axis. A double peak in χ'' is evident in a finite field range.

are drastically affected by the lattice discreteness and may vanish between the layers, where the vortices are Josephson-like.

The variation in core energy along the c-axis has been considered by Tachiki and Takahashi[26] and Feinberg and Villard[27], leading to an intrinsic barrier, similar to Peierls-Nabarro barrier[28], for the motion of dislocations across the planes. The barrier is against the kink formation in the vortex core; the kinks should appear at some critical orientation of magnetic field Θ_c away from the Cu-O layers.

To test these ideas, we have probed the anisotropy of the irreversibility line of YBaCuO single crystals, by angle resolved measurements of χ in a magnetic field range up to 7 Tesla[29]. We observe a development of two dissipation peaks in χ'' in a range of magnetic fields which is dependent on the orientation of the crystal with respect to the field. We suggest that the appearance of two peaks indicates the presence of two distinct components of critical current density J_c (T), and argue that they arise from vortices conforming to the intrinsic layered crystal structure by kink formation in a "staircase" pattern, as suggested by Feinberg and Villard[27].

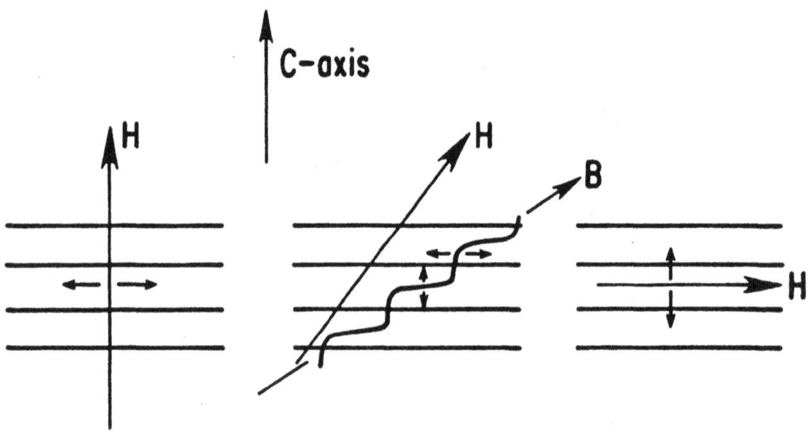

Fig. 5. Schematic drawing of vortex motion due to a.c. excitation. For the intermediate field orientations Θ, the two channels of dissipation arise from the motion of the in-plane segment of the vortex and of the kink crossing the a-b plane.

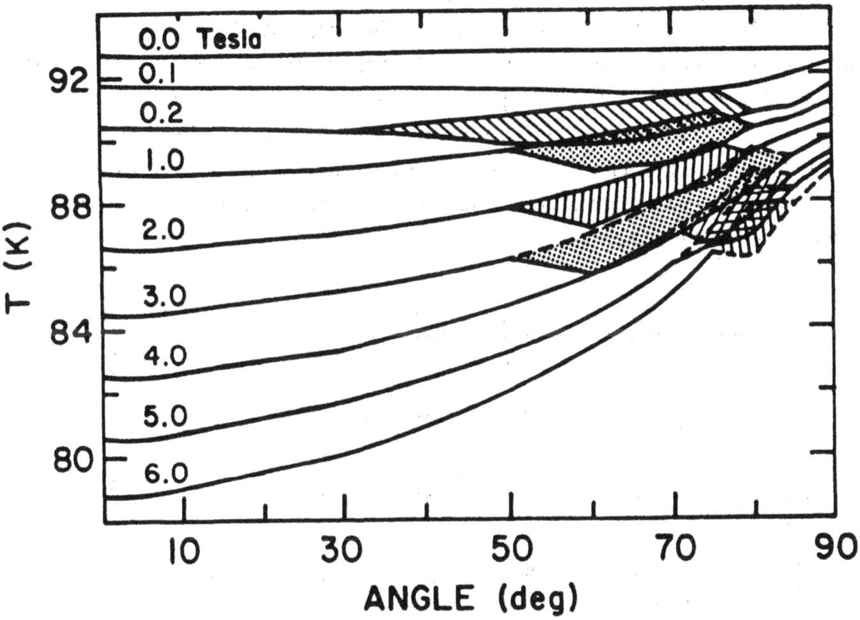

Fig. 6. Temperatures of the peaks in χ'' as a function of the field orientation Θ for different values of H_{dc}. The shaded areas indicate the finite range of fields and angles at which the double dissipation is observed.

As we discussed earlier, the dissipation peak in χ'' is used as an indicator of the onset of the hysteretic behavior in the principal orientations, i.e. with the applied field either parallel or perpendicular to the a-b plane. However, in the intermediate orientations the behavior is more complex. Figure 4 shows χ'' for several dc magnetic fields at an intermediate angle of $\Theta = 70\,^\circ$, where Θ is measured from the c-axis. At 0.1 Tesla, χ'' is a single peak, as seen in the principal orientations. As the field is increased, in addition to the shift towards lower temperatures and broadening of the whole feature, which also occurs in the principal orientations, a double peak is emerging. This double peak exists over a range of magnetic fields, which depends on Θ, and disappears again at high enough fields (Fig. 4 (d)). The temperature locations of the peaks obtained from both single- and double-peaked χ'' as a function of the field orientation Θ for various fields are shown in Fig. 5. With the applied field either parallel (Θ = 0 °) or perpendicular (Θ = 90 °) to the c-axis, the anisotropy of irreversibility line is ~5 and mimics that of H_{c2}. The shaded areas for each applied field show the extent of the double dissipation peak region and we note again that it occurs in the *finite range* of fields and angles.

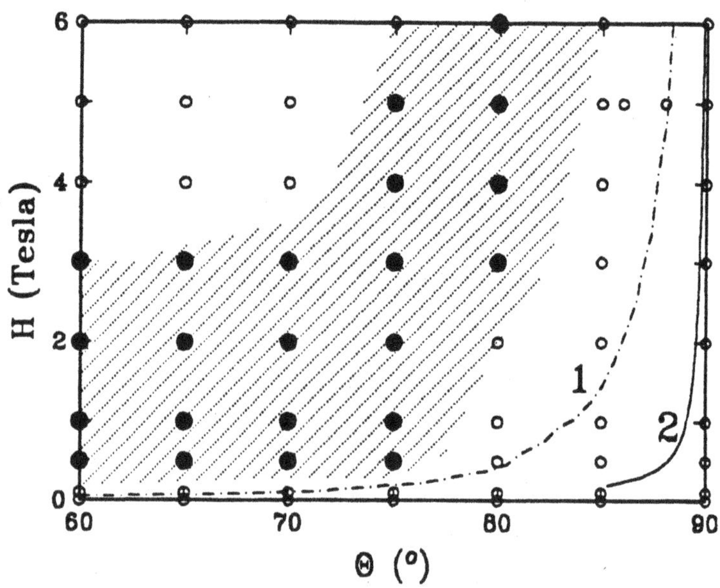

Fig. 7. Phase diagram in H-Θ plane. The shaded area indicates where the double peak is found (pluses are the actual data points). The single peak data are shown by the solid dots. The prediction of Feinberg and Villard (dot-dash line 1) for the lock-in boundary and of Bulayevskii (solid line 2) are indicated.

Double peaks in χ'' have been observed in granular materials, where both intergrain and intragrain currents contribute to dissipation[30]. There, however, the peak associated with the weaker intergranular currents moves very rapidly towards lower temperatures when relatively small dc field is applied, because of long Josephson penetration length. We do not see any effects of granularity in our single crystals. We argue that the two peaks correspond to two components of J_c; one which is parallel to Cu-O planes and another, crossing the planes as shown schematically in Fig. 5. The two channels of dissipation correspond to losses due to the motion of two (albeit coupled) vortex segments: one in the ab-plane and another one corresponding to a kink crossing the ab-plane. This new vortex structure is present in a finite region of T-Θ and H-Θ space, as shown by the hatched areas in Fig. 6 and Fig. 7, where the data for the intermediate angles are summarized. Feinberg and Villard describe such "staircase" vortices above a critical boundary, below which a perfect lock-in of flux lines parallel to the layers and on their full length is predicted[27]. The region of strong intrinsic pinning is confined by a "lock-in boundary" given by[27]:

$$\cos \Theta_c = \frac{1}{\pi} [2\alpha \frac{H^\star}{H} \{1 + \varepsilon^2 \frac{H^\star}{H}\}]^{1/2} , \qquad (1)$$

where, $H^\star$ is of the order of H_{c1}, ε is the anisotropy parameter and $\alpha \simeq 0.5$. This is represented by boundary 1 in Fig. 7, where we take $\varepsilon = 5$. At angles above this boundary, kinks in vortex core form and a "staircase vortex" will appear with the average B-field away from the direction of applied field. This boundary is within less than 5 ° of the onset of the shaded region where we observe two peaks. This is roughly our resolution limit; although we can vary dc field orientation to better than 1 °, the width of χ'' prevents us from fully resolving the peaks.

Such remarkable effect should be easier to observe in Bi or Tl based copper oxides, which are considered to be more anisotropic[31,32], than in YBaCuO[33-35], and perhaps two-dimensional. Indeed, single crystals of BiSrCaCuO show two distinct peaks in the mechanical oscillator response as well as in ac susceptibility[36]. There, two separate and independent boundaries in H-T plane are proposed, which are completely determined by the field component along the c-axis, H cosΘ. We do not find such H cosΘ scaling in YBaCuO, which is apparent from Fig. 6; the two J_c components appeared to be not decoupled, as it would be in the 2D case at low enough fields. Recently, Bulayevskii[37] proposed a two dimensional description taking into account 2D pancake structure of vortex cores, which also predicts a boundary for intrinsic pinning:

$$\cos \Theta_p = \frac{\Phi_0 \alpha}{4\pi\lambda_{\|}^2 H}. \tag{2}$$

Here, $\lambda_{\|}$ is penetration depth for fields in ab-plane and α is of order unity. This boundary is labeled 2 in Fig. 7 and it appears at angles much closer to the ab-plane than "lock-in boundary"[27] and is further separated from our data bounded in the shaded region. Indeed, our results indicate that a 2D description is less likely to be applicable to YBaCuO. At high fields, and closer to the c-axis, the interaction between the vortices will stiffen them, and one expects 3D Ginzburg-Landau description to hold. Indeed, we do not observe a double peak feature in that regime.

4. REDUCED PINNING AT LOW FIELDS - THERMAL SOFTENING OF CORE PINNING

So far we have discussed the a.c. response in high d.c. fields. Now we will describe a remarkable behavior we discovered recently in at low d.c. fields. We have found a *sudden collapse of the irreversibility line*[38] near T_c above the lower critical field H_{c1}. Observed in ac susceptibility measurements, the irreversibility line undergoes an abrupt jump from the usual power-law observed at high fields into a lower field regime near T_c, where it follows a different power-law, and where the *reversible* region is considerably enlarged. This observation represents experimental evidence for the existence of the *thermal softening boundary*, below which the collective pinning force is substantially reduced due to delocalization (on the scale of coherence length) of vortex cores by thermal fluctuations[39,40]. In Fig. 8 we show in detail the irreversibility line below 0.6 Tesla and near T_c for dc field parallel to c-axis ($H_{dc}^{\|}$). A sharp (and reproducible on cycling) step in $H_{irr}(T)$ initiates at $H_{dc}^{\|} \simeq 0.1$ Tesla and completes, within $\Delta T \sim 0.020$ K, at $H_{dc}^{\|} \simeq 200$ Oe. We observe this new transition in both twinned and fully untwinned crystals; six crystals were measured, all with T_c between 92.5 and 93.5 K, ΔT_c of ≤ 400 mK and 100% shielding.

We first note that the applied dc field at which the collapse occurs is much above the lower critical field $H_{c1}^{\parallel}$ [41]. At these fields the flux lattice spacing is smaller than magnetic penetration depth λ and hence, it is unlikely that this new effect is due to the change in the nature of the vortex-vortex interaction (i.e. from logarithmic at higher fields to exponential near H_{c1}). We argue now that the

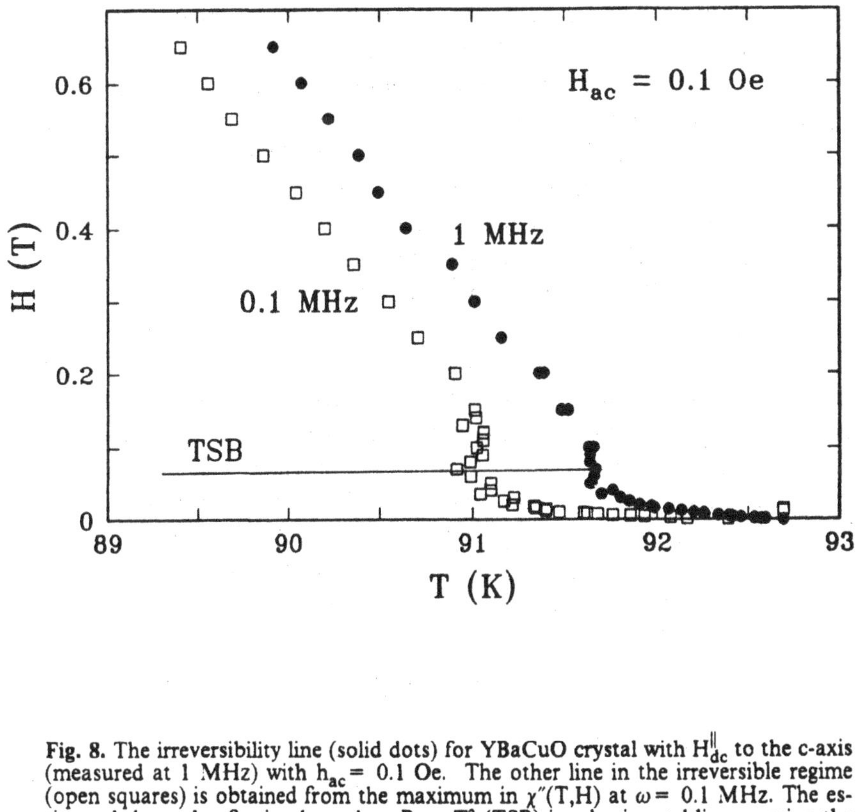

Fig. 8. The irreversibility line (solid dots) for YBaCuO crystal with $H_{dc}^{\parallel}$ to the c-axis (measured at 1 MHz) with h_{ac} = 0.1 Oe. The other line in the irreversible regime (open squares) is obtained from the maximum in $\chi''(T,H)$ at ω = 0.1 MHz. The estimated thermal softening boundary $B \propto T^2$ (TSB) is a horizontal line crossing the other two contours.

irreversibility collapse is due to the thermal softening of the core pinning of the vortices, proposed recently by Feigel'man and Vinokur[39]. The essence of it is, that the pinning barriers become unimportant when the thermal displacement $\sqrt{<u^2>}$ of the vortex core is of the order of the coherence length ξ, if the pinning centers are of the atomic origin[42].

Feigel'man et al.[39] predict a parabolic boundary $B \propto T^2$ in H-T plane below which pinning is vastly reduced. This boundary crosses the usual upward-curving irreversibility line. We propose that the collapse edge articulates where the irreversibility and thermal softening lines meet. This means of course, that the anomaly we observe in the irreversibility line should appear in the a.c. response at lower frequencies, which probe further into the nonlinear regime. And indeed, the line obtained from the position of the maximum in χ'' measured at 0.1 MHz shows this anomaly as well. The estimated *thermal softening boundary* (TSB) is also shown in the Fig. 8 and there is a remarkable correspondence. We should like to point out, that our technique is able to probe only a very narrow temperature window near T_c and a true critical current may not yet have developed in this range. For a more complete mapping at low temperatures, however, transport and M(H) hysteresis loop measurements are being pursued.

One still has to explain the abruptness of the collapse; it may be controlled by the temperature dependence of J_c, which is suppressed strongly by the thermal fluctuations of the vortex lines and shows a fast power-law or exponential decay with increasing temperatures[39,40]. This thermal softening is quite different from the dislocation melting in the Lindemann sense[39,43] which occurs when the thermal displacement of the vortices is about 10% of the lattice spacing, and at low enough fields the vortex lattice will first depin by thermal agitation and then possibly melt at higher temperatures[39]. Of course, it will be important for the existing theories of vortex-glass (or lattice) melting to address this new complexity of the magnetic phase diagram of high temperature superconductors.

5. CONCLUDING REMARKS

In summary, a.c. magnetic measurements provide a considerable insight into the macroscopic magnetic behavior of high temperature superconductors in both linear and nonlinear response regime. In addition to describing here a simple procedure for determining the temperature dependence of critical currents, we have discussed some of the interesting new pinning behaviors apparent from such measurements. This is clearly just a beginning of our unraveling of complexity of magnetic phase diagram of the oxide superconductors and exploring many novel phase transitions and pinning crossovers.

REFERENCES

1. A.P. Malozemoff in **Physical Properties of High Temperature Superconductors**, edited by D. Ginsberg (World Scientific, Singapore, 1989), p.71.
2. C.P. Bean, Phys. Rev. Lett. **8**, 250 (1962).
3. U. Welp et al., Phys. Rev. Lett. **62**, 1908 (1989).
4. T.K. Worthington, F. Holtzberg, and C.A. Feild, Cryogenics **30**, 417 (1990).
5. R.H. Koch, V. Foglietti, W.J. Gallagher, G. Koren, A. Gupta, and M.P.A. Fisher, Phys. Rev. Lett. **63**, 1511 (1989).
6. K.A. Müller, M. Takashige and J.G. Bednorz, Phys. Rev. Lett. **58**, 408 (1987).
7. Y. Yeshurun and A. P. Malozemoff, Phys. Rev. Lett. **60**, 2202 (1988).
8. A. P. Malozemoff, T.K. Worthington, Y. Yeshurun, F. Holtzberg, and P.H. Kes, Phys. Rev. B **38**(10), 7203 (1988).
9. T.K. Worthington, Y. Yeshurun, A.P. Malozemoff, R. Yandrofski, F. Holtzberg, and T.R. Dinger, J. Phys. Colloque (C8 Suppl.), **12**, 2093 (1989).
10. Y. Yeshurun, A.P. Malozemoff, T.K. Worthington, R. Yandrofski, L. Krusin-Elbaum, T.R. Dinger, F. Holtzberg, and G. V. Chandrasekhar, Cryogenics **29**, 258 (1989).
11. P.L. Gammel, D.J. Bishop, G.I. Dolan, J.R. Kwo, C.A. Murray, L.F. Schneemeyer, and J.V. Waszczak, Phys. Rev. Lett. **59**, 2592 (1987).
12. J.R. Clem, H.R. Kirchner, and S.T. Sekula, Phys. Rev. B **14**, 1893 (1976).
13. L. Civale, T.K. Worthington, L. Krusin-Elbaum, and F. Holtzberg, Proc. of the Workshop on Magnetic Susceptibility of Superconductors and other Spin Systems, Coolfont, May 19-22, (1991), to be published by Plenum Publishing.
14. V.B. Geshkenbein, V.M. Vinokur, and R. Fehrenbacher, Phys. Rev. B **43,** 3748 (1991).
15. T.K. Worthington, F. Holtzberg, and C.A. Feild, Cryogenics **30**, 417 (1990).

16. S. Martin, A. T. Fiory, R. M. Fleming, G. P. Espinosa, and A. S. Cooper, Phys. Rev. Lett. **62**, 677 (1989).

17. P. Gammel, L.F. Schneemeyer, J.V. Waszczak and D.J. Bishop, Phys. Rev. Lett. **61**, 1666 (1988).

18. Y. Yeshurun, A.P. Malozemoff, T.K. Worthington, R.M. Yandrofski, L. Krusin-Elbaum, F. Holtzberg, T.R. Dinger, and G.V. Chandrasekhar, Cryigenics **29**, 258 (1989).

19. see for example **Physical Properties of High Temperature Superconductors**, edited by D. Ginzberg (World Scientific, Singapore, 1989).

20. D.R. Nelson, Phys. Rev. Lett. **60**, 1973 (1988);

21. S.P. Obuchov and M. Rubinstein, Phys. Rev. Lett. **65**, 1279 (1990).

22. V.G. Kogan, Phys. Rev. B **24**, 1572 (1981).

23. W.E. Lawrence and S. Doniach, in **Proceedings of the 12th International Conference on Low Temperature Physics**, edited by J.G. Daunt, D.V. Edwards, F.J. Milford, and M. Jaqub (Plenum, New York, 1965), Part A, p. 566.

24. L.N. Bulayevskii, Zh. Eksp. Teor. Fiz. **64**, 2241 (1973) [Sov. Phys. JETP **37**, 1133 (1973)]

25. J.R. Clem, Phys. Rev. B **43**, 7837 (1991).

26. M. Tachiki and S. Takahashi, Solid State Comm. **70**, 291 (1989).

27. F. Feinberg and C. Villard, Phys. Rev. Lett. **65**, 919 (1990).

28. J. Friedel, in **Dislocations** (Pergamon Press, 1064).

29. L. Krusin-Elbaum, L. Civale, T.K. Worthington, and F. Holtzberg, to be published in Physica C.

30. Q.H. Lam, Y. Kim, and C.D. Jeffries, Phys. Rev. B **42**, 4846 (1990).

31. S. Martin, A.T. Fiory, R.M. Flemming, L.F. Schneemeyer, and J.V. Waszczak, Phys. Rev. Lett. **60**, 2194 (1988).

32. D.E. Farrell, S. Bonham, J. Foster, Y.C. Chang, P.Z. Ziang, K.G. Vandervoort, D.J. Lam, and V.G. Kogan, Phys. Rev. Lett. **63**, 782 (1989).

33. L. Krusin-Elbaum, A.P. Malozemoff, Y. Yeshurun, D.C. Cronemeyer, and F. Holtzberg, Phys. Rev. B **39**, 2936 (1989).

34. G.J. Dolan, F. Holtzberg, C. Feild and T. Dinger, Phys. Rev. Lett. **62**, 2184 (1989).

35. U. Welp, W.K. Kwok, G.W. Crabtree, K.G. Vandervoort, and J.Z. Liu, Phys. Rev. Lett. **62**, 1908 (1989).

36. J. Yazyi, A. Arribére, C. Durán, F. de la Cruz, D.B. Mitzi and A. Kapitulnik, preprint.

37. L. N. Bulaevskii, preprint.

38. L. Krusin-Elbaum, L. Civale, F. Holtzberg, A.P. Malozemoff, and C. Feild, to be published Phys. Rev. Lett.

39. M.V. Feigel'man and V.M. Vinokur, Phys. Rev. B **41**(13), 8986 (1990).

40. A.E. Koshelev and V.M. Vinokur, Physica C **173**, 465 (1991).

41. L. Krusin-Elbaum, A. P. Malozemoff, Y. Yeshurun, D. C. Cronemeyer, and F. Holtzberg, Phys. Rev. B **39**, 2936 (1989).

42. L. Civale, A. Marwick, M. W. McElfresh, T. K. Worthington, A. P. Malozemoff, F. H. Holtzberg, J. R. Thompson and M. A. Kirk, Phys. Rev. Lett. **65**, 1164 (1990).

43. A. Houghton, R.A. Pelcovitz, and A. Sudbo, Phys. Rev. B **40**, 6763 (1989).

WHY THE CRITICAL CURRENT DENSITIES OF HIGH T_c's VARY ANOMALOUSLY WITH TEMPERATURE, FIELD AND TIME

S. Senoussi, C. Aguilllon-Levillain and M. Mosbah
Laboratoire de Physique des Solides, Bâtiment 510, 91405 Orsay Cedex, France.

and

P. Régnier
SRMP/DTM/CEREM, CE Saclay, 91191 Gif sur Yvette Cedex, France.

ABSTRACT

The irreversible magnetization (M_{irr}), the associated critical current density (J_{mag}) and the transport current (J_{tr}) of high-T_c oxides (HTSC) decrease extremely rapidly with temperature, time, field (in some H-T region in the last case) and to a lesser extent as function of the dimension of the sample. All of these effects are reviewed here together with other several features of the hysteresis cycle not fully understood presently:

(1) The cycle exhibits a central peak near H = 0, the amplitude and the width of which depend only weakly on the radius R of the sample.

(2) As T approaches T_c the general shape of the cycle becomes dominated by the equilibrium magnetization.

(3) The cycle acquires a "butter fly" shape in some special conditions: (i) a limited T-H region (typically $T_c/2 \leq T \leq 2T_c/3$, $H \leq 10$ kG, (ii) H oriented far from the (a-b) planes, (iii) It is always present in the M(H) curves of large single crystals of $YBa_2Cu_3O_7$, $La_{2-x}Sr_xCuO_4$ and hardly seen in standard granular materials, (iv) this anomaly is generally much more pronounced in the two oxides above than in Bi-based HTSC.

A three currents model developed recently allows us to reproduce the whole hysteresis cycle. Here we shall discuss the microscopic origins of the different currents associated with the various regions of the hysteresis cycle. It will be shown that correct interpretation of these anomalies requires elasticity and collective theory.

1 Introduction

The question addressed by the organizers of this conference at Warsaw concerns the microscopic mechanisms which limit the critical current densities in high T_c superconductors (HTSC). Naturally this question is of fundamental importance from both the theoretical and the practical points of view. Nevertheless we prefer to pay here more attention to the "mechanisms" which presently limit our ability to describe our experiments correctly (in particular transport and magnetic data). We believe indeed that this is a very important aspect of the HTSC problem in view of the complexity of their microstructure and the huge number of publications devoted to these systems. Many experimental data are interpreted in the framework of the very useful Bean picture and more generally in the "linear" flux creep model of P.W. Anderson [1]. However, there are a number of experimental situations where these descriptions are insufficient and sometimes even incorrect. The purpose of this paper is to try to clarify some of these problems.

In experiments (concerning the hysteresis cycle, the transport current, the imaginary part of the ac-susceptibilitie, ac-method of Bean [2], that of A.M. Campbell [3], torque...) what is measured is not the local current density, which generally depends on all the quoted local parameters enumerated previously, but some macroscopic current which can differ significantly from one experimental technique to another (Here $\theta_B(r)$ is the angle of B(r) with c-axis; it depends on r for inclined field H). This is because the measured current always represents some average of the local currents in time (t), space (r), field (B(r)) and angle $\theta_B(r)$. One of the reasons of this is that in real systems we do not readily controle the geometry of the vortices (i.e. B(r) and $\theta_B(r)$) within the sample as it depends on the self field which in turn depends on many experimental conditions, particularly the shape of the sample. Moreover the average value of J or M is not unique but can depend on the experimental technique. Therefore, the most natural question to ask first is that of the physical meaning of the measured macroscopic quantities (J_{mag}, J_{tr}) and their relation with the local current density $J(t,r,T,B(r),\theta_B)$. We believe that the easiest way to answer this question is to consider some physically plausible current distribution, $J(r,B,T,\theta_B)$, to calculate the corresponding macroscopic quantities (hysteresis cycle, χ'', J_{tr}...) and then to compare the theoretical results with the experimental data. Next question concerns then the microscopic origin of the current density $J(r,B,T,\theta_B)$ used in the calculation (this is the procedure followed in ref. [4]). A very similar and related problem is that of the physical meaning of the effective potential barriers acting on the vortex lines deduced from various experimental data. Here too the result always represents some macroscopic average of the local potential barriers U(r). It turns out that there is no simple answer to this problem which is in fact related to the old puzzling question of how to sum out the pinning potentials in the collective pinning model. Indeed, as for the apparent critical current the effective pinning potential U_{eff} depends on the experimental method and represents a combination of at least three different averages (across the material) corresponding to different characteristic volumes V_b,V_c and V. Here V_b is the volume of the usual flux bundle (entering the flux creep theory of P.W. Anderson [1]), V_c is the correlation volume of the collective pinning theory [5,6,7] defined as the minimum volume over which the flux line lattice (FLL) keeps its periodic structure (i.e. within which the elastic distortion of the (FLL) is less than the coherence length : $|u(r)|<\xi$) and V is the volume of the whole specimen. Note that : (I) according to [6,7] we have $V_b>V_c$ especially at high enough temperature, (II) both V_c and V_b depend directly upon the nature of defects under considerations (weak, strong..) as well as on the elastic properties (or rigidity) of the (FLL). A schematic illustration of the situation is shown in Fig.1.

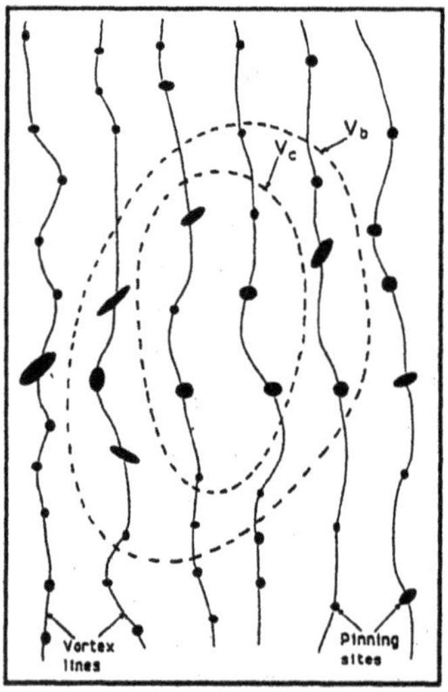

Figure 1: Schematic drawing of the vortex bundle volume V_b and the correlation volume V_c. In general $V_b>V_c$ because V_b causes much more elastic distortions in the vortex lattice as it is defined dynamically (flux jump) and as the jump costs compressional elastic energy via the large elastic modulus C_{11}. We see that the vortex lines are distorted by the pinning centers. This is quite similar to phenomena observed in ferromagnetic materials where the domain walls meandre through the pinning centers. The final state of the vortex results from a compromise between this energy gain (due to defects) and the interactions between vortex lines. Anisotropy helps the tilting of the vortices at low applied field.

The question of how to average of $U_p(r)$ in a given experiment is probably one of the most difficult situation faced by experimentalists. Another important point is that of the dependences of the characteristic fields and currents and other fundamental parameters such as λ, ξ, T_c upon the concentration of defects. In the case of conventional superconductors, this dependence is generally simply related to the value of the electron mean free path of the conduction electrons l, compared to λ and ξ. However, what is the real meaning of l in the superconducting state ? Formally, we assume that, l refers here to the mean free path of a fictious normal state which would exist at the field and temperature of measurements. It turns out that for high quality single crystals and thin films of High Temperature Superconductor (HTSC), we have $l \approx 14$nm at $T \geq T_c$ so that $\lambda >> l >> \xi$ near but not very close to T_c (since λ and ξ diverge there and since l increases with decreasing T, like $1/\rho_n$ where ρ_n is the 'normal' resistivity). However the situation could be radically different in the case of heavily irradiated materials in which l could approach $\xi(T)$ particularly near the transition temperature as ξ diverges at T_c. This shows that when comparing the magnetic and transport properties of various materials it is important to be very careful and check whether or not the condition $l >> \xi$ is satisfied. The situation is sketched in Figure 2.

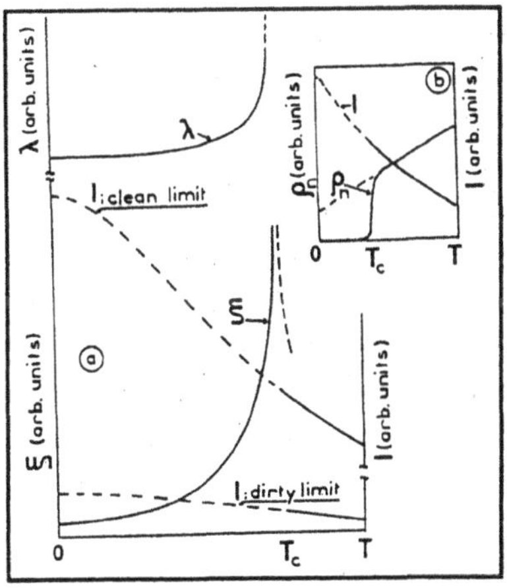

Figure 2: (a) Sketch of the variation with temperature of the electron mean free path in the clean limit and in the dirty limit. ξ is the order parameter which is in principle sligthly different in the dirty and in the clean limits (this difference is neglected here). Figure 2(b) suggests that the resistivity $\rho_n(T)$ entering the flux flow resistivity and other parameters of the superconducting state can be obtained by interpolation from the normal phase.

We emphasize however that this figure is a simple suggestion since we do not know how to define really the 'imaginary normal state' and the associated ρ_n and l in the region $T<T_c$, $H<H_{C2}$. The situation is probably much more complicated in the case of HTSC than in that of conventional superconductors where ρ_n is determined essentially by the residual resistivity ρ_{no} due to defects which is, in a first approximation, independent of T and H (in the limit of high $\kappa=\lambda/\xi$ materials). Thus, the aim of this short discussion is to address the question of the role of ρ_n rather than to answer it. We shall see that ρ_n enters also the flux-flow resistivity as well as in the so called attempt frequency ν_o entering the flux creep model of Anderson. The ρ_n of the flux flow state is the resistivity associated with the 'normal electrons' of the vortex cores and thus has a clear signification. However it is not clear wether or not this ρ_n is the same as the one defining the mean free path l. The latter has no direct relation to the vortices but enters the values of the characteristic parameters λ, ξ,..(in the dirty limit) because of the fact that the relationship between the microscopic current at some point r and the vector potential at another point r' contains an attenuation factor $e^{-|r-r'|/\lambda}$ (see for example [9]). In fact it is to be emphasized that the usual representation of an Abrikosov's vortex with a core of radius ξ made of normal electrons is an

approximation [10-12] which is probably questionable in the case of HTSC (this point will be discussed further below).

For pedagogical reasons and for the sake of clarity, the paper is organized roughly as a 'talk' with as many illustratrive examples as possible. Also the physical arguments presented here are generally qualitative and often demonstrated by means of some appropriate experimental examples. More quantitative considerations together with many experimental aspects not considered here can be found in a more general review paper reported by one of us [4]. These two articles have unavoidably some points in common but are rather complementary.

The outline of the paper is as follows. First we examine in details some problems encountered by us in the interpretation of our magnetic and transport data. This concerns in particular the role of anisotropy, heterogeneities, granularity, self field, time relaxation effects etc. We shall then try to explain some of the very characteristic features of the hysteresis cycle such as the central peak and the butter-fly shape exhibited by the hysteresis loop in some experimental conditions. These experimental features are tentatively correlated with the local (semi-microscopic) currents $J(r,H,T_c, \theta_B)$. The V-I characteristic and the comparison between transport and magnetic data are also considered. Finally we shall try to summarize as clearly as we can some of the most characteristic and most important information that can be simply deduced from transport and magnetic measurements having in mind the above qualitative approach.

2 Preliminary discussion and definitions

High Tc superconductors differ from conventional ones because of their (i) high critical temperatures and their high H_{C2} (due to the short coherence length), (ii) layered structures (anisotropy) and (iii) tendency to form granular materials (ceramics). This has important consequences on many properties particularly on the vortex structure we wish to consider briefly below.

2.1 On the nature and shape of individual vortices in HTSC

Most of the physical problems we are concerned with here deal with the properties of the vortex line structure. This concerns in particular (1) their mutual interactions (controlled by the elastic properties of the (FLL), (2) their individual geometrical shapes, (3) the way they appear and move through the sample, (4) their interaction with defects (usual pinning properties) (5) the interaction with the crystalline structure (anisotropy) and (6) with the shape of the whole specimen (demagnetizing and self fields).

Abrikosov's and Josephson's vortices in conventional superconductors

A remarkable property of classical superconductors is the fact that vortices may penetrate into the sample or leave it only cross the surface boundary with either vacum or insulator. Moreover a vortex line can be generated by only two possible mechanisms: either an external magnetic field or a biasing applied current flowing in the specimen. In addition it is accepted that in conventional superconductors vortices can exist only in the form of Abrikosov lines (Fig.3a,b,c) or (occasionally) in the shape of Josephson vortices (Fig.4) of length comparable to the sample dimension (in perfect samples and in both cases). Beside these classical vortices have rather simple geometrical shapes, often cylindrical.

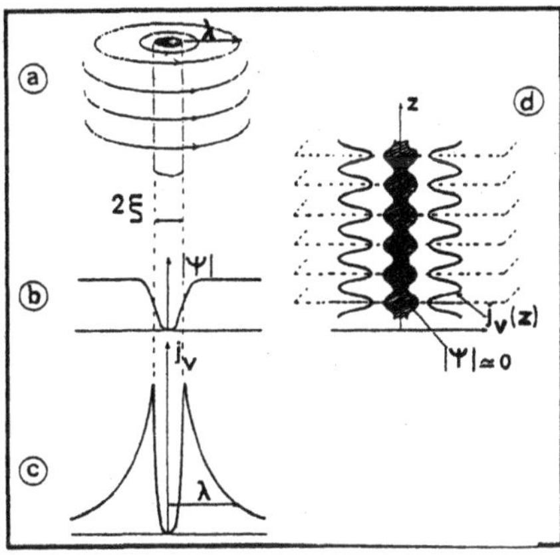

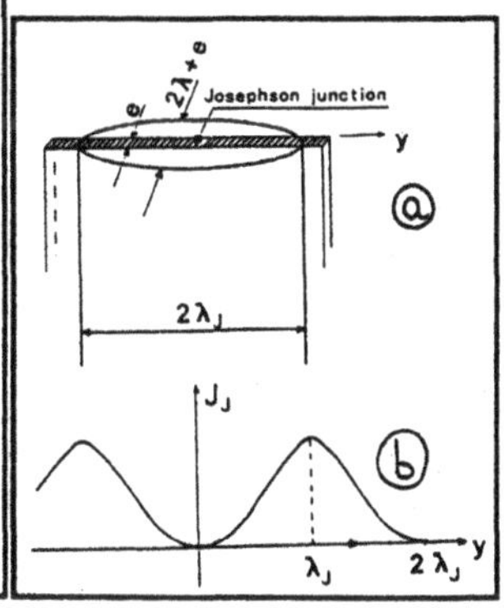

Figure 3 (left): a,b,c, profiles of the order parameter |Ψ| and the supercurrent J_v around a cylindrical Abrikosov's vortex. Here |ψ| and J_v are invariant along the direction (z) of the vortex. d : Schematic representation of the profiles of |ψ| and J_v along the direction of a vortex parallel to the c-axis in HTSC. The outer sinusoidal curves of Fig.d would represent the variation of J_v as a function of z at fixed distance $r_\perp = \sqrt{x^2+y^2}$ from the line whereas the dashed region would correspond to the vortex core where $|\psi| \approx 0$. The radial variation is expected to be monotonical and the same as for more classical line of Fig.3c,d.
Figure 4(right): a) Intended to sketched a Josephson's vortex in a conventional "short junction". b) Within the function (i.e // to y) the current of a Josephson's vortex is thought to build up gradually on a distance of about λ_J and then decreases to zero over the same length scale. A Josephson's vortex is said to have no core and is characterized by a unique length scale λ_J.

Let us first consider the case of an Abrikosov's vortex as sketched in Fig.3a,b,c. We see that the order parameter Ψ is zero at the center of such a vortex and tends to its equilibrium value at distances greater than ξ from the axis. This defines the core of the vortex as a region where the electrons can be considered as normal [10, 11]. Strong persistent supercurrents J_v flow around the core. As it is clear from the figure 3 J_v is equal to zero at the center (r=0) passes by a sharp maximum at $r \approx \xi$ and then drops again down to zero over a distance of about λ. These currents produce a magnetic field $h_v(r)$ (directed along the vortex axis) extending over the same distance λ. In conventional superconductors $|\psi|$, J_v and h_v are constant along the vortex direction and vary monotonically with r in the perpendicular direction. As depicted qualitatively in Fig.3d (for B parallel to the c-axis) and in Fig.5 (for B parallel to the a-b planes) the situation is expected to be radically different in layered structures.

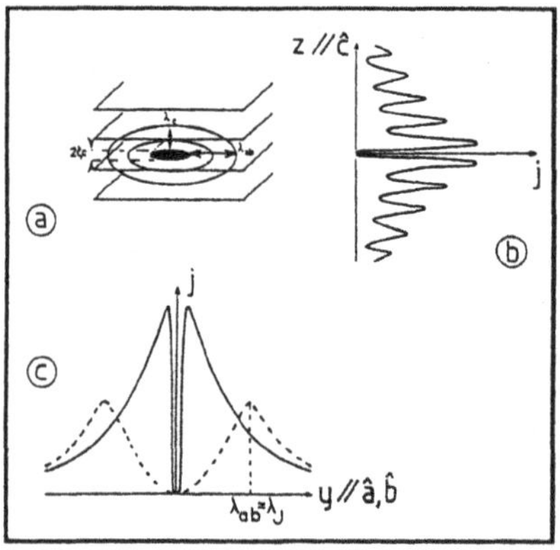

Figure 5: (a) and (b) are an attempt to represent the field and current profiles in a vortex parallel to the a-b planes of HTSC. In the c-direction the curent J_v exhibits minima and maxima the amplitude decreases as a function of y and z. (c) The profile of Jv parallely to the basal planes is expected to be different according to the nature of the vortex (Josephson ou Abrikosov). If the region between Cu-O sheets behaves like a thin weak link J_v would vary as the dashed curve. However, such a variation would probably lead to some anomalies in the hysteresis cycle when the distance between lines along the a-b direction becomes comparable to λ_J. However, as will be discussed later this requires very high degree of alignment of H with the a-b planes, very difficult to realize experimentally.

The physical meaning and the geometrical shape of Josephson vortices are less clear than in the case of Abrikosov lines. However it turns out that the most characteristic features of Josephson vortices are as follows: (i) they have no core (see below) [14,15,16] and (ii) the screening currents J_v in the junction build up very slowly, over a distance of the order of the Josephson penetration depth λ_J (instead of ξ in the above Abrikosov's case) and then drops towards zero over the same distance λ_J. At this point we must add that usually λ_J is quite large compared to λ.

Abrikosov's and Josephson's vortices in high-Tc superconductors

It is believed that because of their layered structure and due to the large thermal fluctuations in HTSC Abrikosov's vortices can now be kinked [17], meandre through the sample [18,19] and, at high enough temperature, can even exist in the form of decoupled thin slices (called 'pan cakes') of thickness approximatively equal to the spacing between Cu-O sheets. In addition, vortex loops can be generated spontaneously by thermal activation processes. This is because the thermal agitation energy (k_BT) can overcome the energy barriers for direct nucleation of small and stable vortex loops in HTSC. It turns out that spontaneous nucleation of vortices by thermal fluctuations can occur even in the Meissner state ($H<H_{C1}$) [18] and as a consequence

leads to energy dissipation (i.e. to a non zero resistivity) in this regimee. We shall not consider thermally activated nucleation of vortices any more in this paper.

An interesting question is that of the physical conditions under which Josephson vortices can exist in HTSC. On the one hand it is commonly accepted that Josephson vortices can be generated naturally in the weak link structure of granular materials. On the other hand some people believe that even in single crystals the vortices lying between Cu-O sheets should have a very pronounced Josephson character (since the order parameter is very depressed there). If true the last property will have important consequences on the critical current density along the c-axis ($J_{c,ab}$ see below). The reason is that in the case of Josephson vortices the core condensation energy would contribute very little to the pinning forces whereas it is often the principal source of pinning for Abrikosov lines (see [20-22]). In fact the question concerning the nature of vortices in the weak link structure of ceramics or between Cu-O sheets in single crystals is not simple at all and still controversal. For instance according to Likharev [14b, p.121] "Abrikosov vortices cannot exist under any circumstance in S-N-S weak links".

Other deviations from Abrikosov's description

The accepted concept of normal electrons in the core of the vortex is based on the fact that the Fermi energy E_F of such electrons is considerably higher than the energy Δ of the superconducting gap ($E_F \gg \Delta$). Caroli, de Gennes and Matricon [10] were the first to show that there is a density of low lying excited states localized in the vortex core with eigenvalues of order $\mu.\Delta^2/E_F$. Here μ is a numerical factor of order unity ($\mu=1/2$ for the first excitation energy). In conventional superconductors the ratio $\frac{\Delta^2}{E_F}$ is of order $10^{-4}\Delta$ so that these "Landau-like" levels within the vortex cores are populated through thermal excitations at least in usually accessible temperature. In HTSC Δ/E_F is probably of order 10^{-1} whereas Δ is of order 20 meV or more. As a result it is probable that the condition $\Delta^2/E_F \ll k_BT$ is hardly satisfied at low enough temperatures.

The energy spectrum of the electrons in the core was also investigated by J. Bardeen et al [10] and can be considered as a part of a more general problem known as Andreev reflexions [11,12]. Such energy spectrum should lead to large deviations from the usual predictions concerning specific heat, thermal conductivity, the flux flow resistivity, sound attenuations, NMR, EPR...in the presence of a magnetic field greater than H_{C1} (to have vortices). Clearly these very interesting questions are behind the scope of the present paper and need more experimental and theoretical developments. In the following part we shall accept the usual assumption that the vortex has a core of normal electrons.

Consequently the new high-Tc can be characterized by a number of new effects among which we can quote the following : (1) The appearance of three different pinning forces (which will be defined later). (2) The possible decomposition of vortices (particularly for H // $\hat{c}$) into elementary "pan cakes" of thickness $\sim\xi_o$ (3). The onset of "phonon like collective thermal fluctuations" in the (FLL) in addition to the more usual activated jumps. (4) The possibility of a melting transition in the (FLL). (5) The pinning of vortex lines via the kinetic energy of the screening currents by defects located in the regions of space where J_v passes by a sharp maximum (see fig.3 and 4). (6) The possible breakdown of many other usually accepted results in particular the classical expression of the flux flow resistivity we want to consider now briefly

Flux flow resistivity

Let us consider a given classical vortex induced either by the external field or by an applied current. In both cases the vortex is submitted to a Lorentz force density (per unit length in SI units)

$$\vec{f} = \Phi_o (\vec{n} \times \vec{J}) \tag{1}$$

Here $\vec{J}$ is the average current density at the core of the vortex of direction $\vec{n}$. If the vortex is created by the current flowing within the specimen (in this case the vortex forms a closed loop or a ring), the Lorentz force $\vec{f}$ pushes the vortex in the direction normal to the current and to $\vec{n}$. Then, in the absence of strong enough pinning centers this force will tend to establish a continuous flow of additional vortices across the sample. As a result we have a non zero dc-voltage and thus some amount of heat dissipation. This is a simple consequence of the Lentz law $e = -\frac{d\Phi}{dt}$ combined with the fact that each moving vortex carries with it a quantum of flux Φ_o. The resulting flux flow resistivity $\rho_{FF}=\rho_n B/B_{C2}$ has been established long time ago by Barden and Stephen [23] and is easy to calculate if we accept that it is simply equal to that of the normal electrons of the core of the lines (see [24] for a qualitative justification of this assumption). In this case indeed the fractional volume V_n of such normal electrons is equal to the vortex density (which is B/Φ_o by definition) times the volume $\pi\xi^2$ of the core (we assume that the vortex line is of length unity). Thus, $V_n = \frac{B\pi\xi^2}{\Phi_o} = \frac{B}{B_{C2}}$ since we know that $H_{C2} = \frac{\Phi_o}{\pi\xi^2}$. Consequently the calculated resistivity is $V_n\rho_n=(B/H_{c2})\rho_n = \rho_{FF}$. It is known that this linear flux flow resistivity is not well obeyed by conventionel superconductors for which more general theoretical treatments [25] reveal that ρ_{FF} should follow a concave-upward curve as a function of B rather than the linear relationship of Barden and Stephen. As we have noted above the deviations should be probably more dramatic in the case of HTSC because of the low lying energy excitations in the core of the line which could prevent some of the electron scatterings responsable for ρ_{FF}. To our knowledge, as far as HTSC are concerned, this very important topic has not received many attention up to now but will probably be the object of many investigations in the near feature.

2.2 Characteristic magnetizations, fields and currents

Even in the simplest example of a homogeneous and isotropic material with J independent of H, the hysteresis cycle reflects different characteristic magnetizations corresponding to different distributions of magnetic flux and currents within the material. These are defined qualitatively in Fig. 6 for ideal materials and in Fig. 7 for more realistic materials.

Characteristic magnetizations

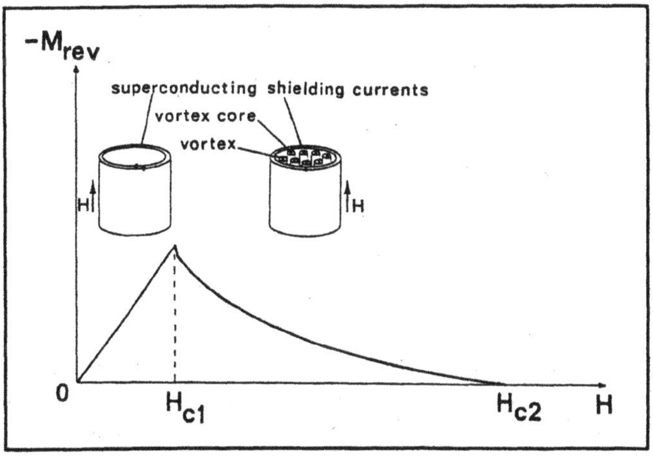

Figure 6: The equilibrium magnetization of an ideal isotropic type II superconductor as a function of H. Left insert: below H_{C1}, the fields and currents are restricted to the London penetration depth λ. Right insert : The current is still restricted to the depth λ because flux enters the sample uniformly on the form of vortex line.

Fig.6 represents the equilibrium magnetization $M_{eq}(H)$ of an ideal superconductor (i.e. without defects) determined first by Abrikosov. Note that $M_{eq}(H)$ is associated with surface screening currents ($J_{L\text{-}A}$) distributed in the London penetration depth λ both in the Meissner ($H<H_{C1}$) and in the mixed states. Therefore the specimen carries no bulk or volume currents if $R>>\lambda$ and hence it is of no use for technical applications. For most applications we need defects to pin the vortices in order to have a flux gradient ($\partial B(r)/\partial r\neq 0$) and thus a current (via Maxwell eq. $\mathrm{curl}H(r) = J$) within the material.

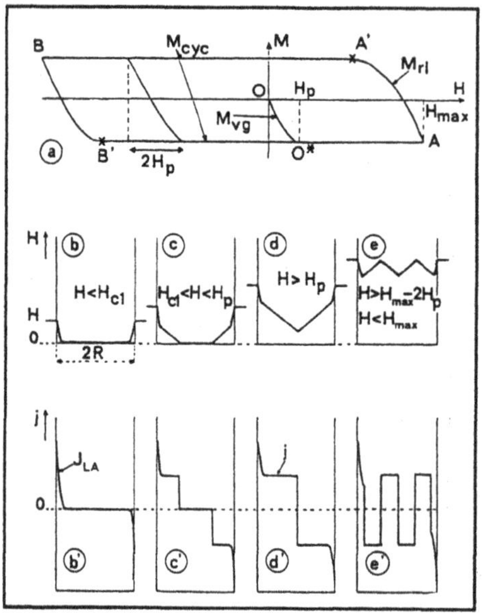

Figure 7: Definition of the various irreversible magnetizations for hard superconductors. The initial ZFC branch OO* defines the virgin magnetization M_{vg}.BB' and AA' refer to the reversal magnetization M_{rl}. A'B and B'B corrrespond to the cyclic or critical magnetization M^{+}_{cyc} and M^{-}_{cyc}. Also, the intersection point O* between M_{vg} and M^{-}_{cyc} defines the field H_p of full flux penetration. The other figures show the field and current profiles associated with M_{vg}, M_{rl}, and M_{cyc}. (b),(c),(d),(e) field profiles corresponding to the various branches of the cycle. b',c',d' and e':current profiles within the sample. The surface discontinuities are due to λ.

In Fig.7a we define schematically the various magnetizations associated with different branches of the hysteresis cycle. Also shown in Fig.7 are the profiles of fields and currents (across the specimen) associated with these branches respectively. We see from these figures that in all cases the measured magnetization includes the equilibrium magnetization $M_{eq}(H)$ which is associated with the surface currents represented on the figure as discontinuities. Beside it is clear that $M_{eq}(H)$ influences the various irreversible magnetizations (M_{vg}, M_{rl}, M_{cycl}) via the screening field produced by the equilibrium currents J_{LA}.. For a long cylinder with negligible classical demagnetizing factor N, this screening field is produced by the currents flowing in a surface region restricted to the thickness λ where $J=J_{LA}$. This gives a discontinuity $\Delta H \sim M_{eq}$ in H which is associated with J_{LA} by $J_{LA} \approx \Delta H/\lambda$ via Ampere's theorem. We now wish to define rapidly these characteristic magnetizations.

-M_{eq} is the London-Abrikosov's magnetization generally assumed to be independent of the defect structure at least in the clean limit $l >> \xi$ where l is the normal electrons mean free path.

-M_{vg} is the virgin or initial magnetization. It is measured just after the sample had been cooled in zero applied field (ZFC magnetization). M_{vg} corresponds to a vortex state where the sample is not yet fully penetrated by fields and currents.

-M_{rl} : this magnetization is characterized by the coexistence of positive and negative currents (i.e currents flowing in opposite directions).

-$M^{\pm}_{cycl}$: here the sample is fully penetrated by fields and currents which in addition have a constant sign through the sample positive for one branch and negative for the other one (note that J_{LA} is independent of the magnetothermal history of the material and has the same sign for a given field). We call this state a cyclic state and we define the irreversible magnetization below.

$$M_{irr} = \frac{M^{+}_{cycl} - M^{-}_{cycl}}{2} \qquad (2)$$

This magnetization is to a first approximation independent of $M_{eq}(H)$ (in the usual limit that J is not very dependent on H). We also define a "reversible" experimental magnetization as follows.

$$M_{rev} = \frac{M^{+}_{cyc} + M^{-}_{cycl}}{2} \qquad (3)$$

In the literature on HTSC, M_{rev} is generally identified as M_{eq}. However it is to be emphasized that M_{rev} depends in general on the shape of the sample and on J(H) through M_{irr}. It tends towards $M_{eq}(H)$ only in the limit $|M_{irr}| < 10|M_{eq}|$ typically. This condition is often difficult to satisfy in the case of single crystals of $YBa_2Cu_3O_7$ for which $M_{irr} \approx 2.10^4$ emu/cm^3 while $|M_{eq}| < H_{C1}/4\pi \approx 20$ emu/cm^3 at 4.2K typically (in CGS units).

Because of the anisotropy of HTSC we shall also need to introduce the following magnetizations :

$M_L = \vec{M}.\vec{H}/H$ (longitudinal component) (4)

$M_T = \left| \vec{M} x \frac{\vec{H}}{H} \right|$ (transverse component) (5)

$M_C = M(\theta_H = 0)$ (M for H oriented along the c-axis) (6)

$M_{a\text{-}b} = M(\theta_H = \pi/2)$ (M for H oriented along the a-b planes) (7)

where θ_H is the angle of **H** with the c-axis.

Characteristic fields

Shown schematically in Fig.8 are some of the most characteristic fields encountered in HTSC materials. Other typical fields are also defined below.

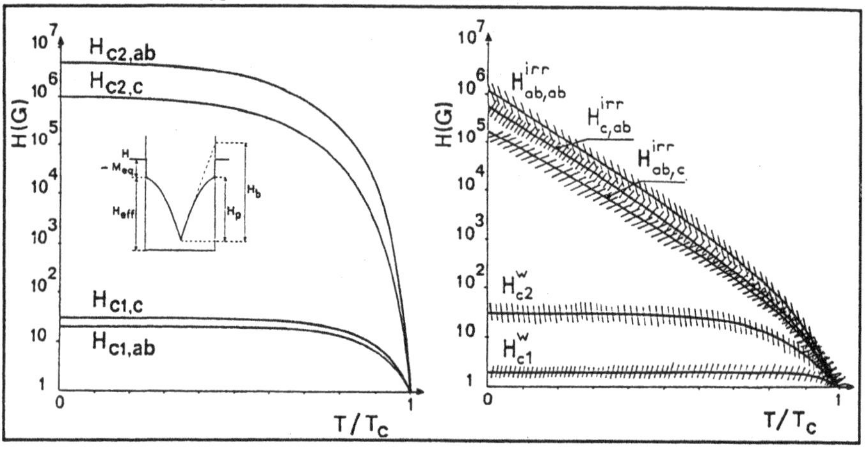

Figure 8: Schematic representation of the temperature variation (log(H) -T scales) of the first critical fields $H_{C1,c}$ and $H_{C1,ab}$, (parallel and perpendicular to the c-axis respectively), the upper critical fields $H_{C2,c}$ and $H_{C2,ab}$ and the irreversibility lines ($H^{irr}_{ab,ab}$, $H^{irr}_{ab,c}$, $H^{irr}_{c,ab}$); The first index defines the current direction whereas the second one stands for the direction of the vortex lines (or equivalently that of B). See Fig 16 for more details. The lower dashed lines correspond to the field H^{w}_{C1} of the first penetration of vortices within the weak links. H^{w}_{C2} stands for the Josephson decoupling field of the same Junctions. Insert : The thick curve represents the field profile within the sample (assumed to be cylindrical) and defines the fields M_{eq}, the field of complete penetration H_p and the Bean field H_b defined by the extrapolation of the linear region of the field profile (dashed curve).

$H_{C1,c}$ and $H_{C1,ab}$ are the first critical fields of individual grains and single crystals for H parallel and perpendicular to the c-axis respectively.

$H_{C2,c}$ and $H_{C2,ab}$ are the upper critical fields defined as above.

H_J is the first critical field of a single Josephson junction.

H^{w}_{C1} is the apparent first critical field of the weak link structure as deduced from magnetic measurements. This field is defined rather phenomenologically and is generally significantly higher than the true H_J because of irreversibility effects.

H^{w}_{C2} is the apparent Josephson decoupling field between adjacent grains. As for H^{w}_{C1} defined above this field is generally badely defined from experiments.

H^{irr}_{jk} : These fields represent the depinning lines (in the T-H plane) defined mathematically by the equation $J_{jk}(H,T) = 0$ (i.e. the vanishing of the critical current densities at the sensitivity of the experimental technique). This line is also called irreversibility line and is equally ill defined.

In fact we believe that in most experimental conditions it is more correct to call these lines constant J or constant E (E is the electric field). This is because the irreversibility line is defined by some (more or less) arbitrary experimental condition such as $E=10^{-5}V/cm$ (in transport data) or J=0 to within some experimental uncertainty ΔJ in magnetic measurements. As a consequence the measured irreversibility line will depend strongly on the sensitivity of the experiment and on the experimental technique used as well. In particular it has no much meaning in transport measurements if one uses the usual electric field criterion E=constant. Of course in this case it is a constant E line in our opinion.

$H_{eff,s}$: It is the effective field existing at the surface of the specimen (including the demagnetizing contribution H_d for samples of arbitrary shapes).

H_{eff} : It is the effective field seen by the vortex lines at the surface of a very long cylindrical specimen: because of the field induced by the equilibrium currents (J_J or J_{LA}) circulating at the surface of the specimen H_{eff} is different from the applied field even for a cylindrical sample with no classical demagnetizing field.

B_o (or H_o) and T_o : They define some characteristic scaling field and temperature respectively governing the variation of $J= J(B/B_o,T/T_o)$ with B and T. In other words, J is assumed to depend on T and H through the reduced variables B/B_o and T/T_o (in the simplest situation).

H_p : It is equal to the applied field at which the first flux lines just arrive and meet at the center of the specimen. It is a shape dependent field. Note that because of anisotropy effects we shall need to introduce three different penetration fields when discussing these effects.

H_b is the usual Bean field defined in the case where the critical current density J is assumed to be independent of H (as in the original Bean model). It coincides with H_p when J is independent of B. Here too we shall need to introduce three different H_b.

$H_{//}$, $H_{\perp}$: These fields enter in the transport measurements of the resistivity ρ and of J_{tr}. They mean that the applied field is either parallel or perpendicular to the mean direction of the macroscopic current (More on these parameters in the appropriate sections).

Characteristic fields in the experimental measurements

In real experiments J depends on H and as a consequence the hysteresis cycle exhibits various anomalies associated with characteristic fields and currents of the kind just defined. This is illustrated in Fig.9 and Fig.10 which show the evolution of the hysteresis cycle of a granular (but very compact) $YBa_2cu_3O_7$ as a function of the maximum applied field H_{max} (Fig.9) and as a function of temperature (Fig.10).

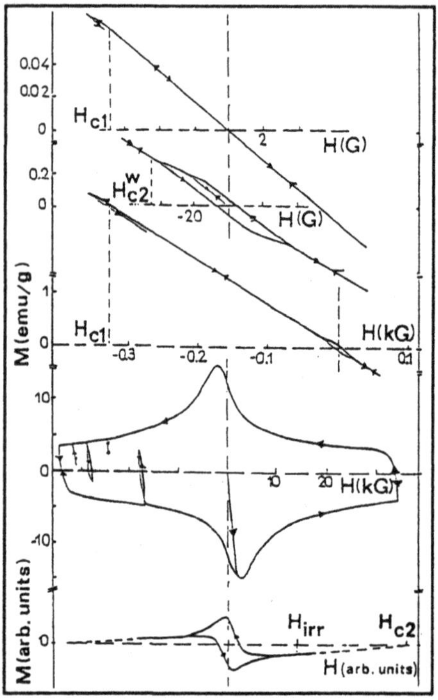

Figure 9: Typical magnetization loops in a granular, highly compact (>95%) $YBa_2Cu_3O_7$ at different field scales extending from H~1G to near H_{C2} (the last cycle defining H_{C2} is only schematic). Also defined here are other characteristic fields discussed in the text.

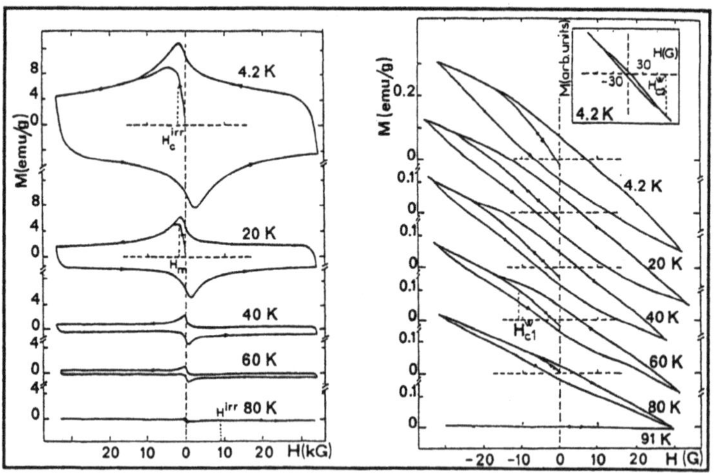

Figure 10: This figure illustrates how the hysteresis loops of standard $YBa_2Cu_3O_7$ ceramics evolve with temperature.

It is of great interest to discuss the physical meaning of the various curves of Fig.9 in some details.

(a) It is seen that at very low H, the magnetization is linear (i.e. $M = \chi_1 H$) and reversible (to within our experimental precision of about 1%). Beside, it exibits perfect screening if $R \gg \lambda_{J,eff}$, (see below the definition of $\lambda_{J,eff}$) after correcting for the macroscopic demagnetizing factor N : $\chi_1 \approx -1/(1-N)$.

(b) As the field is further increased we see the gradual onset of a low irreversibility regime connected with intergrain currents $J_{WL}(=J_{inter})$.

(c) For still higher field we find a second quasi reversible and quasi linear region ($M=\chi_2 H$) with a smaller slope ($|\chi_2|<|\chi_1|$) and thus reduced screening (after correcting for the macroscopic demagnetizing factor (Fig.9d). Here we have exceeded the Josephson decoupling field ($H>H_{J,C2}\approx H^{W}_{C2}$) and hence destroyed all of the intergrain currents. In this case the magnetization is given only by the London currents circulating around the individual grains in the London penetration depth λ.

(d) At $H > H_{C1}$(grains) vortices penetrate into the grains and M(H) becomes strongly irreversible and presents various typical behaviours depending on the exact J(H) relationship.

(e) Finally for high enough field such as $H> H^{irr}(T)$), M becomes again reversible (mixed state) and goes down to zero at $H \geq H_{C2}$.

Fig.(11a) shows the same kind of plot (i.e at various field scales)) as Fig.9 but for a melt textured sample now [26]. Here we see no detectable low-H irreversibilities in agreement with the idea that the low-H cycle of Fig.(9a,b,c) is due to the weak links network. Of course we see no low-H irreversibility in single crystals either, Fig.(11b). Fig.(11c) is the same as Fig.(11b) but for a diluted aligned granular sample (about 10% in volume of superconducting material in ~90% of epoxy resin). Here too we have no trace of the low-H cycle, but the physical argument behind this behaviour is different : It is because of the fact that the grains are now so dispersed in the epoxy that Josephson coupling is virtually zero within our experimental precision. This means that we are always above H^{W}_{C2} here. Note that when the distribution of the Josephson coupling energy is very broad as it is the case for insufficiently reacted or multiphased materials the field H^{W}_{C1}, H_{C1}, ... as well as the low-H cycle are ill defined. Figure 11c compares the cycle of a granular sample and a single crystal.

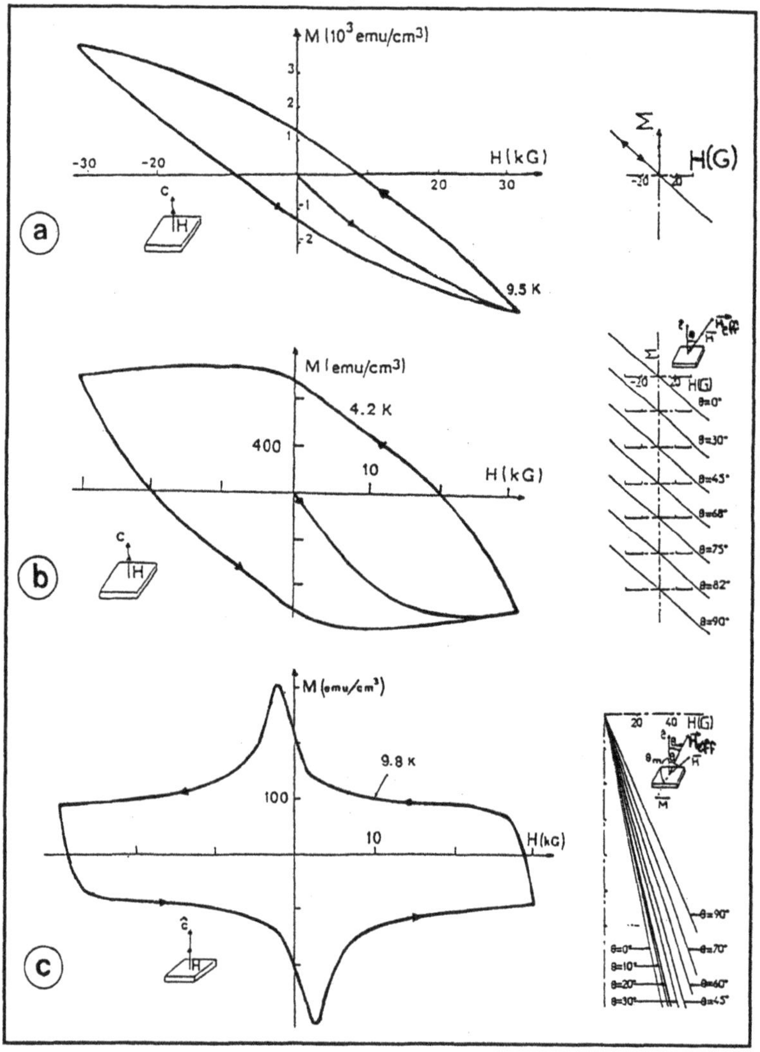

Figure 11: Evolution of the cycle at 4.2K as a function of the maximum applied field, i.e the same kind of plot as in figure 9 : a) melt textured, b) single crystal (R≈1mm) and c) oriented grains (r_g ≈ 10μm) finely dispersed in epoxy resine. No low-H cycle is detected in any of the three examples.

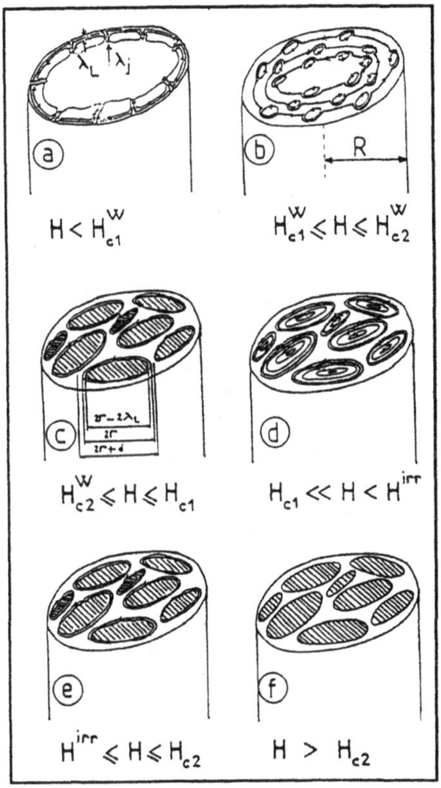

Figure 12: a) The current is made of macroscopic loops exclusively circulating at the periphery of the sample either within the grains in the London penetration depth λ or in the Josephson depth λ_J.
b) Corresponds to the low hysteresis regime where M is induced both by macroscopic current loops associated with irreversible penetration of flux through the junctions and reversible current loops around the individual grains.
c) Here the grains are decoupled and M is due London reversible currents alone (as far as $H \leq H_{C1}$).
d) Corresponds to strong irreversible regime due to intragranular currents.
e) Intragranular current is equal to zero, and M $\approx$ Meq (H). Reversible or equilibrium currents persist in the effective London penetration depth. H_{irr} defines the irreversibilty line. Here the sample is assumed to be isotropic, homogeneous, cylindrical with N=0 (no demagnetizing field) and with J independent of H.

Shown in Fig 12 is the evolution of the current profiles corresponding to the hysteresis cycles of Fig.9. Fig.13 illustrates how the low-H cycle of ceramic materials depends on the external dimension of a granular sample [27], in particular the influence of the demagnetizing factor N : Here we see that after correcting for the macroscopic demagnetizing field all the cycles coincide with the experimental precision.

The low-H cycle of several hollow granular cylinders [28] prepared from the same material with fixed external shape are presented in Fig.14. Here it was found [28] that at very low-H the signal is independent of the diameter of the internal hole, i.e. it depends only on the external volume. However at $H \geq H_{C2}^{W}$ the signal becomes proportional to the mass of the specimen and thus to the number of grains which compose each pellet. A more systematic investigation of size effects can be found in [29].

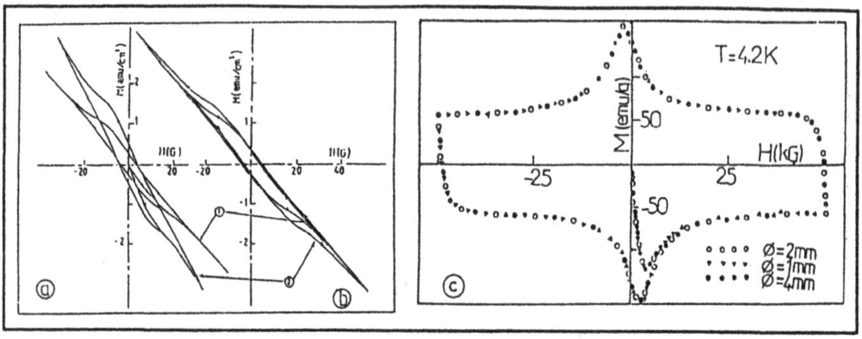

Figure 13: a), b) The low field hysteresis loops of two sintered pellets (R=0.5mm and 2mm) obtained after cycling the field up and down. Cycles labelled (1) and (2) correspond to cylinders with R=0.5mm and 2mm, respectively. c) The high field hysteresis cycles (up to 35kG) of the same samples as in fig. a,b.

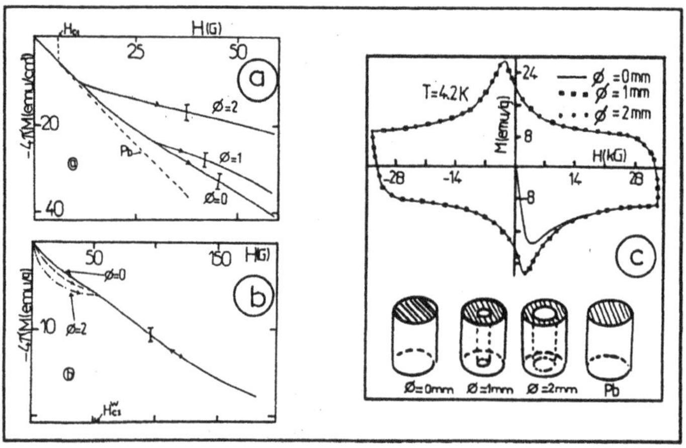

Figure 14: a) A lead and three tubular YBaCuO samples with the same external dimensions 3 mm and with internal diameter 0,1 and 2 mm.a) $M(H_{eff})$ is divided by the external volume ($\pi R^2 h \approx 35 mm^3$). The four samples exhibit the same degree of magnetic screening and the same demagnetizing factor N_S as the lead one for $H<H^W_{C1}$. Data are corrected from the demagnetizing field. b) The measured magnetization is now divided by the masses of the samples and expressed in (emu/g). We see that for $H>H^W_{C2} \approx 60G$, M is independent of both external and internal diameters. c) The high-field hysteresis cycles of three YBaCuO samples (hot pressed materials) differing by their external diameters. M(H) is the same for the three samples (except a very small effect around the hump region).

It is interesting to consider now the particular limit $H \rightarrow 0$ for all of the above figures. It is clear that in this limit the current is made up of macroscopic loops exclusively, located at the periphery of the macroscopic specimen (in some effective penetration depth λ_{eff}) either near the surface of the grains within the London penetration depth λ or at the edge of the junctions within the Josephson depth λ_J. Let d be the average physical thickness of the Junction (that is the region where the order parameter is practically zero) then we can show [27,28] that :

$$\lambda_{J,eff} = \lambda \left[1 - \alpha \frac{2\lambda + d}{r_g}\right] + \alpha \lambda_J \frac{2\lambda + d}{r_g} \qquad (r_g \gg 2\lambda + d) \qquad (8)$$

Here α is a factor of the order of 2 whereas r_g ($r_g >> \lambda$) is the average radius of the grains. Note that the above formula applies also to single crystals but in this case r_g must be replaced by the average distance between the "superficial" Josephson junctions at the surface of such a single crystal. Only when the surface is perfect that one has $\lambda_{J,eff} = \lambda$. Therefore in this field domain we have

$$M \approx \frac{-H}{(1-N)} \left[1 - \frac{\lambda_{J,eff}}{R}\right] \qquad (\lambda_{J,eff} \ll R) \qquad (9)$$

with $\beta \approx 2$ for a long cylindrical sample ($N \approx 0$) and $\beta \approx 3$ for a spherical specimen ($N \approx 1/3$).

Some of the most characteristic currents in HTSC

Sketched in Fig. 15 are some of the most characteristic current densities existing in layered superconductors.

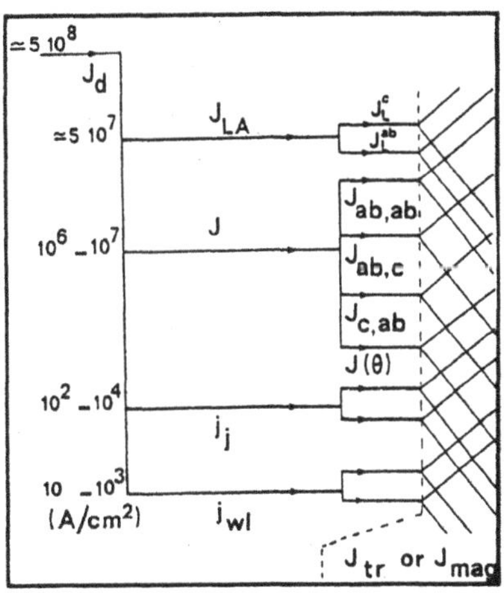

Figure 15: Schematic representation of the most important current densities exhibited by high T_c superconductors and defined in the text. Generally they all contribute to the measured current in a way which depends on the measurement technique used (hysteresis cycle, transport, torque, magnetic susceptibility...), the direction of the applied field with respect to the crystalline axes, the microstructure(granularity...) and the shape of the specimen.

(1) At the very microscopic level is the depairing current J_d. For isotropic materials J_d is of the order of the thermodynamic current ($J_d \approx 2J_{th}$) defined from the condensation energy or the gap energy (Δ) by : $J \approx \alpha.n_S.(e\Delta/m^*)$. Here n_s is the density of superconducting electrons and m^* their effective mass. In general J_d is much higher than any measurable quantity. It is thought that in conventional superconductors $J_d \approx 50J_c$ where J_c is the optimum critical currrent density of such materials ($J_c \approx J_{max}$). It is interesting to note that J_d is of the order of magnitude of the maximum current density near the core of the vortices (Fig.3c). Formally speaking, the thermodynamic current J_{th} can be determined if one is able to realize extremely thin wires having a diameter of the order of ξ [30, p.116]. As far as HTSC are concerned J_d is estimated to be of the order of $10^9 A/cm^2$ in the a-b planes at $T \ll T_c$. The depairing current J_d is lower in the c direction by a factor related to the mass anisotropy ratio ($J_{d,c} \approx \Gamma J_{d,ab}$ with $\Gamma^2 = M/m$).

(2) The second most typical current density always present in superconducting materials is the London-Abrikosov's screening current density J_{LA} flowing in the London penetration depth λ near the surface (Fig.6,7). It is this screening current which gives rise to the equilibrium magnetization M_{eq}. In a first approximation this current is independent of the nature and the density of defects in the material. More precisely, J_{LA} would be constant in the so called clean limit defined by the condition $l \gg \xi$ where l is the electron mean free path in the normal state. We recall that the fundamental parameters λ, ξ, T_c, H_{C1}, H_{C2}... are thought to be roughly independent of defects concentration in the same limit (see [30, p.67 and 261]. For standard single crystals of $YBa_2Cu_3O_7$ the mean free path l is of the order of 150 Å at 100K and increases as T is decreased (Fig.2) so that the condition $l/\xi \gg 1$ is easily realized for good quality single crystals. It is also important to recall that in the same condition the critical current J increases with defect concentration and would decrease in the opposite limit $l/\xi \leq 1$. A second important difference between J_{LA} and J concerns the fact that the former drops very slowly with T (like H_{C1}) whereas the latter falls off exponentially with T and even fastly in many experimental cases. A third distinguishing feature of J_{LA} is that it is independent of time while J generally exibits a logarithmic relaxation. Finally, J_{LA} is expected to be anisotropic with $J_{LA}^{a\text{-}b} \approx \Gamma J_{LA}^{c}$ $(\Gamma = \sqrt{M/m})$.

3) Looking again to Fig.15 we encounter the bulk critical current density J (just mentioned) which is a defect dependent quantity as explained above. The current J is expected to be highly anisotropic in layered superconductors where it is possible to define at least three different critical currents $J_{ab,ab}$, $J_{ab,c}$, and $J_{c,ab}$ depending on the directions of B and J with respect to the crystalline axes (see below).

4) Because of their ceramic nature HTSC often present a granular structure with intergrain currents called J_{WL} (Fig.15 and ref.[31,32]). By analogy with bulk materials and following Josephson we can also define a surface like screening Josephson current restricted to flow near the edges of the junctions in the penetration depth λ_J (Fig.12) and a "bulk like"

intergrain (or intrajunction) current flowing in the interior of the junction. However, it is very important to keep in mind that in general the junctions (i.e. the grain boundaries) are so imperfect that J_{WL} is strongly deteriored by such defects. In other terms there are always much more defects in the junction than needed to optimise J_{WL} and it is in general necessary to improve the quality of the junction in order to enhance this current. In fact it is likely that in general the junction is in the extreme dirty limit where J is severely deteriored by defects.

5) In macroscopic experiments the measured current always reflects some spatio-temporal average of all the currents depicted in Fig. 15. In addition, the macroscopic result may depend on the experimental technique used (transport, hysteresis cycle, ac-susceptibility via χ'', ac-techniques of Bean [2] and of Campbell [3], torque, ...) This is illustrated by different examples in the following sections in which, for the sake of simplicity, we shall concentrate on J_{tr} and J_{mag} principally.

3 Anisotropy and critical current density

Influence of anisotropy on J_{tr} and J_{mag}

We have seen in Fig.15 that layered HTSC are characterized by at least three different critical currents associated with three different pinning forces. These are defined in Fig.16a,b,c where it can be shown that the vortices prefer to be located between the Cu-O sheets where their core energy is the lowest.

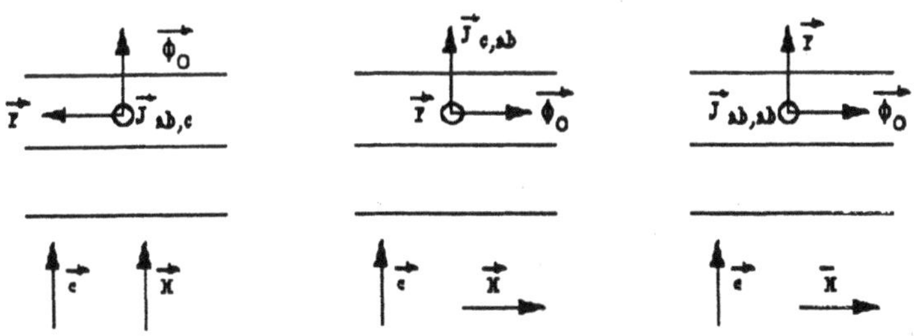

Figure 16: Schematic representation of the three current densities J and the associated Lorentz forces F that can exist in HTSC single crystals. Φ_0 is the unit vector along the vortex line direction whereas H is the applied field.

(a) $\vec{J} \in$ [ab] $\vec{\Phi}_0$ //c $\Rightarrow \vec{F} \in$ [ab] : we mesure $J=J_{ab,c}$

(b) $\vec{J}$ //c $\vec{\Phi}_0 \in$ [ab] $\Rightarrow \vec{F} \in$ [ab] : we mesure $J=J_{c,ab}$

(c) $\vec{J} \in$ [ab] $\vec{\Phi}_0 \in$ [ab] $\Rightarrow \vec{F}$ //c : we mesure $J=J_{ab,ab}$

This leads to an intrinsic pinning barrier when the Lorentz force is directed along the c-direction. In this geometry indeed to displace the vortex, it is necessary to overcome the gap energy at the Cu-O plane. The same arguments show that the pinning force would be the lowest when both the Lorentz force and the vortex are parallel to the a-b planes. At this point, for highly anisotropic materials, we note that the core energy which is assumed to be almost zero between Cu-O sheets cannot contribute significantly to the pinning energy in the configuration of Fig.16b. This suggests that for this configuration the pinning of the (FLL) should be via the energy of the kinetic currents circulating around the cores [14]. This in fact would be particularly true for the so called Josephson-vortices which, as we have noted (Fig.4), have no core at all and which are thought to exist in extremely layered superconductors such as Bi-based compounds. See the sketches of Figs (3-5) which show the distribution of currents around the vortices in different cases ($H//\hat{c}$ and $H\perp\hat{c}$).

The influence of both anisotropy and the equilibrium currents

The apparent anisotropy of macroscopic quantities such as M, J_{tr}, J_{mag}, χ... depends not only on the local anisotropy properties but also on the shape of the specimen. This occurs even in the absence of any usual demagnetizing field (N=0). For this reason the apparent anisotropy can change dramatically with the shape, particularly the aspect ratio w/e (i.e. width over thickness) of the specimen. A striking illustration of this is provided by Fig.17 which shows the apparent anisotropy of J_{tr} and J_{mag}.

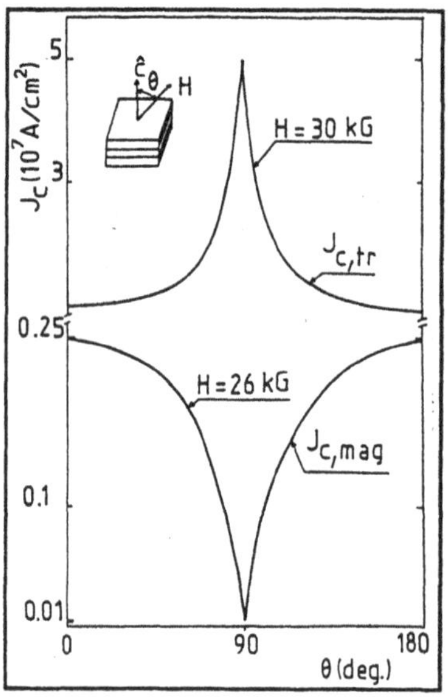

Figure 17: The anisotropy of the apparent critical current density as a function of the angle θ of H with c-axis. The lower curve, J_{mag}, is deduced from magnetic measurements whereas the upper one corresponds to transport data on thin films. Note that here H is higher than the self fields in both cases.

3.1 J parallel to the a-b plane.

We have already mentioned that the measured current can be considered as a combination of the various currents depicted in Fig.(15) especially at low enough field such that $H<H_{sf}$ where H_{sf} is the self field. This is illustrated in the following examples.

At first we see in Fig.18 that in transport measurements the vortices have the form of rings around the current pathways. It is clear that in this case the critical current results from the combination of $J_{ab,ab}$, $J_{ab,c}$ and J_{LA}. The exact contribution of each of these terms depends strongly both on the shape of the sample and on the field value. In the example of Fig.18 (case of very thin film) we expect that the total current density J_{tot} is imposed by $J_{L\text{-}A}$ exclusively for ($H \leq H_{C1}$) and mostly by $J_{ab,ab}$ for $H_{C1}<H<H_{sf}$. At $H \leq H_{sf}$ (i.e. for not very high H) J_{tot} will be determined as depicted in Fig.18b where it is seen that $J_{ab,ab}$ gives the largest contribution. The application of an external field will gradually change the relative contribution of each of the three above current terms to J_{tot} and will finally impose the appropriate J (and the final Lorentz force $\vec{F} = \mu_0 \vec{J} \times \vec{h}$) when the condition $H>>H_{sf}$ is satisfied everywhere in the sample. Therefore we see that both J_{tr} and J_{mag} can have different physical origins depending on the ratio H/H_{sf}, the direction of H and the shape of the specimen.

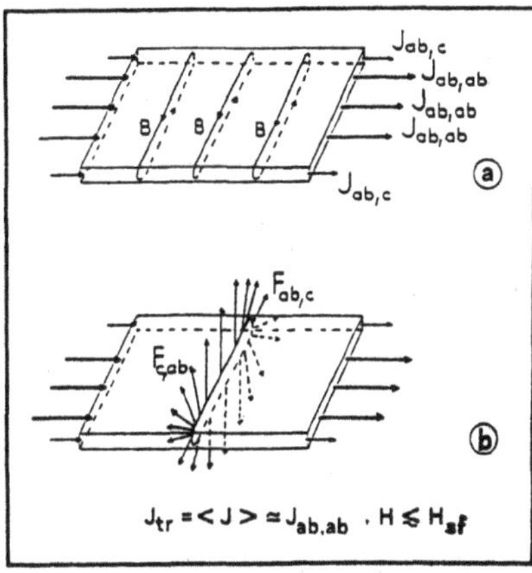

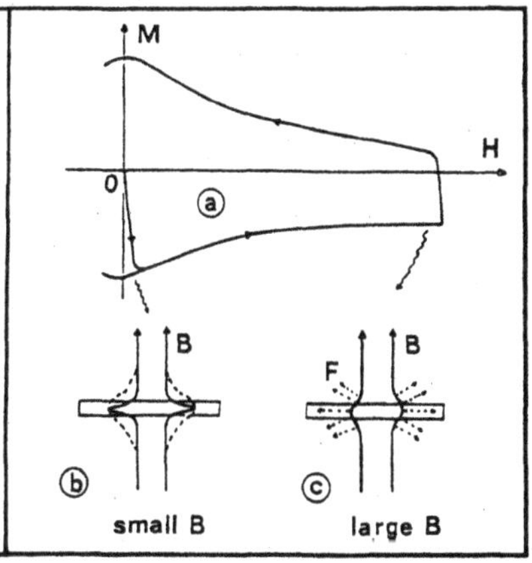

Figure 18 (left): Qualitative representation of the vortex rings (upper part) and the Lorentz forces (lower part) induced by the transport current feeding a thin monocrystalline film and their self fields (negligible applied field). It is clear that the resulting critical current is dominated by $J_{ab,ab}$ in this case compared to $J_{ab,c}$ (here, for simplicity we have not represented the London-Abrikosov's current density J_{LA}).

Figure 19 (right): a) Example of a hysteresis cycle of a thin film of $YBa_2Cu_3O_7$. R~1mm, e~1500Å and J_{mag} estimated to be about $2.10^7 A/cm^3$. b) Suggests that because of the anisotropy of the layered structure the vortices prefere to lie in the a-b planes at low B whereas at large B (fig.c) the interaction betweene vortices favours straight lines hence diminishing the radial field and the associated current. Also shown is the Lorentz force the direction of which varies through the film.

Fig.19 corresponds to the same thin film as in Fig.18 but represents now the field profile corresponding to hysteresis cycle measurements for H applied along the c-axis. Here too, it is clear that at sufficiently low-H the vortex lines are strongly bended near and within the film. In addition, this distortion is strongly reinforced by the layered structure. This is because, at low H, the tilt modulus C_{44} of the (FLL) is divided by the mass anisotropy ratio Γ^2 [19], compared to the isotropic material. As in the transport case of Fig.18, it is seen that the direction of the Lorentz force varies regularly across the film from about $F_{ab,c}$ to $F_{c,ab}$.

As shown in Fig.19b the distortion of the flux lines decreases with increasing H and hence reduces the relative contribution of $J_{ab,ab}$ to the measured magnetization. This can explain the drop of M with H which is illustrated in Fig.19a. In fact as will be reported elsewhere this effect also leads to a variation of J with H having the Kim form $J = J_o/[1+H/H_o]$.

3.2 H in the a-b plane

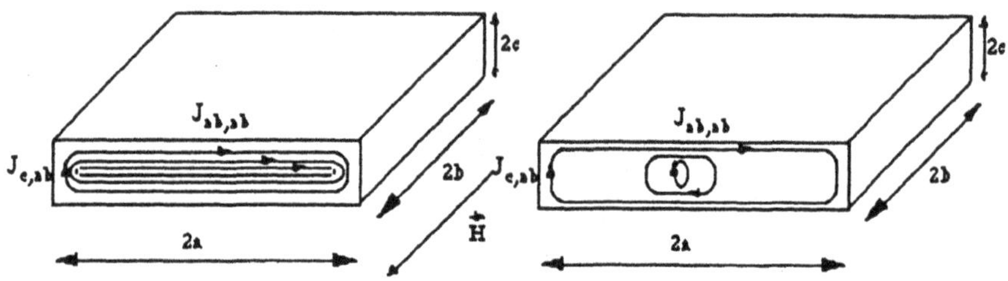

Figure 20: Schematic representation of the current loops developed within a single crystal of HTSC for $\vec{H}$ parallel to the a-b planes. It is clear that the measured current J_{mag} involves the two currents $J_{ab,ab}$ and $J_{c,ab}$. The real contribution of each of the two terms depends both upon the strength of the applied field and the aspect ratio w/e.

Fig.20 shows the development of the current sheets through the specimen as a function of the external field H which is assumed to be directed along the a-b basal planes (see [4] for more details). It is clear that the magnetization (M_{ab}) depends both on $J_{ab,ab}$ and $J_{c,ab}$. However it is easily seen that the saturation value of M_{ab} (or equivalently the cyclic branch of the cycle) depends mostly upon one of the two currents (i.e either $J_{ab,ab}$ or $J_{c,ab}$) : such a saturation is accomplished as soon as one of these two current densities ($J_{c,ab}$ here) reaches the center of the specimen. Therefore the cyclic a-b branch depends essentially on one of the two current densities depending on the dimensions of the sample. More precisely if the aspect ratio w/e is such that w/e « $J_{ab,ab}/J_{c,ab}$ then M_{ab} is imposed by $J_{c,ab}$ alone otherwise it is governed by $J_{ab,ab}$ (we recall here w and e refer to the width and the thickness of the sample respectively).

Now it is important to note that the current ratio $J_{ab,ab}/J_{c,ab}$ is thought to be of the order 200 to 300 in the case of $YBa_2Cu_3O_7$ and up to 10^3 to 10^4 for Bi-based families. This implies that except perhaps in some thin films $J_{ab,ab}$ would be absent in the cyclic (or saturated) magnetization since the aspect ratio w/e hardly exceeds 50 in ordinary experimental conditions. Fortunately the virgin M^{ab}_{vg} and the reversal M^{ab}_{rl} branches of the cycle (Fig.20) depend on both $J_{ab,ab}$ and $J_{c,ab}$ and hopefully can give information on $J_{ab,ab}$. However, these information are safe only if H is strictly parallel to the a-b plane everywhere in the material (a very hard condition to fulfil, as we shall see below). Moreover, because of the low-H peak (or central peak), which is always present in the virgin branch $M_{vg}(H)$, it is generally very difficult to analyse correctly this branch. Nonetheless, there is no such complication (from the central peak) in the returning branch (M^{ab}_{rl}) provided that the maximum applied field is high enough ($H_{max}>2H_p$, see Fig.7,8) or equivalently $B(r)\geq 2B_{C1}$ everywhere in the sample). In this way we can avoid what we have called H_{C1} effect.

3.3 Calculations of hysteresis cycles

calculation of the hysteresis cycle in the limit $J_{ab,ab}$, $J_{c,ab}$ and $J_{ab,c}$ independent of field and H along a symmetry direction.

Shown below are the equations of the magnetizations (i.e the hysteresis cycle) calculated under the following hypothesis (see ref. [4,33] for details)

1) H strictly parallel to the a-b planes.

2) No demagnetizing effects.

3) Both $J_{ab,ab}$ and $J_{c,ab}$ are independent of H and more generally of the position in the sample r.

To perform these calculations we must introduce the following full penetration fields equivalent to Hp :

$$H_{ab,ab} = e\, J_{ab,ab}\,, \quad H_{ab,c} = w\, J_{ab,c}\,, \quad H_{c,ab} = w\, J_{c,ab} \qquad (10)$$

We recall that w and e stand for the width respectively the thickness of the sample. For the analytical calculations we start from the general definition of

$$\vec{M} = \frac{-1}{v} \int [\vec{J}(r) \times \vec{r}]\, dx\, dy\, dz \qquad (11)$$

where the integration is performed over the whole volume v of the sample assumed to be either a long cylinder or a slab with negligible demagnetizing field in both cases (see [4] for practical applications of this equation). Of course J(r) and H(r) are related by Maxwell's equation. Let us first assume that J is isotropic and independent of H (i.e. J(r) =constant and cylindrical symmetry). This is realized approximately if H is parallel to the c-axis since in this case one has $J = J_{ab,c}$ everywhere. Then, the equations describing the whole cycle (we still neglect demagnetizing effects) can be written for positive H as :

$$M_c^{cyc} = -\frac{J_{ab,c}\,R}{3} + M_{eq}(H) \qquad (H \geq H_{ab,c}) \qquad (12)$$

$$M_c^{vg} = -H_{eff}\left[1 - \frac{H_{eff}}{H_{ab,c}} + \frac{1}{3}\left(\frac{H_{eff}}{H_{ab,c}}\right)^2\right] + M_{eq}(H) \qquad (H \leq H_{ab,c}) \qquad (13a)$$

$$\text{with} \qquad H_{eff} = H + M_{eq}(H) \qquad (13b)$$

$$M_c^{rl} = -\frac{J_{ab,c}R}{3} + \Delta H_{eff}\left[1 - \frac{\Delta H_{eff}}{2H_{ab,c}} + \frac{1}{3}\left(\frac{\Delta H_{eff}}{2H_{ab,c}}\right)^2\right] + M_{eq}(H) \qquad (14)$$

$$\text{with} \qquad \Delta H_{eff} = H_{max} - H_{eff} \leq 2H_{ab,c} \qquad (15)$$

We recall that the equilibrium magnetization $M_{eq}(H)$ contributes to the total M in two ways: either directly (i.e. terms M_{eq} in the above formule) or via its screening field which enters H_{eff} above. These equations describe the cyclic, the virgin, and the reversal branches of the hysteresis cycle respectively.

The general equations of the magnetization for H parallel to the a-b planes (i.e. of $M_{ab}(H)$) are more complicated than for H parallel to the c-axis (Eq.12-15 above) so that we only give here some particular cases (see [4] for details).

Let us indeed consider the hysteresis cycle M_{ab} (H) for H lying in the a-b basal planes. In this configuration, the intensity of the local current density J(r) is no longer constant (along a given current loop), and varies from $J_{ab,ab}$ to $J_{c,ab}$ (Fig.20). This rises the important question concerning which of these two currents is really at the origin of the magnetization M_{ab}? To answer this question it is necessary to calculate M_{ab} analytically, at least approximately. The final result depends on the ratio $\frac{eJ_{ab,ab}}{wJ_{c,ab}} = \frac{H_{ab,ab}}{H_{c,ab}}$. In general $H_{ab,ab} \gg H_{c,ab}$ for single crystals. In this case we find for the virgin curve with H parallel to the basal planes :

$$M_{ab}^{vg} = -H_{eff}\left[1 - \frac{1}{2}\left(1 + \frac{wJ_{ab,c}}{eJ_{ab,ab}}\right)\frac{H_{eff}}{H_{c,ab}} + \frac{w\,J_{c,ab}}{3e\,J_{ab,ab}}\left(\frac{H_{eff}}{2H_{c,ab}}\right)^2\right] + M_{eq}(H) \qquad (16)$$

We recall that H_{eff} is given by an equation of the form Eq.13b. In the case where $H_{ab,ab}<H_{c,ab}$, the virgin magnetization is again given by the same equation except that $J_{c,ab}$ must be replaced by $J_{ab,ab}$ and vice versa. The cyclic curve is given by :

$$M_{ab}^{cyc} = -\frac{\inf(wJ_{c,ab}\,;\,eJ_{ab,ab})}{2}\left[1-\frac{\inf(wJ_{c,ab}\,;\,eJ_{ab,ab})}{3\,\sup(wJ_{c,ab}\,;\,eJ_{ab,ab})}\right]+M_{eq}(H) \qquad (17)$$

In general (except sometimes for thin films) we have $wJ_{c,ab} < eJ_{ab,ab}$ so that the above equation transforms into

$$M_{ab}^{cyc} = -\frac{w\,J_{c,ab}}{2}\left[1-\frac{w\,J_{c,ab}}{3e\,J_{ab,ab}}\right] + M_{eq}(H) \qquad (18)$$

It is to be noted that we can easily recover equations (12-15) of the hysteresis cycle for H parallel to the c-direction from the above equations (16-18) simply by assuming $J_{c,ab}=J_{ab,ab}=J_{ab,c}$ in eq.(12-15)

Moreover we can also recover the equation of both cylindrical (w=e) and platelet samples (w>>e) of Bean [2] from these equations assuming again that the material is isotropic ($J_{c,ab}=J_{ab,ab}=J_{ab,c}$)

H out of the symmetry directions: the longitudinal and the transverse magnetizations

In the previous example of Fig.20 we have assumed that the applied field H was strictly parallel to a symmetry direction everywhere in the material. As we can expect the situation becomes exceedingly complicated when H is rotated out off the a-b planes (but still kept far away from the c-axis). This is illustrated in figures 21 and 22 which show how the hysteresis cycle of a single crystal of $YBa_2Cu_3O_{7-\delta}$ evolve with θ_H, the angle of H with the $\hat{c}$-axis.

The origin of this complicated behaviour is probably due to the fact that both the amplitude and the direction of the vector $\vec{B}$ vary across the specimen. If θ_{BH} is the angle between $\vec{B}$ and $\vec{H}$ then this means that B and θ_{BH} vary through the specimen in a way imposed by the applied field. In other words now we have $B=B(H,\theta_H,r)$ and $\theta_{BH} = \theta_{BH}(H,\theta_H,r)$. By virtue of Maxwell's equation $\vec{M} = -\vec{H} + \vec{B}/\mu_o$

(where $B = \int B(r)dv/v$ is the magnetic induction averaged over the whole sample) the three magnetic vectors $\vec{M}$, $\vec{H}$, $\vec{B}$ become misaligned. In terms of currents this implies that now all the currents $J_{ab,ab}$, $J_{ab,c}$ and $J_{c,ab}$ contribute to $\vec{M}$

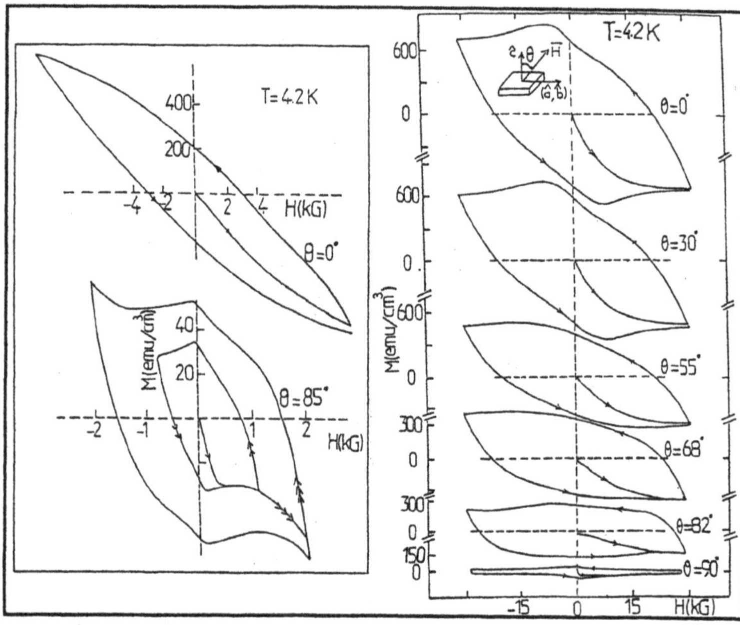

Figure 21 (left): Low-H loops of a single crystal of $YBa_2Cu_3O_{7-\delta}$. The behaviour is highly anomalous for H oriented close to the a-b planes ($\theta \approx 85°$).
Figure 22 (right): The hysteresis cycle of the same crystal as in Fig.21 but at much higher field and for various angles θ. Note that for $\theta \geq 50°$ the initial branches deviate strongly from the monotonical conventional behaviour.

To give some supports to this claim it is useful to decompose the magnetization vector $\vec{M}$ into a longitudinal (M_L) and a transverse (M_T) components defined as in eq. (4-5) can be written as below :

$$M_L = - H + B\cos\theta_{BH} \tag{19}$$

$$M_T = B \sin\theta_{BH} \tag{20}$$

Only M_L is probed by means of usual magnetic measurements as in the examples of the above figures. With torque techniques we have access to M_T. We suggest here that the anomalous behaviours for very tilted lines (Fig21-22) is first of all due to the variation of θ_{BH} as H is increased (at fixed direction θ_H).

At this point it is important to recall that in the case of layered superconductors an isolated (i.e a lonely vortex) always prefers to lie between Cu-O sheets where it minimizes its free energy.

Since at low enough field ($B \approx \mu_0 H_{C1}$) the interaction between vortex lines is exponentially small ($\sim \exp[-\sqrt{\frac{\Phi_0}{B\lambda^2}}]$) the vortex tends to be parallel to the a-b planes for this field domain. What is interesting to emphasize now is that such a situation of isolated lines always occurs in the virgin curve of the cycle. By definition of M_{vg} there is a region of the sample defined by $r \leq r^* \approx R[1-H/H_p]$ such that $B(r<r^*) \approx 0$. Therefore there is a region of the sample where the interaction between lines is vanishingly small, where the vortices tend to be parallel to the a-b planes. These arguments suggest that if the field H is tilted with respect to the c-direction and for sufficiently thick sample (or equivalently for very large H_p) we will always have a fraction of the vortices directed along the a-b planes.

The same arguments can be repeated for the reversal curve of the hysteresis cycle if the maximum applied field is smaller than $2H_p$ ($H_{max} \leq 2H_p$). This means that M_{rl} would present some anomalous effect if the maximum measuring field H_{max} is smaller than $2H_p$ but M_{rl} would be rather regular at much higher measuring field such that $H_{max} > 2H_p$. Therefore the physical origin of the anomalies in the hysteresis cycles near H_{C1} (i.e. for low H) is probably connected with this fact that tilted vortex lines tend to aligne with the a-b planes. On the other hand we expect that at high enough field they will be close to the $\vec{H}$ direction. This suggests that θ_{BH} would vary rapidly in this $r \approx r^*$ region such that $B(r) \approx B_{C1}$. Such a domain is crossed unavoidably when H varies from zero to H_p. We shall call this anomalous behaviour "anisotropic H_{C1} effect", the term "anisotropic" is intended to distinguish the present phenomena from the so called low-H peak (or central peak) effect which is also connected with anomalies in the pinning forces for B close to H_{C1} (see [4] and below). From the above discussion we expect that the influence of "anisotropic H_{C1} effect" would be present in the virgin curve especially near the a-b planes. However, as long as M_{rl} is concerned this will depend on whether the ratio $H_{max}/2H_p$ is greater or smaller than 1 (to a first approximation). This is exactly observed experimentally, as illustrated in the above figures: Figure 21 shows the low-H loop ($H \ll H_p$, T=4.2K) of a single crystal of $YBa_2Cu_3O_{7-\delta}$ at $\theta_H=0$ and $\theta_H \approx 80°$. It is quite clear that the loop $\theta_H=0$ is fairly regular whereas at $\theta_H \approx 80°$ both the virgin and the reversal branches of the loop are very anomalous. Fig.22 shows a selection of several cycles (T=4.2K) in relatively high field ($H_{max} > 2H_p$) and for different θ_H. In accordance with figure (21) we observe that for θ_H in the range 60-88° the virgin branch exhibits an anomalous behaviour whereas the reversal one is now quite regular. Therefore both Fig.21 and Fig.22 seem to confirm the above discussion that as H is increased (at fixed direction) M_{vg} is to a large extent governed by the variation of the angle θ_{BH} of B with H while M_{rl} is principally controlled by the amplitude B of the magnetic induction (if $H_{max} > 2H_p$). Of course for a more rigorous interpretation we must measure M_L and M_T simultaneoulsy.

Displayed in Fig.23 from [33] is the evolution of M_{vg} of oriented grains embedded in epoxy resin ($r_g \approx 10\mu m$, concentration ~1% in volume).

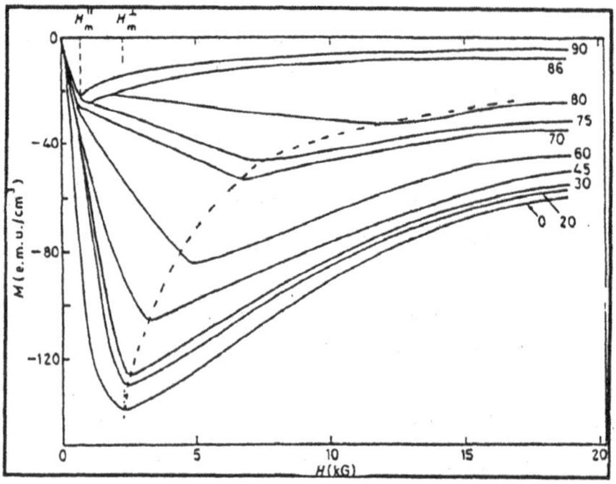

Figure 23. M vs H for different angles θ (= 0 to 90°) between the applied field and the c axis. The dashed curve represents the singular points which vary roughly as $H_m^{\perp}/\cos\theta$ where θ is defined by the numbers labelling the curves. The sample consists of oriented powder of YBaCuO.

We see that we have qualitatively the same anomalies as for the single crystal. Indeed as for the single crystal of Fig.22 we find that, when θ_H is far away from the c-axis and out of the a-b planes, each curve presents two singular points. There are arguments [33] according to which the first of these anomalies is connected with the effect we have ascribed to the anisotropy of H_{C1} (i.e. the tendency of isolated lines to lie between a-b planes) whereas the second one is related to the isotropic H_{C1} effect (i.e. to the vanishing of the elastic constants c_{ij} (this question will be discussed further in a forthcoming paper). In fact, it is found [33] that to a first approximation the anomalous points on the curves of Fig.23 (i.e those for which $\vec{H}$ is out of the symmetry directions or equivalently the curves such that $\theta_H \neq 0$ and $\theta_H \neq 90°$) can be deduced from the peaks on the curves $\vec{H} /\!/ \hat{c}$ (or $\theta_H=0$, uppermost curve) and $\vec{H}$//(a-b) (or $\theta_H \approx 90°$, lower most curve) by a projection procedure outlined in ref.[33]). In particular, according to this projection model the two singular points for curves such that $0<\theta_H<90°$ are related to the fields $H_{m,c}$ and $H_{m,ab}$ of the peaks of curves $\theta_H=0°$ and $\theta_H \approx 90°$ by $\frac{H_{m,c}}{\cos\theta_H}$ (see dashed curve) and $\frac{H_{m,a\text{-}b}}{\sin\theta_H}$ respectively. However we find that this projection model works much better for oriented grains samples than for single crystals. This is probably because of the fact that in the former case the applied field is relatively low (because H_p is small) and thus much closer to the average field $\langle H \rangle \approx B/\mu_o = H+M(H)$ seen by the vortices (the difference between H and $\langle H \rangle$ increases roughly as R if $M \approx J.R/3$). This splitting procedure can be justified qualitatively if we accept the idea presented previously according to which inclined vortices are generally not straight lines but have kinked shapes with negligible coupling between the vertical and the horizontal sections of the line. It is also probable that this projection model does not hold at very high field such that $H>H_{m,c}/\cos\theta_H$. On the other hand this description should be justified in the infinite anisotropy limit ($M/m\rightarrow\infty$) since in this case the currents are constrained to flow in the a-b planes exclusively whatever θ_H so that $\vec{M}$ is necessarly perpendicular to the a-b planes (see eq.(11) and imposed by the projection of H on the c-axis alone. This should be approximately the case for Bi based HTSC.

3.4 The link between the anisotropy of the magnetization and that of the critical current.

In this section we wish to examine the relationship between the measured magnetic anisotropy ratio $M(\theta_H\approx 0)/M(\theta_H\approx\pi/2)$ and the real anisotropy of the critical current densities. However this question is very subtle and involves three or four basic ambiguities discussed briefly below.

(I) The first one is related to the ambiguity about the microscopic origin of the magnetization $M(\theta_H\approx\pi/2)=M_{ab}$ for H parallel to the basal planes which depends on two different currents $J_{ab,ab}$ and $J_{ac,ab}$ via the shape of the specimen (see Fig.20).

(II) The second one arises because of the extreme difficulty to ensure that H is readily in the a-b plane. This is due to the accumulation of several unfavorable factors. First of all because both single crystals and aligned grains always have internal stresses and substructures which induce fluctuations of several degrees in their crystalline orientations(and even more for Bi based compounds). A second unavoidable cause of misalignment of the local field H(r) results from the self field which forms closed loops around the specimen thus inducing a small B component along the c-direction. Finally the condition that H is everywhere parallel to the a-b planes depends also on the spatial homogeneity of the applied field, hence on the homogeneity of the magnet producing such a field, the exact position of the sample within the cryostat and so on.

(III) A third and non independent cause of uncertainty in the definition of $M(\theta_H=\pi/2)$ is the fact that $M(\theta_H)$ varies extremely rapidly with θ_H close to a-b planes (i.e. for $\theta_H\approx\pi/2$). As a result a small error $\Delta\theta_H$ in θ_H can lead to huge variations in M_{ab}.

(IV) Apart from the above problems the link between the anisotropy ratios $M(\theta_H=0)/M(\theta_H\approx\pi/2)$ and $J_{ab,c}/J_{c,ab}$ or $J_{ab,c}/J_{ab,ab}$ involves the geometrical factor (w/e) and in addition depends on H and probably continue to vary with H (in some cases) up to very high field close to H_{C2}.

An interesting and striking example is that of Bi based systems which, according to many examples of the literature exhibits an apparent magnetic anisotropy ratio $M(\theta_H=0)/M(\theta_H=\pi/2)$ of the order of 10 typically. This is very surprising since there are other indications (from transport) data) that the anisotropy of the currents of Bi systems is probably greater than 10^3. However this apparent magnetic anisotropy can be understood quite naturally in the framework of the projection model developed in ref[4,33] and recalled above. This model becomes quite simple in the very anisotropic limit such that $J_{c,ab} \approx 0$ which is probably valid for Bi based compounds. Let us call $\Delta\theta_H$ the average error angle of $\vec{H}$ with the (a-b) planes. Then if the projection procedure of ref .[4,33] is valid over the whole field domain of the experiment the apparent magnetization $M_{a\text{-}b} = M(H,\pi/2-\Delta\theta_H)$ would be related to the magnetization $M_c = M(\theta_H = 0)$ for H parallel to the c-axis by the formula

$$M_{a\text{-}b} = \vec{M}_c(H_z).\vec{H}/H \approx M_c(H\Delta\theta_H)\,\Delta\theta_H \qquad (21)$$

valid for $\Delta\theta_H$ small. Thus the apparent value of $M_{a\text{-}b}$ is simply equal to the longitudinal component of M_c at $H = H_z = H\cos(\pi/2-\Delta\theta_H)) \approx H\,\Delta\theta_H$.

A striking illustration of the complexity of anisotropy measurements was presented in Fig.17 which shows the experimental critical current anisotropy of $YBa_2Cu_3O_7$ in two different situations.

4 Factors influencing the macroscopic critical current.

4.1 Influence of time effects

Influence of the experimental characteristic time scale on J_{tr} and J_{mag}.

The transport critical current density J_{tr} is usually defined by a condition on the electric field of the form $E_{tr} \approx 10^{-5}$ Vcm^{-1} where E_{tr} is the electric field within the sample carrying an external current $I_c = J_{tr}.\pi.R^2$ defined as the critical current (sometimes I_c is defined not with E_{tr} but with some critical resistivity value ρ_{cr}). This is to be compared with the electric field E_{mag} associated with flux creep in magnetic relaxation experiments [1]. Because of relaxation effects J_{mag} depends on the sweep rate of the applied field (see Fig. 24) and thus somehow on the associated induced electric field E_{mag}.

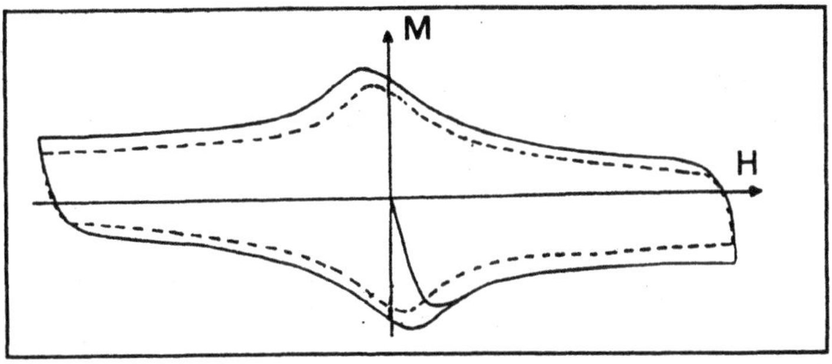

Figure 24: Illustrates relaxation effects during the recording of the hysteresis cycle. This example corresponds to Bi (2212) for B varying from -3 to 3 Tesla at 5K and at two sweeping rates of the applied field : the dashed cycle is for 2mT/s whereas the continuous one is for 80 mT/s.

Using Lentz law ($e=-d\Phi/dt$) and assuming cylindrical symmetry we can easily determine E_{mag}. On the one hand the electric field E is related to the E.M.F. e by (assuming again cylindrical geometry):

$$e = \int E(r)dr = 2\,\pi\, r\, E(r) \tag{22}$$

On the other hand from Lentz law one also has

$$-e = \partial\Phi/\partial t = 2\pi \int [\frac{\partial B(r)}{\partial t}\, r\, dr \approx \pi\, r^2 \frac{\partial \langle B \rangle}{\partial t} \approx \mu_o\, \pi\, r^2 \frac{\partial M(r,t)}{\partial t} \tag{23}$$

with

$$M(r,t) = M_o(r,t=0)\ \frac{1 - \dfrac{kT}{U_{eff}} \ln \dfrac{t+\tau}{\tau_o}}{1 - \dfrac{kT}{U_{eff}} \ln \dfrac{\tau}{\tau_o}} \tag{24a}$$

$$M_o \approx J_o\, r/3 \tag{24b}$$

After few rearrangements of the above equations we obtain :

$$E_{mag} \approx \frac{\mu_o\, J_o}{6}\ \frac{\dfrac{kT}{U_{eff}}}{1 - \dfrac{kT}{U_{eff}} \ln \dfrac{\tau}{\tau_o}}\ \frac{r^2}{t+\tau} \tag{25}$$

Here τ is some correlation time introduced by Senoussi in [4]. It depends on the sweep rate (dH/dt) of the applied field and on some characteristic field (not completely known). We propose here the following ansatz : $\tau \approx \alpha J_{max} R/(dH/dt)$ where J_{max} is the critical current density in the most critical state at the measuring T whereas α is a dimensionless coefficient which is estimated qualitatively in [4] see also [34] and which is expected to depend on T as it would tend to zero at T«Tc. As usual, t is the time counted just after the variable field H(t) had been stopped. It is of interest to remark that contrary to E_{tr} which is by definition fixed to some value (generally assigned by practical conditions), E_{mag} depends on J, r, T, B, U_{eff}, t and the sweep rate of H(t) via the correlation time τ. Moreover E_{mag} varies across the wire from zero at the center (r=0) to a maximum value at the periphery (r=R).

However it turns out that in general we have $E_{tr} >> E_{mag}$ and therefore $J_{tr} > J_{mag}$. Let us have an order of magnitude estimate of E_{mag} in some typical conditions for $YBa_2Cu_3O_7$ at 4.2K.

(1) $J_0 \approx 2.10^{10} A/m^2$, R = 1 mm, $kT/U_{eff} \approx 10^{-2}$ and $t + \tau \approx 50s$

Putting in eq.(25) gives $E(R)_{mag} \approx 10^{-8}$V/cm. This value will be reduced to about 10^{-9} to 10^{-10} V/cm at 77°K (in the same crystal)

(2) Keeping the same values as above except for $R \approx r_g \approx 10\mu m$ now gives $E_{mag}(r_g \approx 10\mu m) \approx 10^{-12}$V/cm at 4.2 K and 10^{-13} to 10^{-15}V/cm at ~77K. Note that E_{mag} decreases with T because $J_o(T)$ drops exponentially with T while the factor kT/U_{eff} varies rather slowly.

The consequence on J_{mag} of the weakness of E_{mag}.

First of all we stress that as far as magnetic measurements are concerned the exact relationship between E_{mag} and J_{mag} is not clear for us and is probably not unique. Accepting the same correlation between E_{mag} and J_{mag} as between E_{tr} and J_{tr} we would like to show that the influence of E_{mag} on J_{mag} depends strongly on whether we are in the usual flux creep regime (and thus very far from the depinning line) or in the so called diffusive or TAFF(thermal activated flux flow) regime close to this line. At first we want to show that in the flux creep region very large variations in E have very little incidence on J. To prove this claim we write the J-E characterisitic in the usual phenomenological form below :

$$E = E_1\left(\frac{J}{J_1}\right)^m \tag{26}$$

Here $E_1 \approx 10^{-5}$V/cm is the "initial" criterion associated with the "initial" critical current density J_1. Let us change the electric field E_1 defining the critical current density and take a new electric field E_2 and let J_2 be the corresponding critical current to be determined. From eq.26 we can write :

$$J_2 = J_1\left(\frac{E_2}{E_1}\right)^{1/m}$$

It seems that in the flux creep limit (not very far from J_{max}) the exponent m is generally very large 15 to 50 typically. Taking m=20 and assuming 10 orders of magnitude variations in E (i.e $E_2/E_1=10^{-10}$) results in a factor of only 3 between the two critical currents ($J_2 \approx J_1/3$).

However the situation is probably radically different near the depinning line (or in the TAFF domain) where the exponent m is probably of the order of one while at the same time E_{mag} should be extremely low. These two effects contribute simultaneously to make $E_{mag} \ll E_{tr}$ in this diffusive domain. The two limits are illustrated schematically in Fig.25.

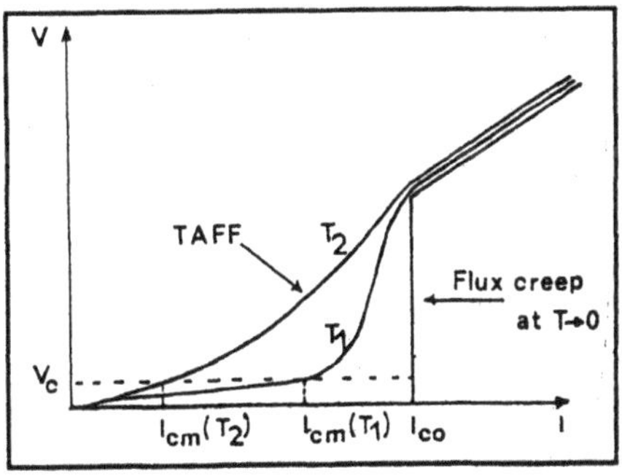

Figure 25. Schematic illustration of the V-I characteristic in the two extreme limits of flux creep at very low T and the diffusive or TAFF regime close to the depinning line. We predict that the transport and magnetic currents would be very different in the diffusive limit.

In conclusion of this section we can say that as far as time effects are concerned we find that J_{tr} should be slightly higher than E_{mag} in the flux creep regime. In the case of $YBa_2Cu_3O_7$ this difference should be in general less than a factor of two except close to the so called irreversibility lines where the relaxation of M can be very rapid leading to extremely low E_{mag}. Concerning $YBa_2Cu_3O_7$ the last region is generally hardly accessible experimentally (with laboratory available fields) except very close to Tc. However in the case of Bi-based HTSC the TAFF region is much more extended experimentally and the difference between E_{mag} and E_{tr} would be large as a general rule (except for $T < \approx 15K$). Lastly we would like to emphasize again that the exact relationship between E_{mag} and J_{mag} is more complex than in transport measurements and needs more investigations.

4.2 Influence of the spatial average on J(r) across the sample on the apparent critical current. Comparison between J_{tr}, J_{mag} and J_{ac}.

We have just seen that because of the difference in the characteristic time scales we generally have $J_{tr}>J_{mag}$. Here we wish to consider the effects of spatial average on J and show that if, for any reason, J varies across the sample (i.e $J=J(r)$) then this will lead to a difference between J_{tr}, J_{mag} and J_{ac} (where J_{ac} is the critical current density measured by the ac-method of Bean [2b]). This can be seen quite simply if we write down the mathematical expressions defining these quantities. For simplicity let us assume that the material is isotropic and that the sample is a very long cylinder with no demagnetizing field. Then by definition the transport critical current density is given by summing over a transverse section of the wire :

$$J_{tr} = \frac{\sum J(r)ds}{S} = \frac{\int J(r)ds}{S} = \frac{2\int J(r)rdr}{R^2} \tag{28}$$

(for cylindrical symmetry) whereas J_{mag} is related to the normalized magnetization per unit volume by :

$$M(\text{SI unit}) = \frac{\sum s\, dI}{v} = \frac{\int J(r)r^2dr}{2\int rdr} \tag{29}$$

with $dI = JLdr$ and $s=\pi Lr^2$ for a a cylindrical sample of length L. This equation can be used to define J_{mag} as :

$$J_{mag} = 3\frac{M}{R} = \frac{3}{2}\frac{\int J(r)r^2dr}{R\int rdr} \tag{30}$$

First of all we note that Bean formula ($J=3M/R$) is deduced immediately from eq.30 if one assumes J=cte. Secondly, we see that we also have $J_{mag} = J_{tr}$ if J=cte, otherwise these two currents are always different from each other.

For comparison we recall that in the a-c method of Bean J_{ac} is essentially equal to the critical current density average, not over the whole sample as in J_{mag} and J_{tr}, but only over a small penetration depth δr proportional to the amplitude h_o of the alternating field : $\delta r \approx \frac{h_0}{J(r)}$. Now we want to consider few concrete examples in order to illustrate the above claims.

(1) The first and perhaps the most interesting example is that of surface pinning defined here by assuming that J is restricted to a very thin layer $\Delta R<<R$ in which $J(r)$ is different from zero, that is $J(r) = J_s$ for $R\geq r>R-\Delta R$ and $J(r)=0$ for $r<R-\Delta R$.

In these conditions the calculation gives the following results :

$$J_{mag} \approx 3\, J_S \frac{\Delta R}{R}, \qquad J_{tr} \approx 2 J_S \frac{\Delta R}{R}, \qquad J_{ac} \approx J_S \quad (\Delta R << R) \qquad (31)$$

Thus for surface currents restricted to a surface zone of depth $\Delta R \ll R$ the relation between the currents depends on the ratio $R/\Delta R$ and is given by :

$$J_{mag} = \frac{3}{2} J_{tr} = J_{ac} \frac{3\Delta R}{R} \qquad (\Delta R << R) \qquad (32)$$

In conclusion, when J(r) increases with r as for surface pinning one has $J_{mag} > J_{tr}$ whereas the inverse is true when J(r) decreases with r. In particular surface pinnings advantages J_{mag} while surface weak links have the opposite effect.

4.3 Influence of self field on J_{tr} and J_{mag}.

This section is devoted to the influence of the self-field on both transport and magnetic critical current densities. In view of the complexity of the subject and for the purpose of clarity we make the simplifying hypothesis that the superconducting material is homogeneous (i.e J does not depend explicitely on r) and isotropic. We further assume that time effects are negligible and that the applied field $H \approx 0$ or at least $H << H_{sf}$(H_{sf} is the self field). We recall that all of these points were considered previously except the influence of the field dependence on J which will be the object of the next section. Under these conditions the possible difference between J_{tr} and J_{mag} comes from the self field only because of which J can depend implicitely on r via the relation $J = J(H_{sf}(r))$. To illustrate this point we consider the three different J(H) relationships below and calculate in each case the associated magnetic and transport currents as a function of the radius of the sample assumed again to be a very long cylinder (to avoid classical demagnetizing effects) :

$$J(H) = \frac{J_0}{1+|H|/H_0} \qquad \text{(Kim model)} \qquad (33)$$

$$J(H) = J_0\, e^{-|H|/H_0} \qquad \text{(exponential model)} \qquad (34)$$

$$J(H) = J_0 \left[1 - |H|/H_0\right] \qquad \text{(linear model)} \qquad (35)$$

$$R_0 = H_0/J_0 \qquad (36)$$

J_o and H_o are dependent on the temperature (respectively critical current density and field).

As will be seen just below it is very interesting to define a characteristic length $R_o = H_o/J_o$ (Eq.36) and then investigate J_{tr} and J_{mag} as a function of the reduced radius R/R_o. This is done in Fig.26a where it is clear that for $R<R_o$ we have $J_{mag} \approx J_{tr} \approx J_o$ (for all of the three laws considered) whereas for $R>>R_o$ the inequality $J_o>>J_{mag}>J_{tr}$ holds for all the above relationships (this result is just the opposite of that due to time effects). It is now important to make the following remarks

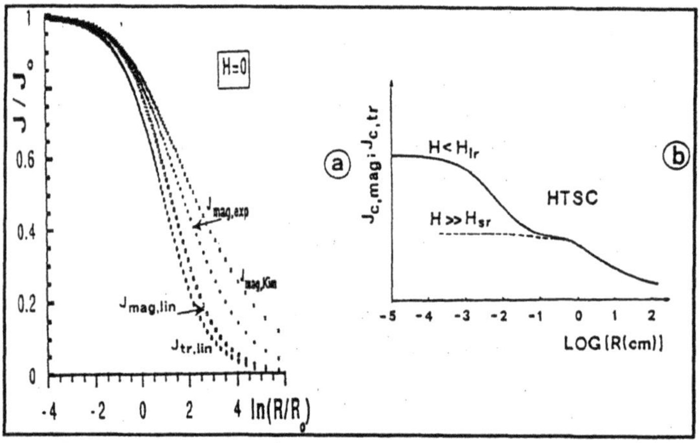

Figure 26: (a) Variation with the radius R of the sample of the reduced transport (J_{tr}/J_0) and magnetic (J_{mag}/J_0) currents as a function of the reduced radius $r = R/R_0$. This is qualitatively valid for intergrain currents. b) expected behaviour for intragrain currents. See text and eq.(33-36) for details.

(1) As we shall see later [4,35] the above equations alone cannot describe the field dependence of the magnetization and the associated currents in HTSC. As a matter of fact, we must include the equilibrium magnetization and in addition we need at least two such functions and hence two characteristic lengths R_{lr} and R_{sr}. The consequence of this on the effective critical current density is illustrated in Fig.26b where $J_{eff}(R)$ exhibits two different plateaux centered around R_{lr} and R_{sr}. Note that because of the reversible current J_{LA} the transport critical current density should present a third plateau for $R \lesssim \lambda$, not represented here.

(2) The estimated numerical values of the characteristic lengths R_o 20 to 200µm for intergrain currents, while for intragrain currents in standard $YBa_2Cu_3O_7$ $R_{lr} \approx$10-30µm and $R_{sr} \approx$1-5mm.

(3) The role of the self field always becomes negligible in the limit $H>>H_{sf}$. The latter is proportional to the product J.R in both transport and magnetic data.

5 Relationship between the macroscopic magnetization and the local current densities. Influence of field and temperature.

Going back to Figs.(8-11) we see that M(H) exhibits a variety of anomalous effects that cannot be described by the usual Bean model at any temperature. As shown in refs [4,35] the Bean model never applies for the low-H loops of intergrain currents nor for the central peak of intragrain currents. The model proposed in ref.[4,35] reproduces all of the features displayed at low-T as well as near T_C (Fig.8-11). For the sake of simplicity we will consider here separately the low-H cycles (connected with intergrain current J_{WL} threading the weak links) and the high H cycles induced by intragrain currents circulating inside the decoupled particles.

Large single crystals and decoupled grains.

According to Senoussi et al [4,35] to describe the magnetic properties of HTSC superconductors correctly we need at least two scaling laws similar to that of given by Eqs (33-35) together with the equilibrium magnetization, that is:

$$J_{tot}=J_{sr}(H/H_{sr}) + J_{lr}\,(H/H_{lr}) + J_{LA}(r,H) \tag{37a}$$

We recall that J_{LA} is the London Abrikosov's current (cf Fig.(7,11)); H_{sr} is a scaling field connected with short range (sr) pinnings potential of short range of order ξ (core pinning); H_{lr} is a scaling field connected with long-range (lr) interactions between vortices (here the spatial scale is of order λ). Physically, H_{lr} is due to the fact that this interaction tends to zero exponentially in the field region $H \approx H_{C1}$. In fact it is the current J_{lr} which is responsable for the central peak of all the previous Figures 7-16. We believe that this peak has the same physical origin as the peak near H_{C2} observed in some conventional superconductors. As in this case the elastic modulus C_{11} and C_{66} go down to zero rapidly (exponentially here) as $H \to H_{C1}$. In addition the modulus C_{44} is reduced by a factor of the order of the mass anisotropy ratio M/m as compared to the isotropic case [19]. Fig.27 illustrates how H and J would vary through the sample as a function of both its radius R and the maximum applied field H_{max}.

In fact it was demonstrated in [4] that within the limit of validity of the usual critical state picture the pinning force would vary as

$$B\,J = \left[1 + \mu_0 \frac{\partial M_{eq}}{\partial H_{eq}}\right] P_v \tag{37b}$$

Thus the critical current would exhibit a peak near H_{C1}. From the above discussion and comments we expect that both the shape and the amplitude of the central peak should depend on the radius of the specimen in a complicated way that has nothing to do with Bean model. We see from the above figure that the Bean model would apply only when the radius of the sample is smaller than the characteristic radius $R_{lr} \approx H_{lr}/J_{lr} \approx 10\ \mu m$ (In $YBa_2Cu_3O_7$).

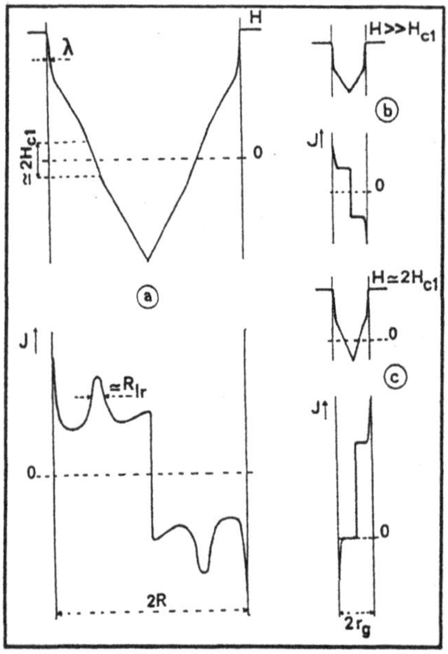

Figure 27. Predicted fields (a) and current (b) profiles within large single crystals (left) and fine oriented grains (right). The rapid variations near the surface is due to London-Abrikosov's currents (J_{LA}) whereas the peak around H_{C1} is due to J_{lr} (see text). This peak effect is more difficult to see in small grains (right figures) because the conditions $H \gg H_{C1}$ everywhere in the sample (c) is more easily realized there. d : When $H \leq H_{C1}$ everywhere in the material the current is dominated by J_{lr} alone and the peak is not visible either.

We tried different laws (see Eqs.33-36) for the field and the temperature variations of J_{lr} and J_{sr} and we performed extensive calculations of M(H) taking the function below (see [35] for more details) :

$$J_{sr} = J_{o,sr}\,[1 - \frac{T}{T_c}]^n\, e^{-|H|/H_{sr}}\, e^{-T/T_o} \qquad (38)$$

$$J_{lr} = J_{o,sr}\,[1 - \frac{T}{T_c}]^n\, e^{-|H|/H_{lr}}\, e^{-T/T_o} \qquad (39)$$

$$J_{LA} = M_{eq}(H)\,/\lambda \qquad R-\lambda \leq r \leq R \qquad (40a)$$

$$J_{LA} = 0 \qquad r < R-\lambda\ ,\ \ R \gg \lambda \qquad (40b)$$

Here $M_{eq}(H)$ is given by a curve like that of Fig.6 (Abrikosov's curve). Of course these equations must be completed by Maxwell's equation $\mathrm{curl}H(r) = J_{tot}(r)$ which couples the various currents written above via their respective self effects.

The central peak and the hysteresis cycle of decoupled grains and single crystals.

We have already noticed that the magnetization of crystallits exhibits a variety of unusual behaviours as a function of H, T and the shape of the specimen, especially its radius. To calculate the hysteresis cycle and reproduce the experimental data we used formula (29) (together with Maxwell's equation curlH(r) J(r). Beside the variable J(r) was replaced by the total critical current density below.

$$J_{tot}(r) = J_{sr} + J_{lr} + J_{LA} \qquad (43)$$

For the field and temperature variations of J_{sr}, J_{lr} and J_{LA} we used Eqs.(37-40) respectively (see [33] for more details).

The results of the calculations model and the comparisons with experimental data are presented in figures 28 to 30 below.

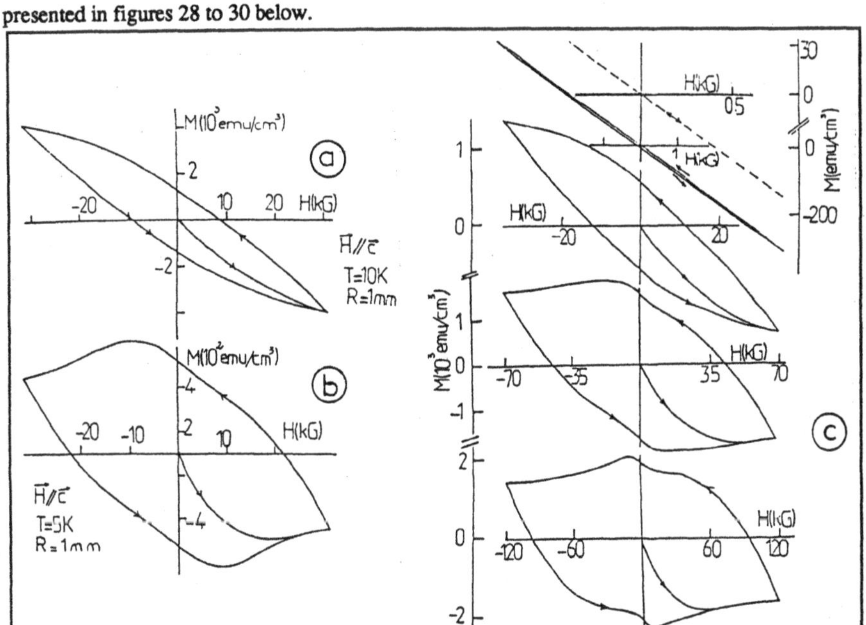

Figure 28. (a,b) Measured hysteresis cycles of two single crystals of $YBa_2Cu_3O_7$ (H parallel to the C axis) a: R≈1mm, b: R≈0.1mm. (c) Development of the calculated cycle (T=4K) of a single crystal as a function of the maximum cycling field H_{max}, $J_{01}=J_{02}=10^{10}A/m^2$, r_g=0.5mm, H_{01}=400kG.

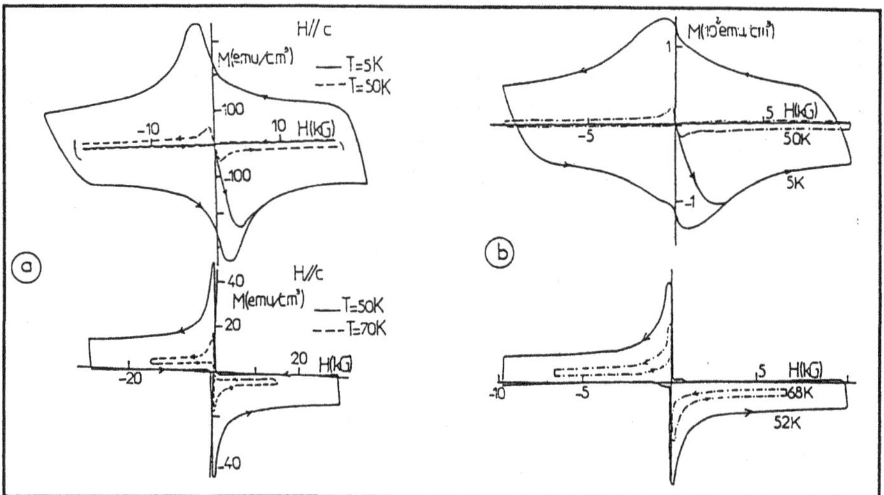

Figure 29. (a) M versus H for oriented grains diluted (~1%) in epoxy resin. H parallel to the c axis, $r_g \sim 5$ to 10mm. (b) Calculated cycle at the same reduced temperature and the same parameters as in (a), $J_{01} \sim 1.7 \times 10^{10}$ A/m², $J_{02} \sim 2J_{01}$, $r_g \sim 10$mm, $H_{01} \sim 200$kG, $H_{02} \sim 3$kG, n=1 and m=2.

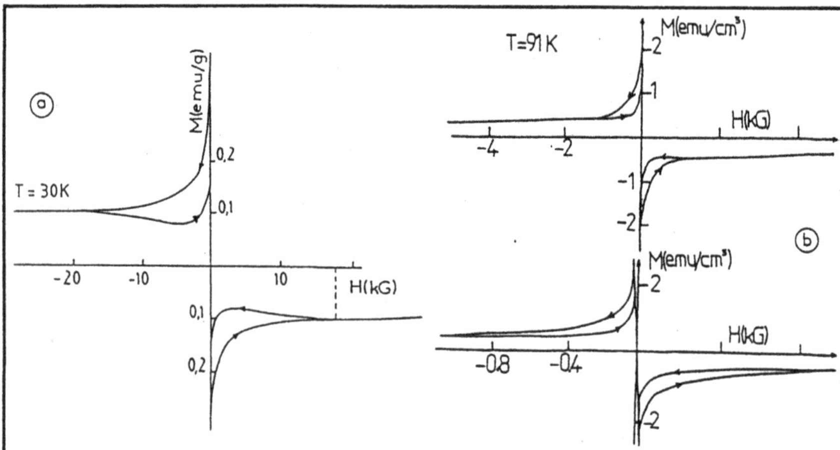

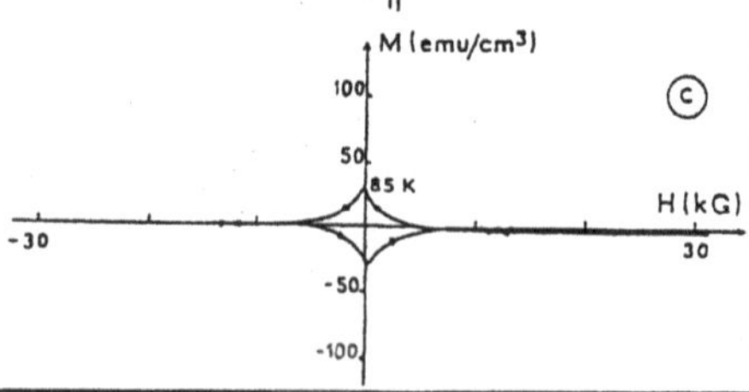

Figure 30: (a) Typical magnetic cycle of $La_{1.85}Sr_{0.15}CuO_4$ ceramics close to T_c. Similar curves are observed with other HTSC at temperatures very close to T_c. (b) Calculated cycle for T=91K at different field scales with approximately the same parameters as in (a). (c) Calculated cycle but here the scaling field H_{sr} is assumed to vary much rapidly than in (b), a situation often encountered in single crystals.

"Fish" or "butter fly" shaped cycle.

With the intragranular properties we have to mention another effect. In the model considered above we did not account for the so-called "fish" or "butter fly" shaped cycle exhibited by single crystals and (more rarely) by decoupled grains in some experimantal conditions. A typical example of such a cycle is shown in Fig.31

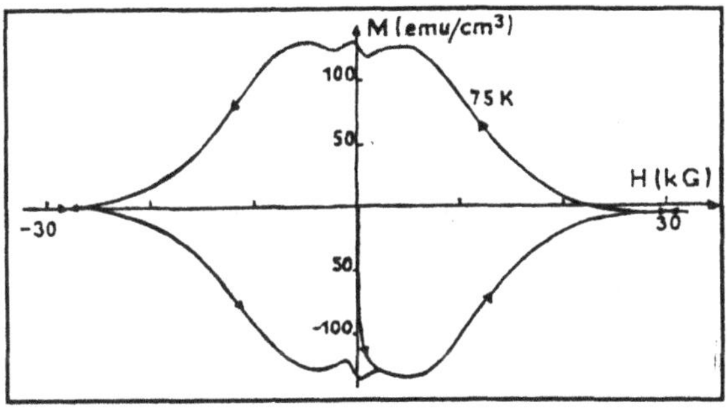

Figure 31: Example of "butter-fly" cycle for YBaCuO single crystal for which R=0.5mm.

Of course this unusual cycle can be reproduced by adding to Eq.(43) a fourth current term of the form $J_{bf}(\frac{|B-B_{1,bf}|}{B_{2,bf}})$ which reaches by a maximum at some field $B_{1,bf}$ and then decreases to zero on a field scale of order $B_{2,bf}$ (here the subscript 'bf' stands for 'butter fly'). However, we find this procedure not interesting and not physical if we do not know the physical origin of the butter fly current. It turns out that the butter fly shaped cycles exist only when the pinning is low (i.e for low J). Since collective behaviour of the vortex lattice lines is more likely applicable in the limit of low pinning disorder we suggest that the butter fly is associated with collective pinning effect. Indeed, according to ref. [7] there exists a domain in the T-B plane such that $H_{C1} \ll H \ll H_{C2}$ and $0 \ll T \ll T_c$ where J increases with B as $B^{3/4}$ and then drops exponentially with both B and T. According to [7] this interval is defined by :

$$T > \max(T_L^*, T_y^*) \qquad (44)$$

with $T_L^* = \phi_o^{3/2} \frac{\sqrt{B}}{(2\pi\kappa)^2}$ and

$T_y^* = [\gamma C_1/6]^{1/3}$

Here $\kappa=\lambda/\xi$, γ a pair correlation function characterizing the disorder potential in the limit of large wave lengths distortions (or of small wave vector q) and $C_1=(\Phi_o/4\pi\lambda^2)$ $[1-\ln(q\xi)]$ is the vortex tension ($\sim C_{44}$ of isolated lines).

This is consistent with the experimental data of Fig.31 which show indeed that J(B) increases rather slowly with B in the region $B<B_{1,bf}$ and then drops fastly at higher B.

Note that the theory of ref.[7] neglects anisotropy and assumes $B>>H_{C1}$ as well as weak pinning (i.e weak γ). This problem will be discussed in more details elsewhere.

Intergranular contribution in the model

We consider the grains network as a set of filaments of very high J and high H_{C1} materials embedded in a second material (weak links) of low $J(=J_{WL})$ and low H_{C1} $(=H_J)$. Then the total magnetization is the sum of three terms sketched in Fig.32 below:

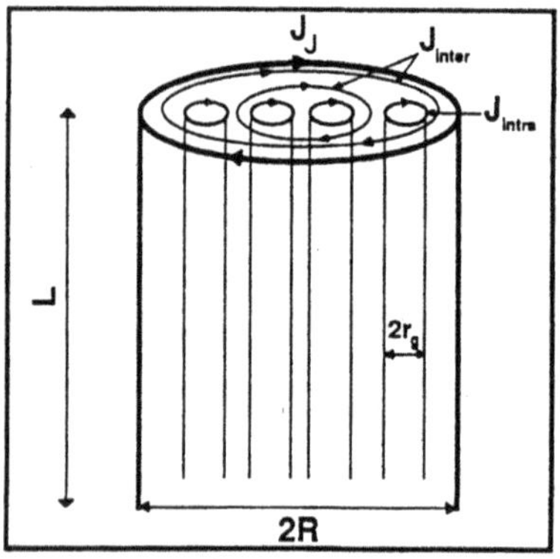

Figure 32: Sketch of the three current densities used in the model (see text) for the calculation of the low-H husteresis cycle of ceramics. J_J is the Josephson reversible current restricted to flow in a depth λ_J, J_{inter} is the current inside the junctions and J_{intra} is the current of the individual grains. For simplicity we assume $H<H_{C1}$(grains) so that J_{intra} is reduced to J_{LA} exclusively.

Fig.32 is a schematic illustration of the three current densities contributing to the total magnetization. J_J is the Josephson reversible current (the analogue of J_{LA} for perfect materials) restricted to flow in a depth λ_J, J_{inter} is the current inside the junctions and J_{inter} is the current of the individual grains. The latter is limited to J_{LA} only if $H<H_{C1}$.

$$M_{tot} = M_{inter} + M_{intra} + M_{J,eq} \tag{41}$$

or in terms of currents :

$$M_{tot} = \frac{1}{R^2} [\int J_{inter}(r)\, r^2 dr + \int J_{intra}(r)\, r^2 dr + \int J_J(r)\, r^2 dr] \tag{42}$$

It is to be recalled that M_{inter}, M_{intra} and $M_{J,eq}$ are not independent but coupled together via their self fields. To determine this coupling we assume that the sample is very long (L>>R) so that a given filament sees only its own self field, plus the field produced by the large loops (of J_{inter} and J_J) surrounding it, plus the applied field. For the same reasons J_J is imposed by the external field exclusively (since it is not surrounded by any other current as it flows at the edges of the sample in the depth λ_J) but it influences the other inner loops (of J_{inter} and J_J). For concrete calculations these equations need to be completed by Maxwell's equation relating J(r) and H(r) and by some assumptions about the field variation of each current terms. It was found in ref.[4,33] that an exponential relationship of J as a function of H of the kind given by Eq.(34) reproduces correctly the experimental data. Finally it is to be added that to reproduce correctly the high field hysteresis cycle of decoupled grains and single crystals we must also take into account (in Eqs.41,42) the fact that J_{intra} itself includes three terms we have called J_{LA}, J_{sr} and J_{lr}. The details of all these calculations can be found in [33]. Here we shall only compare some experimental and theoretical data.

The low-H cycle of granular samples

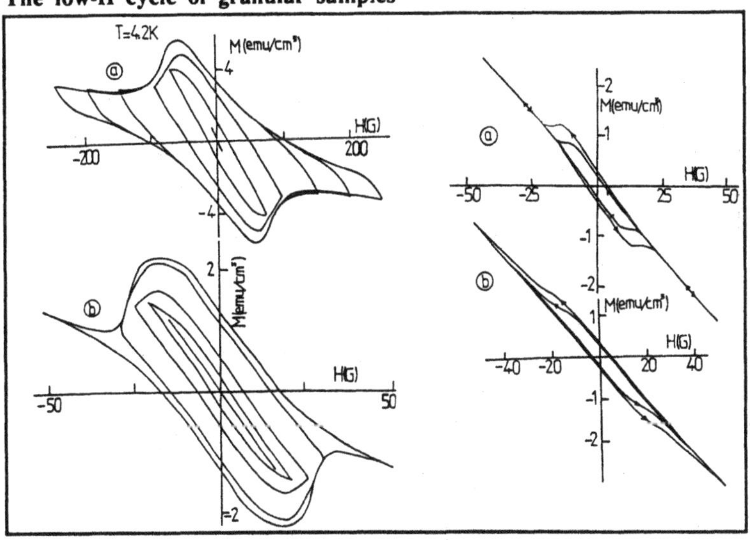

Figure 33 (left): Development of the low-H cycle of a granular pellet as a function of the maximum applied field H_{max} for R = 3mm, r_g = 1,5μm, and H^W_{C1} = 2oe, a : experimental, b : calculated.

Figure 34 (right): Influence of the radius R of the pellet on the low-H cycle : (a) calculated, (b) experimental. R=1 and 3 mm for the inner and outer cycles respectively.

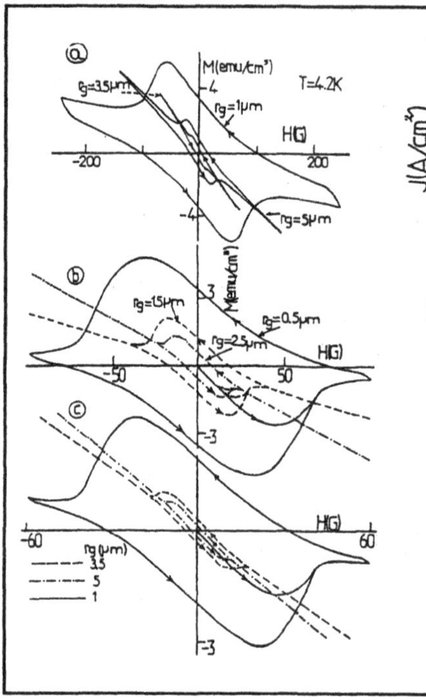

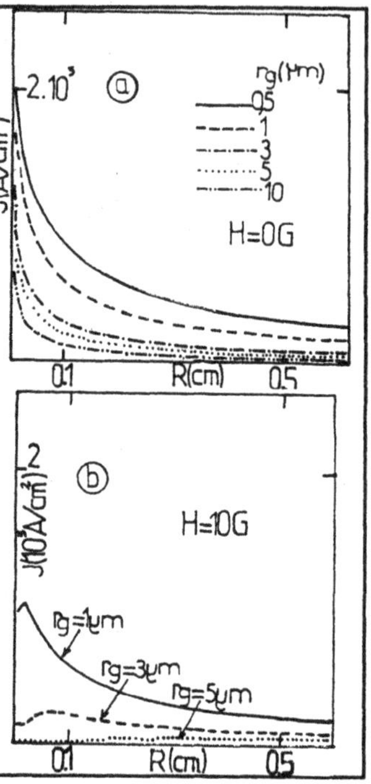

Figure 35(left): Influence of the radius of the grains (r_g) on the low-H cycle for different R of the pellet; a: experimental, b,c : calculated.
Figure 36(right):Calculated critical current density as a function of the radius R of the pellet at H=0 and H=10oe, for the indicated grains radius r_g.

The series of cycles displayed in figure 33-36 allows to compare some experimental and calculated data as a function of H, R, r_g, H_{max} and T. Here we have assumed that the applied field is lower than H_{C1} of decoupled grains and single crystals so that J_{intra} is reduced to the London current J_{LA} only. It is clear from these figures that the above model reproduces the most characteristic features of the experimantal data for several orders of magnitude of all these parameters (namely : R varies fro 1 µm to 1mm, H_{max} from zero to 200kG, T from zero to T_c). This is achieved with a minimum number of fitting parameters (J_o,T_o H_{sr}, H_{lr}, J_{wl} and J_J).

5 The I-V characteristic comparison between theoretical and experimental results.

We have already discussed the J versus E characteristic of superconductors from a phenomenological point of view (Eq.26) when we have compared the electric field criterions defining J_{tr} and J_{mag}. In fact these I-V (or equivalently the J -E) characteristics have been investigated extensively in the past in the framework of several theoretical models. Here we would like to discuss briefly three common approaches corresponding to increasing degrees of complexity. Indeed, as in the previous examples dealing with the interpretation of magnetic and transport data our purpose is to examine the experimantal constraints and other problems which can limit the applicability of these theories to the experimantal data. We have already examined some of these problems in particular those connected with anisotropy, self-field, spatial average, granularity ...We are going to examen now other difficulties for the interpretation of data.

(1) We recall again that the V(I) function is often written in the form of a power law $V=kI^m$. This power law is rather phenomenological and has no real justifications, especially dealing with the properties of fundamental quantities such as the pinning energy.... However, it is useful for rough characterization of the material, particularly when its shape is ill defined or in the presence of other experimental problems as those enumerated just above. In addition, the parameters m and k depend in general on T, H, the shape of the sample and sometimes on its magnetothermal history.

(2) A second and much more general description of the E(J) relationship is the famous flux-creep model of P. Anderson [1] which predicts the following relationships between the electric field E and the current density J (in this section we replace E_{tr} by E and J_{tr} by J to simplify the notation) in transport measurements.

$$E_{creep} = E_o \exp\left[-U_p(1-J/J_{max})/kT\right] \qquad (U_p/kT >> J_{max}/J), \quad (45)$$

$$E_{TAFF} = 2E_o \frac{JU_p}{J_{max}k_BT} \exp\left[-U_p/k_BT\right] \qquad (U_p/kT << J_{max}/J), \quad (46)$$

$$E_o = XB\nu_o \quad (47)$$

Here the notations 'creep' and 'TAFF' mean flux-creep and thermal assisted flux flow limit respectively, ν_o is an attempt frequency discussed briefly previously, which generally depends on both B,T (via the normal resistivity ρ_n) and the defect structure. X is the jump distance of the flux bundle. In principle, X should also depend on T and H. To compare with experiments it is important

(i) to make clear the limits of the validity of such equations,

(ii) to discuss the real meaning of the prefactors ν_o, X ... (usually these terms are not precisely defined in the literature) and

(iii) to define the experimental procedure, most appropriate for comparison with the above predictions.

First of all we note that equations (45-47) assume tacitly that J is uniform within the material and that the pinning potential U_p is independent of T up to about $T \approx T_c/2$. It is to be recalled that in conventional superconductors it is believed that U_p is function of $H_{C2}(T)$ and thus depends on temperature via this field principally. The situation is certainly very different for HTSC since the depinning line is very far from $H_{C2}(T)$ curve there.

5.1 On the physical meaning of the prefactors v_o, X, ρ_n.

In the literature on HTSC, v_o is always refered to as an "attempt frequency" of the order of $10^{10}s^{-1}$. However it turns out that v_o depends explicitely on $\rho_n(T,B)$, T, B and on the defect structure [4]. In addition, we have seen (when we discussed the structure of vortices in HTSC) that ρ_n is the normal state resistivity at the temperature of the measurements (the exact meaning of which is not yet very clear). It is expected that ρ_n depends both on T (as it tends towards the residual resistivity as $T \rightarrow 0$) and perhaps on B because of the effect of ordinary magnetoresistance which can contribute significantly to ρ_n particularly whenever ρ_n becomes very low. Neglecting magnetoresistance effects and assuming that ρ_n has the same physical meaning as the normal state above $H>H_{C2}$ (T) we can write

$$\rho_n = \rho_o + \rho_{exc}(T/T_c)^n \qquad (48)$$

Here ρ_{exc} is the resistivity due to the collisions of the 'normal electrons' within the core of the vortices with phonons and other possible thermally induced excitations. The exponent n is unknown. However measurements of the magnetoresistivity [36] in very high pulsed field (45 Tesla) as a function of T and interpolation of the data to low temperatures suggest that ρ_n would have the form
$\rho_o + \alpha T$ for $T<T_c$. It is to be recalled that the true resistivity can be much more complicated in reality, particularly at $T \ll T_c$, if the vortex core is not totally normal but has an internal structure (discrete energy spectrum) as already discussed [9-13].

We have insisted on the great difficulties of how to define v_o, X, ρ_n which enter Anderson's equations (45-47) to show that it is often a non sens to try to deduce quantitative conclusions (From I-V measurements) about those prefactors without having in mind these problems. It is interesting to note that as far as v_o is concerned the situation is somewhat simpler in magnetic than in transport measurements, especially in relaxation experiments. This is because

v_0 enters flux-creep equations such as Eq.(24) only logarithmically term instead of linearly in transport equations (45-47).

5.2 Influence of the experimental procedure (zfc or fc) on the measured I-V characteristic.

In the previous section we insisted on the great difficulties to define some prefactors (v_0, X, ρ_n) of the theoretical flux creep equations. The I-V characteristic is generally determined following the standard procedure outlined below. The sample is cooled in zero applied current (and also in zero field) down to the temperature of measurements. The V(I) curve is then plotted at constant T starting from I=0 and then increasing the current up to some maximum value $I_{max}>I_c$.

By analogy with magnetic measurements we shall refer to this procedure as zero field cooled (ZFC) since cooling in a finite current is equivalent to cooling in a field gradient given by curlH=J.

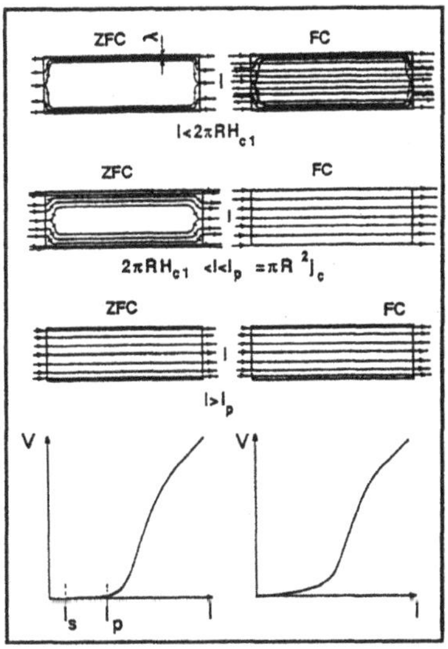

Figure 37: The top figures illustrate how the applied transport current penetrates the sample gradually in the zero field cooled method (left) and almost uniformly in the field cooled case (right). The terms ZFC and FC are justified because the transport current is always accompanied by its self field. (bottom) the V-I characteristics should be different for FC and ZFC processes.

Fig.37 shows that the I-V characteristic determined this way is not related in any simple way to flux creep equations (45-47). These equations assume indeed that I is always uniformly distributed across the wire for any value of the external current. It is clear that this is the case only at large I. In fact the most natural way to get the most uniform distribution of J in the material is to cool it in the measuring current from above Tc (or at least from the depinning line) down to the temperature of interest and then determine V or E ($=E_{FC}$) in this cooling current. This is in fact the usual procedure followed in resistivity measurements. The difference between ZFC and FC data are expected to be very large in the flux creep regime where the V(I) relationship is extremely non linear. Of course in the linear diffusive regime the difference should be negligible. Lastly, it is to be emphasized that if the effective field ($H+H_{sf}$) is lower than H_{C1} the measured critical current (whatever the experimental procedure used ZFC or FC) includes a London reversible term, the contribution of which grows as 1/e until $e \approx \lambda$ (recall that e is the thickness of the sample whereas λ is the London penetration depth). Moreover, in many cases the critical current density J decreases much more rapidly with T than the London current $J_{L\text{-}A}$ does. This also implies that the relative contribution of the London current to J_{tot} grows equally with T. We finally recall that flux creep equations (45-47) have no simple relation with the I-V curves of intergranular currents, as I is highly heterogeneous in these cases (see [4] and Fig.(26) for more details on this point).

6 Summary of the various information that can be deduced from the hysteresis cycle of superconducting materials.

Despite the complexity of the problems outlined above it is nevertheless possible to generalize the Bean model to account for some of these complicated situations and get many qualitative information quite rapidly from the hysteresis cycle.

In this section we wish to summarize the most characteristic information that are directly related to the shape of the hysteresis cycles as those displayed in Fig.9

At first from the initial slope of the M(H) curve (in the limit $H \rightarrow 0$) we can get both the demagnetizing factors $N_{//}$ and $N_{\perp}$ (for H parallel to the principal axes of the sample (assumed to be an oblate ellipsoïd) and an order of magnitude estimate of the fractional superconducting volume V_p :

$$N_{//} = \frac{\chi_{//}}{2\chi_{\perp}+\chi_{//}} \quad (49)$$

$$N_{\perp} = \frac{2\chi_{\perp}-\chi_{//}}{2\chi_{\perp}+\chi_{//}} \quad (50)$$

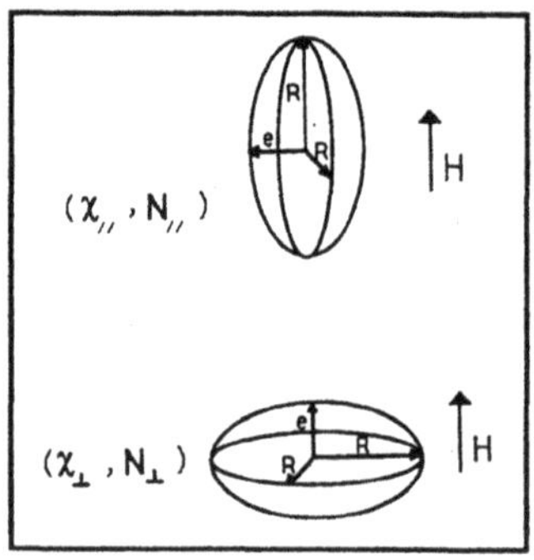

Figure 38: Shows how $\chi_{//}$ and $\chi_{\perp}$ are defined with respect to the principal axes of the sample assumed to have the symmetry of a oblate ellipsoid.

$$V_p = \frac{2}{1-2\lambda_{eff}/r_g} \, \frac{\chi_{\perp}\chi_{//}}{2\chi_{\perp}+\chi_{//}} \quad (\text{if } r_g \gg \lambda_{eff}) \quad (51)$$

with $2N_{//} + N_{\perp} = 1$, $\quad N_{\perp} = \frac{1+u^2}{u^3}[u - \arctan(u)]$, $\quad u = \sqrt{\frac{R^2}{e^2} - 1}$, (52)

Here $\lambda_{eff} = \lambda$ (London) for single crystals and λ_J for granular samples (see Eq.(8)) whereas $\chi_{\perp}$ and $\chi_{//}$ refer to the initial susceptibilities for the two directions of the ellipsoid as illustrated in the figures.

Note however that as far as granular samples are concerned Eq.(51) giving V_p is useful only in the limit $V_p \ll 1$ as it is often the case for oriented grains dispersd in epoxy resin. Moreover in this case χ refers to the slope of M(H) in the region $H^{W}_{C2} < H < H_{C1}$.

(2) The magnetic intergrain current density is directly related to the low-H hysteresis cycle of Fig.39 (field region $H < H^{W}_{C2}$) by the approximative equations below [4,35] :

$$J_{WL,mag} \approx 1.5 \frac{P}{\Delta P} \frac{M^+ - M^-}{R} \quad (53)$$

or (if $\lambda \ll \langle r_g \rangle$)

$$J_{WL,mag} \approx \frac{1.5}{\beta . R} \frac{M^+ - M^-}{\langle 2\lambda / r_g \rangle} \quad (54)$$

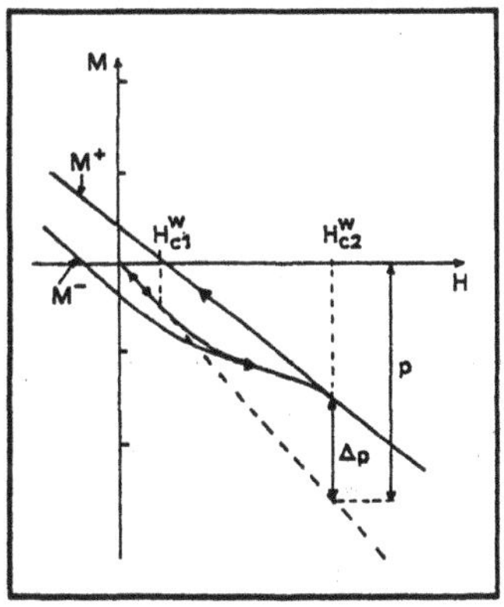

Figure 39. Illustrates the low-H cycle due to the weak-link network and defines the parameters factors p and Δp entring eq.(53, 55) defining the magnetic intergranular current. $H \le H_{C1}$ here.

Here R is the macroscopic radius of the pellet whereas r_g is the radius of the grain. β is a factor of the order of 1 to 1.5 in most cases.

(3) At still higher field, such as $H^{W}_{C2} \le H \le H_{C1}$, the slope dM/dH is related to the average value of r_g entering the ratio λ / r_g (see again Fig.39). If the sample is highly compact (V_p >0.85 to say) and in the limit $\langle r_g \rangle \gg \langle \lambda \rangle$ we can write

$$\frac{\Delta P}{P} \approx \beta \left\langle \frac{2\lambda}{r_g} \right\rangle \quad (\beta \approx 1 \text{ to } 1.5). \quad (55)$$

Therefore if we know λ we can estimate the average radius defined by $<r_g^{-1}>^{-1}$ from low field data. On the contrary if we have a reference sample with a known r_g we can deduce λ.

(4) For much higher applied field such that the grains are fully penetrated by the field ($H>2H_p$) we can estimate the intragrain critical current density from the Bean equation :

$$J_{mag} = 1.5 \frac{M^+ - M^-}{<R>} \qquad (56)$$

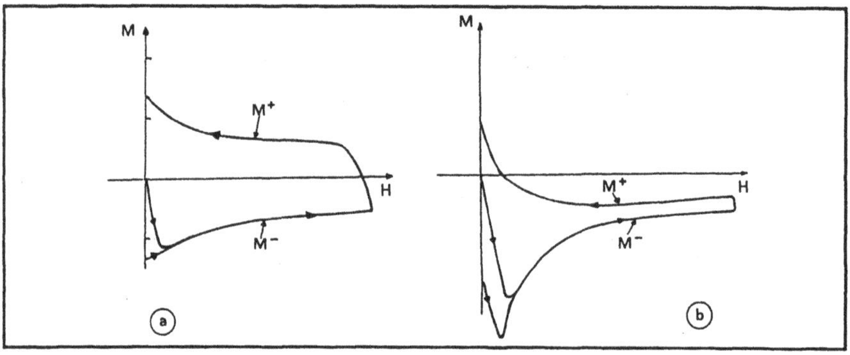

Figure 40: (a) Defines M^+ and M^- entering eq.(56) defining the magnetic intergrain current. Here $H_{max}>H_p$ necessarly. This cycle, which is symetrical with respect to the field axis, is obtained with large single cristals. (b) The cycle recorded with very small crystals ($R \approx 1\mu m$) is highly non symetrical with respect to the field axis.

Now R refers either to the average radius of the single crystal or to that of the grains, for granular samples. At this point we note that both formula (54-56) and depend on r_g(for granular materials) but through different average values. In fact it can be shown that the inequality $<r_g> \geq <r_g^{-1}>^{-1}$ is always true. Despite this inequality we can assume that $<r_g> \approx 2<r_g^{-1}>^{-1}$ and thus use equation (55) (i.e low-H cycle of granular sintered materials) to estimate $<r_g>$ to be put in eq.(56). We propose the following qualitative rule for estimate the intragranular current:

$$J_{mag} \approx \frac{M^+ - M^-}{\lambda} \frac{\Delta P}{3P} \qquad (57)$$

(5) Other useful information related to the symmetry of the high field cycle.

As we know the measured magnetization can be decomposed into an irreversible and a reversible magnetization as follows :

$$M_{irr}^{+} = \frac{M^{+} - M^{-}}{2} = J_{mag}R/3 \tag{58}$$

$$M_{irr}^{-} = - M_{irr}^{+} \tag{59}$$

$$M_{rev}^{+}=M_{rev}^{-}=M_{eq} \tag{60}$$

$$M_{eq}(H) = M_{rev}(H) + M_{irr}(H)\frac{R}{4R_0}\exp(-\frac{B_{eff}}{B_o}) \tag{61}$$

Here the signs + and - refer to increasing and decreasing field respectively. Figure 41 shows the apparent M_{rev} and the apparent H_{C1} as deduced from experiments

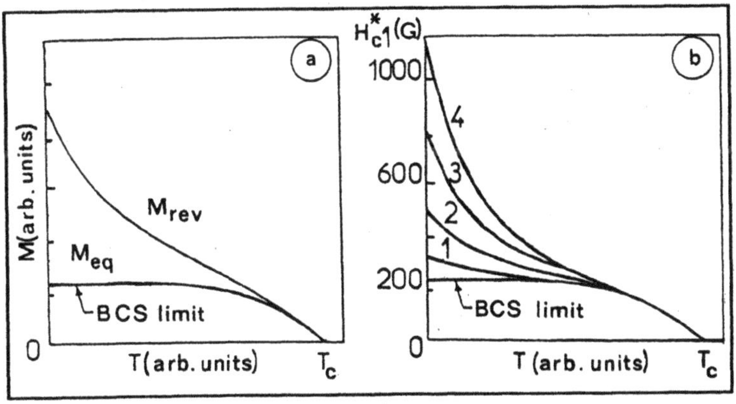

Figure 41: (a) It compares the equilibrium magnetization of London -Abrikosov used in the calculation of the total magnetization (eq(42, 61)) and the reversible magnetization defined from the same cycle by $M_{rev}=(M^{+} + M^{-})/2$. Here the radius R is assumed to be $\approx 200\mu m$ whereas $J_o \approx 2x10^{10} A/m^2$. (b) It shows that the error in H_{c1} becomes also considerable for large radius R of the sample. The labels are such that (1,2,3,4 correspond to R= 1, 10, 100, 500μm) and $J_c=510^6 A/cm^2$ for 1 and 2; $J_c=210^6 A/cm^2$ for 3 and $J_c=810^6 A/cm^2$ for 4.

From equations (58 and 59) we see that $M_{irr}(H)$ is symmetrical with respect to the field axis and is proportional to R and to J_{mag}. On the contrary M_{rev} is to a first approximation independent of R, reversible, negative (for H>0) and symmetrical with respect to the origin.

From these remarks we can easily conclude that a hysteresis cycle which is symmetrical with respect to the H axis (as in fig.40) implies large grains (Fig.40a) while a hysteresis cycle like the one displayed in Fig 40b at T=4.2K implies small effective grains radii. Note that since J decreases extremely rapidly with T whereas M_{eq} decreases rather slowly in the same temperature range we expect that at high enough tempertaures the cycle will always become dissymetrical with respect to H axis (fig.40b). It seems that in the case of usual good quality $YBa_2Cu_3O_7$ ceramics we have $|M_{irr}| \approx |M_{rev}|$ with $r_g \approx 1\mu m$ at T≈4.2K. As we have mentioned all the above equations and the associated quantities represent rather qualitative values which are generally different from their microscopic counterparts.

REFERENCES

[1] P. W. Anderson, (a) Phys. Rev. Lett., **9** (1962), 309.

[2] C. P. Bean, (a) Phys. Rev. Lett. **8** (1962) ; (b) Rev. Mod. Phys. **36** (1964), 31.

[3] A. M. Campbell, J. Phys. C **2** (1971), 1492 ; Ibid, **4** (1971), 3136.

[4] S. Senoussi to be reported in J. de Physique III (1992).

[5] R. Labusch, Phys. Stat. Sol. **19**, (1967), 715; Cryst. Lattice Defects **1**, (1969) 1; Phys. Lett. **22** 1969) 9.

[6] A. I. Larkin, Yu. N. Ovchinnikov, J. Low Temp. Phys. **34** (1979), 409.

[7] M. V. Feigel'man, V. M. Vinokur, Phys. Rev. B **41**, (1990), 8986;
M. V. Feigel'man, V. B. Geshkenbein, A. I. Larkin, V. M. Vinokur, Phys. Rev. Lett. **63** (1989) 2303.

[8] V. B. Geshkenbein, A. I. Larkin, Zh ETF **95** (1989), 1108.

[9] P. G. de Gennes, "Sperconductivity of Metals and Alloys, published by Addison-Wesley Publishing Compny' INC, p. 20-25.

10] C. Caroli. P.G. de Gennes, and J. Matricon, Phys. Lett.**9**, (1964) 307.

[11] J. Barden, R. Kùmmel, A.E. Jacobs and L. Tewordt, Phys. Rev. **187**, (1969) 556.

[12] A. F. Andreev, Sov. Phys. JETP **19** (1964) 1228.

[13] R. Nicolsky, Cryogenics **29** (1989) 388.

[14] (a), K.K. Likharev, Sov. Phys. JETP. **34** 906(1972); (b), Rev. Mod. Phys. **51**, (1979) 101.

[15] M.Y. Kypriyanov, K.K.Likharev, L.A.Maslova, Proc. of the 14th Int. Conf. on Low Temp. Phys. **4** 104 (1975).

[16] I.O. Kulik and I.K. Yansen, Josephson Effect in Superconducting Tunnel Structures Moscaw (1971).

[17] D. Feinberg, preprint (1991); G.Blatter, V.Geshkenbein, preprint 1991).

[18] D.S. Fisher, M.P.A. Fisher and D.A. Huse, Phys. Rev. B **43** (1991) 130.

[19] E. H. Brandt, J. of Mod. Phys. B **5** (1991)751; PhysicaC **162-164** (1989) 1167.

[20] E. Zeldov, N. M. Amer, G. Koren, A. Gupta, M. W. Mc Elfresh and R. J. Gambino, Appl. Phys. Lett. **56**, (1990), 680.

[21] E. Zeldov, N. M. Amer, G. Koren and A. Gupta, Appl. Phys. Lett **56** (1990), 1700.

[22] P. Manuel, C. Aguillon and S. Senoussi, Physica C **177** (1991) 281.

[23] J. Barden and M.J. Stephen, Phys. Rev.A **140** (1965) 1197.

[24] This can be justified qualitatively by saying that a moving vortex does not drag with it any electron just like a phonon wave paquet does not carry with it any atom on the scale of the sample. In this way we feel that as the vortex crosses a sample submitted to some current the electrons which carry such a current become normal during the time they are traversed by the core of the vortex and can thus be scattered by defects suffring a resistance.

[25] C.R. Hu and R.S. Thompson, Phys. Rev. B (1972) 110.

[26] C. Aguillon, D. G. Mc Cartney, P. Regnier, S. Senoussi, G.J. Tatlockt J. Appl. Phys. **69** (1991) 8261.

[27] S. Senoussi, S. Hdjoudj, R. Maury, An. Fert, Physica C **165**, (1990), 364.

[28] S. Senoussi, S. Hadjoudj, C. Weyl and P. Fonder Physica C **165**, (1990), 199.

[29] M.A.K.Mohamed, J.Jung, to appear in Phys. Rev. B (1991).

[30] M. Tinkham, "Introduction to Superconductivity", Ed. Robert E. Krieger Publishing Comp. Malabar, Florida .

[31] S. Senoussi, M. Oussena and S. Hadjoudj ; J. Appl. Phys. **63** (1988), 4176.

[32] S. Senoussi, M. Ousséna, J. B. Bieri, J. Arabski, and R. Reich ; Progress HTS, Vol. **9** ed.

[33] S. Senoussi, and C. Aguillon Europhys. Lett. **12**, (1990) 273. C. Aguillon and S. Senoussi, J. of Less Comm. Met. **164-165** (1990), 1061.

[34] The proportionality factor a is expected to go to zero at T=0 and then to increase monotonically around 1 with increasing temperature.

[35] S. Senoussi, C. Aguillon and P. Manuel, Physica C **175** (1991) 202 ; S. Senoussi, C. Aguillon and S. Hadjoudj, ibid, 215.

[36] M. Oussena, S. Senoussi, G. Collin, J.M. Broto, H. Rakoto, S. Askenazy, J.C. Ousset, Phys. Rev. B **36**, 4016 (1987).

THE EFFECT OF THE INTRAGRANULAR IRREVERSIBLE MAGNETIZATION ON THE INTERGRANULAR AND INTRAGRANULAR CRITICAL CURRENTS

Ginzburg S.L., Khavronin V.P., Luzyanin I.D.

Leningrad Nuclear Physics Institute, Gatchina 188350, USSR

ABSTRACT

Intragranular irreversible magnetization effect on the transport and magnetic properties of ceramic high-T_c superconductors is studied theoretically and experimentally.

The ceramic high-T_c superconductors are known to be a set of grains linked to one another with weak Josephson coupling. In the system like this under the fields below the lower critical field of the grains themselves, $h_{c1\ gr}$, a macroscopic superconductive properties (transport currents, magnetic shielding, etc.) are governed by intergranular links and described by the critical state equation with critical current $j_c(H)$. In this case, Meissner intragranular shielding is taken into account through the introduction of effective magnetic permeability μ_{eff}. This range of magnetic fields is the subject of low-field electrodynamics (see [1,2]). In the fields of $h > h_{c1\ gr}$, intragranular magnetization becomes irreversible and the field penetration into grains is described by the intragranular equation of critical state with $j_{cgr}(h)$. Since $j_{cgr}(h) \gg j_c(h)$ always, under certain conditions this irreversible magnetization can strongly affect the properties of intergranular links, thus affecting the macroscopic electric and magnetic properties. In particular, it results in a hysteresis of the local magnetic field $\vec{h}(\vec{r})$ and, in consequence, in a hysteresis of critical transport current. The equations describing of conventional superconductor are insufficient to describe of these phenomena, therefore they must be supplemented with the equations where intragranular irreversible magnetization is taken into account.

Given the grains, it is easy to show that the total set of electrodynamics equations appears as follows:

$$\mathrm{rot}\vec{h} = 4\pi\vec{j},$$
$$\mathrm{div}\vec{h} = 4\pi\rho,$$
$$\rho = -\mathrm{div}\vec{M} \tag{1}$$

where $\vec{j}$ is intergranular currents density, $\vec{M}_{eff}$ is Josephson medium magnetization dependent on local field $\vec{h}(\vec{r})$:

$$\vec{M}_{eff}(\vec{h}) = f_s\langle\vec{M}_{gr}(\vec{h})\rangle, \tag{2}$$

where f_s is the part of volume filled by superconductor, $\langle\vec{M}_{gr}(\vec{h})\rangle$ is average irreversible magnetization of the grains. Incidentally, $\vec{M}_{eff}(\vec{h})$ is dependent not only on the value of the local field $\vec{h}(r)$ but also on its prehistory. Eqs. (1)-(2) together with the equation of critical state

$$|\vec{j}| = j_c(h); \tag{3}$$

and boundary conditions compose the total equations set provided $j_c(h)$ and $\vec{M}_{eff}(h)$ are known in the explicit form.

Due to the fact that $\vec{M}_{eff} = 0$ in conventional superconductors, combined equations $\mathrm{rot}\vec{h} = 4\pi\vec{j}$ and eq. (3) are general enough to describe them. The conventional critical state theory is based on this equations set. In granular superconductor, indeed, there is granule-generated magnetization $\vec{M}_{eff}(h)$ besides, therefore magnetic density ρ must be taken into account. A total set of the eqs. (1) - (3) determines the generalized critical state theory for granular superconductors. As it will be seen below, the properties of such superconductors can significantly differ from ones of the conventional superconductor.

It is convenient to rewrite eqs. (1) as follows:

$$\vec{h}(\vec{r}) = \vec{h}_1(\vec{r}) + \vec{h}_{dip}(\vec{r}) + \vec{H}$$

$$\vec{h}_1(\vec{r}) = \mathrm{rot}\int \frac{d\vec{r}'}{|\vec{r} - \vec{r}'|}\, \vec{j}(\vec{r}')$$

$$\vec{h}_{dip}(\vec{r}) = \mathrm{grad}\int \frac{d\vec{r}'}{|\vec{r} - \vec{r}'|}\, \mathrm{div}\vec{M}_{eff}(\vec{r}') \tag{4}$$

where $\vec{H}$ is an external field, $\vec{h}_{dip}$ is a dipole field of the grains, and $\vec{h}_1$ is a field created by intergranular shielding supercurrents. Apparently, $\vec{h}_{dip}$ equals zero in conventional superconductors. Let us note that boundary conditions are automatically taken into consideration in eqs.(4). As can be see from eqs. (2) - (4) the field $\vec{h}$ (involved in $\vec{M}_{eff}$ and j_c) is generated by three sources, i.e. supercurrent $j_c(h)$, magnetization of the grains $\vec{M}_{eff}(\vec{h})$ and external field $\vec{H}$.

First of all, we note the following condition can be easily realized:

$$h_{dip} \gg h_1, \tag{5}$$

which means, in this case, intragranular magnetization to make a essential contribution to different phenomena. Therefore it is important to understand the criterion of fulfilment of this condition. It is clear from the expression (4) for h_1 that $h_1 \sim j_c d$, where d is a sample size; and it will be seen below (see eqs. (11), (13)) that $h_{dip} \sim n j_{c\ gr} \langle R_{gr} \rangle$, where n is a demagnetizing factor of the sample and $\langle R_{gr} \rangle$ is a characteristic size of the grains. Although d is usually several orders higher than $\langle R_{gr} \rangle$, but $j_{c\ gr}$ is several orders as much j_c too. Therefore condition (5) can be easily fulfilled in actual samples.

Further, in conventional superconductor there is a single source of the irreversibility, i.e. ambiguity of the current determination in eq. (3). In granular superconductor, there is an additional source of the irreversibility, i.e. ambiguity of function $\vec{M}_{eff}(\vec{h})$, which appears in the region of flux trapping into grains. In particular, this results in hysteresis of $j_c(h)$ and therefore in transport current hysteresis. Let us demonstrate the possibility of hysteresis of $j_c(h)$ by the simple example where condition (5) is met. Incidentally, it is $\vec{h}_{dip}$ that gives a main contribution in eq. (4) and $\vec{h}_1$ can be calculated from perturbation theory. Then the local field $\vec{h}$ is determined by fields $\vec{h}_{dip}$ and $\vec{H}$ only while $\vec{h}_{dip}$ can be obtained from eqs. (2)-(4) assuming $h_1 = 0$. In this case, apparently, $j_c(\vec{h}) = j_c(|\vec{H} + \vec{h}_{dip}|)$ appears hysteresis ambiguous function of H due to hysteresis nature of h_{dip} and M_{eff}. The same situation remains when h_1 is of the order of h_{dip}.

Thus, occurring situation in principle differs from the situation for conventional superconductors. This difference consist in the two point. First, in granular superconductor magnetic field generated by intragranular magnetization can be much higher than the fields out of intergranular shielding currents. Secondly, intragranular irreversible magnetization is a new source of the irreversibility. As a result, many new interesting phenomena occur, some of which were observed, i.e. hysteresis of critical current [3-5], an inverse of the hysteresis loop for the field within hollow cylinder [6,7], etc.

In the present work studied theoretically and experimentally is a number of the phenomena which are determined by the three combined factors, i.e., first, the presence of intragranular irreversible magnetization; secondly, demagnetizing factor value being finite; thirdly, the sample being doubly-connected (hollow cylinder is taken for this purpose in our experiments). Finite sizes of the sample are essential because of $h_{dip} = 0$ in the infinite sample (therefore all the phenomena related to flux trapping disappear). The choice of the doubly-connected sample is governed by the fact that, in this case, the sign of the dipole field h_{dip} in the cylinder hole, as it will be seen below, is opposite to the sign of the field generated by shielding supercurrents h_1 and, as a result, three different types of hysteresis loop can be observed as an amplitude of the dc field changes, i.e. the diamagnetic loop, the loop with the intersection of its ascending and descending branches, and purely paramagnetic loop.

In the present work, intragranular irreversible magnetization influence on the macroscopic properties of high-T_c superconductors was experimentally studied from the behavior of nonlinear dynamic susceptibility in dc magnetic fields. Taken as an object of the studies were the ceramic samples of Y-based and Bi-based. These samples were shaped in the form of a ring (the outer diameters were 12.5 and 10.8 mm, the inner diameters were 7.5 and 5.1 mm, the height - 2.4 and 6.5 mm, respectively). The studies were carried out with the technique described in Ref. [2], for the different directions of the dc field H in respect to the direction of the ac field $h = h_0 \cos\omega t$. In one case, axial dc and ac fields were parallel to each other; and in another case the dc field was directed by the cylinder axis, being perpendicular to the axial ac field. We call the first situation as "parallel fields" and to the second one as "perpendicular fields". Similar to the Ref. [2] from the measuring of even and odd harmonics (C_{2k} and C_{2k+1}) of the induction in parallel fields were determined the important parameters such as the explicit form of the dependence $j_c(H)$ for $H \leqslant 2.5$ Oe, the value of critical current j_0 for $H = 0$ and its dependence on temperature. It was stated, in particular, that $j_c(H) = j_0 H_0^2/(H_0^2 + H^2)$ for Y-based ceramics and $j_c(H) = j_0 H_0/(H_0 + |H|)$ for Bi-based ceramics, moreover in both cases the characteristic field H_0 appeared as about 2 - 3 Oe. Figs. 1a and 1b show the plots of nonlinear susceptibility C_3/h_0 against temperature which were obtained in weak ac field in zero dc field. As one can see, these plots have two peaks. According to the

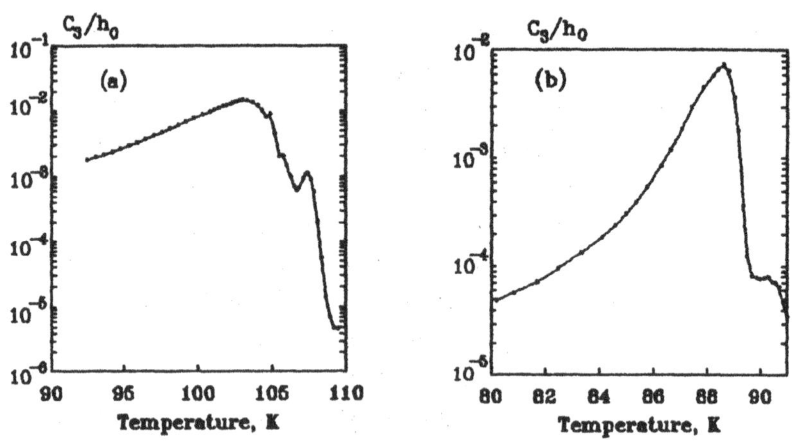

Fig.1a. Bi-based ceramics, $h_0 = 0.8$ Oe;
Fig.1b. Y-based ceramics, $h_0 = 0.27$ Oe, $f = 20$ kHz.

current ideas, high-temperature maximum is due to the field penetration into grains, while low-temperature peak relates to the field penetration to intergranular medium only. However, as it was noted above, the main task of the present paper was to study the effect of dc field penetration into grains upon the macroscopic properties of superconducting ceramics. To that end, it was necessary to choose the temperature range for the measurements, such that $h_{c1\ gr}$ should fall into the dc field range being possible in the given experiments and, besides characteristic field $h_2(H) = 2\pi d j_c(H)$ should be high enough for the condition of ac field h_0 being infinitesimal (i.e. $h_0 \ll h_2(H)$, H (see [2])) to be realized. Then $C_{2k+1} \sim 1/j_c(H)$ and the behavior of function $j_c(H)$ is directly studied in experiment in the case. In addition, for the Y-based sample the dependence of the field in the cylinder center against external dc field was studied with Hall probe. The below data relate to the measurements having been carried out at 90.5 K for Bi-based sample and at 80 K for Y-based sample. Let us note that the phenomena observed in the both samples were qualitatively similar to each other, therefore we shall not give the data obtained for each ceramic sample separately.

Let us consider the phenomena predicted by the generalized theory of critical state and compare them with the results of our experiments. First of all, let us note that since h_{dip} equals zero in infinite sample and $M_{eff}(h)$ does not affect the transport properties, the generalized theory comes to the conventional one. As for real samples, $h_{dip} \neq 0$ and even if demagnetizing factor is small, dipole field may turn out to be higher than the field generated by intergranular currents (see condition (5)), in which case it is necessary for the finite dimensions of the sample to be taken into account. Since the field within superconducting sample is always not uniform, demagnetizing factors, as a rule, cannot be used in the calculating of dipole fields related to finite dimensions of the sample. On the other hand, it is desirable to have the simple formulae in order to be able to compare with the experimental results. To this end, theory of perturbations in demagnetizing factor n, turns out to be conveniently used. Let us consider the hollow cylinder fabricated of superconducting ceramics (with the length L, internal diameter $2R_1$ and external diameter $2R_2$) in the field applied parallel to the cylinder axis z. With $L \Rightarrow \infty$, apparently, there is only z-component of magnetic field $h^{(0)}(r)$, satisfying the equation of critical state

$$\left| \frac{dh^{(0)}(r)}{dr} \right| = 4\pi j_c(h^{(o)}(r)) \tag{6}$$

The expression for the induction $B^{(0)}(r)$ in the sample appears as follows

$$B^{(0)}(r) = h^{(0)}(r) + 4\pi M_{eff}\{h^{(0)}(r)\}, \tag{7a}$$

In asymptotic case when the change of the field H within the sample is small as compared with the field H in itself, the expression for the field in the cylinder hole center $h^{(o)}(0) = h^{(o)}(R_1)$ appears as follows:

$$h_1^{(0)}(0) - H = \pm\ 4\pi(R_2 - R_1)j_c(H) \tag{7b}$$

where negative sign ("-") relates to the ascending branch of the hysteresis loop and positive sign ("+") relates to the descending one. Eqs.(7) are zero-order approximation by the parameter R^2/L^2. Let us note that variation scales for $h^{(0)}(r)$ and $B^{(0)}(r)$ can significantly differ from each other. In fact, the local field hysteresis occurs in the field scales $h_1 \sim j_c d$, whole the hysteresis of $M_{eff}(r)$ occurs in the scale $h_{dip} \sim nj_{c\ gr} \langle R_{gr} \rangle$. When condition (5) is valid, the hysteresis of $B^{(0)}(r)$ occurs in the scale $h_{dip} \gg h_1$.

Now let us calculate these values with the finite value of L. Let us note that it is convenient to use the induction B(r) rather than the field h(r) with the calculations based on perturbation theory. In this case eqs. (1) and (4) give:

$$\vec{B} = \vec{h} + 4\pi\vec{M}_{eff}(\vec{h}),$$

$$\mathrm{div}\vec{B} = 0,$$

$$\mathrm{rot}\vec{B} = 4\pi\vec{j}_{eff},$$

$$\vec{j}_{eff} = \vec{j} + \mathrm{rot}\vec{M}_{eff} \tag{8a}$$

$$\vec{B} = \mathrm{rot}\int \frac{d\vec{r}'}{|\vec{r} - \vec{r}'|}\, \vec{j}_{eff}(\vec{r}') \tag{8b}$$

Eq. (8b) is the same as the last line in the (4) providing for the (7a) and the identity $\mathrm{rotrot}\vec{M} = \mathrm{graddiv}\vec{M} - \Delta\vec{M}$. Let us note that $\vec{M}_{eff} = 0$ and $\vec{B} = \vec{h}$ outside the sample.

After determining $B^{(0)}$ in zero-order approximation by small parameter R^2/L^2 from the eqs. (6) and (7a), let us calculate the value of $j_{eff}^{(0)}$ from the eq. (8a) and substitute it into the right-hand side of the (8b). This results in

$$\vec{B} = \frac{1}{4\pi}\,\mathrm{rot}\int \frac{d\vec{r}'}{|\vec{r} - \vec{r}'|}\, \mathrm{rot}\vec{B}^{(0)}(\vec{r}') \tag{9}$$

With $L = \infty$, the eq. (9) is an identity; while with finite values of L in the first-order approximation by R^2/L^2 for z-components of the fields $B(r) = \vec{e}_z\vec{B}(r, z = 0)$, and $h(r) = \vec{e}_z\vec{h}(r, z = 0)$ we obtain

$$B(r) = B^{(0)}(r) + H_{dip},$$

$$h(r) = h^{(0)}(r) + H_{dip},$$

$$H_{dip} = H_{0\,dip} + h_{dip},$$

$$H_{0\,dip} = \int_{R_1}^{R_2} dr\, \frac{dh^{(0)}(r)}{dr}\, \frac{2r^2}{L^2}$$

$$h_{dip} = 4\pi M_{eff}(R_1)\frac{2R_1^2}{L^2} - 4\pi M_{eff}(R_2)\frac{2R_2^2}{L^2} + 4\pi\int_{R_1}^{R_2} dr\, \frac{dM_{eff}(r)}{dr}\,\frac{2r^2}{L^2}$$

$$M_{eff}(r) \equiv M_{eff}(h^{(0)}(r)) \tag{10}$$

where H_{dip}, the dipole field created by total magnetic moment of superconducting sample, consists of two parts; i.e. H_{Odip} governed by the moment which is generated by shielding transport currents $dh^{(O)}/dr$ and h_{dip} which is governed by the grains moments. It can be demonstrated in the explicit form through the integration of $H_{O\,dip}$ and h_{dip} by parts. This results in

$$H_{O\,dip} = -4\pi n_1 m^{(1)} - 4\pi n m^{(O)}, \; h_{dip} = -4\pi n M^{(O)}$$

$$n_1 = \frac{2R_1^2}{L^2}, \; n = \frac{2(R_2^2 - R_1^2)}{L^2}$$

$$4\pi m^{(O)} = \frac{2}{R_2^2 - R_1^2} \int_{R_1}^{R_2} \{h^{(O)}(r) - H\} r dr,$$

$$4\pi m^{(1)} = h^{(O)}(R_1) - H$$

$$M^{(O)} = \frac{2}{R_2^2 - R_1^2} \int_{R_1}^{R_2} M_{eff} \{h^{(O)}(r)\} r dr$$

From (11) one can see that H_{dip} is not dependent on r in the first - order approximation by R^2/L^2.

Every term in eq. (11) is rather simple in its meaning: $m^{(O)}$ is average moment governed by intergranular currents; $m^{(1)}$ is magnetic moment of the cylinder hole which is resulted from flux trapping in doubly-connected superconductor, $M^{(O)}$ is average intragranular moment, and n and n_1 are demagnetizing factors of the cylinder and its hole correspondingly.

At first, let us consider the case where the field does not penetrate into grains, then $M_{eff}(h) = \chi_{eff} h$, where $\mu_{eff} = 1 + 4\pi\chi_{eff}$ and

$$4\pi(m^{(0)} + M^{(0)}) = \frac{2}{R_2^2 - R_1^2}\int_{R_1}^{R_2}(\mu_{eff}\, h^{(0)}(r) - H)rdr \qquad (12)$$

From (7) and (10)-(12) one can see that, in this case, $4\pi(m^{(0)}+ M^{(0)})$ is of the order of $(h^{(0)}(R_1) - H)$ and, because of n_1, $n \ll 1$, H_{dip} can be neglected and h(r) can be regarded as being determined by $h^{(0)}(r)$ only. Therefore, in the cylinder center $h(o) < H$ in the ascending branch of the hysteresis loop and $h(o) > H$ in the descending one. Such a diamagnetic form of the loop was observed in the event of low-temperature superconducting hollow cylinder; and since, as can see from the (7b), the difference between the ascending and descending branches, in hysteresis loop asymptotic, is $\Delta h(o) = 8\pi j_c(H)(R_2-R_1)$, dependence of critical current against magnetic field has been determined like that from about 1964 forth [8]. Diamagnetic hysteresis loop was also observed in our experiments on the measuring of h(o) in the center of hollow cylinder made of yttrium ceramics, which were carried out in the fields below $h_{c1\ gr}$ (Fig.2a). It is easy to demonstrate that in this case the dependence $j_c(H)$ can also be determined from the difference between ascending and descending branches of hysteresis loop $\Delta h(o)$. However, such a determination of the dependence $j_c(H)$ is valid only in the asymptotic case where the field change over the sample dimensions is less than the field in itself, i.e. for $H \gg \Delta h(o)$. From the $\Delta h(o) - H$ - relation obtained like that (see the inset in Fig.2a), one can see that, beginning from $H > 4$ Oe, $\Delta h(o)$, i.e. $j_c(H)$, is proportional to $1/H^3$. At the same time, according to the results of the studies with parallel ac and dc magnetic fields, the dependence $j_c(H) \sim 1/H^2$ was obtained for this sample. This difference is easy to be accounted for, using the fact that $j_c(H)$ is governed by Josephson contacts and, as a result of the Fraunhofer diffraction, in these contacts under strong fields [9] H has an extra order as compared with weak fields. If, however, flux trapping into grains occurs, h(o) is governed by intragranular critical currents; moreover in this case

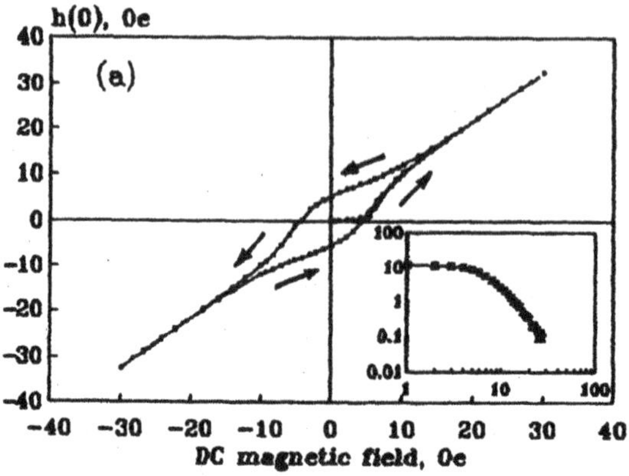

Fig.2a. Hysteresis curve for the field in the center of hollow cylinder h(0), with external field H < h_{c1gr}. The inset shows the difference Δh(0) between ascending and descending branches of the hysteresis loop (Fig.2a). Yttrium ceramics.

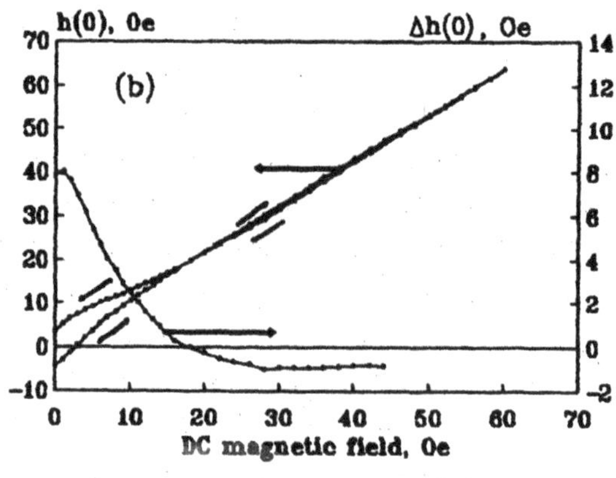

Fig.2b. The same with H > h_{c1gr}.

$$|M_{eff}| = |f_s\langle M_{gr}(h)\rangle| \sim f_s j_{c\ gr}\langle R_{gr}\rangle \tag{13}$$

where $\langle R_{gr}\rangle$ is a mean size of grains. As it was noted above, incidentally, $|M_{eff}|$ can be much more than $m^{(o)}$ and $m^{(1)}$; then, intergramular currents contribution into h(r) being neglected, $h^{(O)} \approx H$ and $h(r) = h$ (i.e. is not dependent on r). In this case, eq.(8) gives:

$$h(0) = h = H - 4\pi n\, M^{(O)} = H - 4\pi n\, M_{eff}(h) \tag{14}$$

$M_{eff} < 0$ in the ascending branch of hysteresis loop, and $M_{eff} > 0$ in the descending one. Making a comparison between eqs. (7b) and (14) one can see that the difference $h(0) - H$ is of opposite signs in both these equations. Of course, this phenomena occurs in conventional superconductors as well. In this case however, the shielding currents in the (7b) and the dipole field in the (14) are determined by the same value of $j_c(H)$ and, since demagnetizing factor is always less than unity, a sum of the dipole field and the field generated by shielding currents appears as such that $(h(0) - H) < 0$ in the ascending branch and $(h(0) - H) > 0$ in the descending one. Thus, hysteresis loop is always diamagnetic in conventional superconductors. In the event of granular superconductor, however, $H_{O\ dip}$ is still governed by the value of $j_c(H)$, and h_{dip} is governed by intragranular currents $j_{c\ gr}(h) \gg j_c(h)$. Therefore, easily be satisfied can the condition $h_{dip} \gg h_1$ which results in the inverse of hysteresis loop. With intermediate fields where h_{dip} is of the order of h_1, hysteresis loop intersection occurs. In fig.2b, this intersection is clearly seen where the sign of the difference between ascending and descending branches of hysteresis loop alters when $H \approx 20$ Oe. Let us note that the hysteresis loop inverse like this was observed before in the works [6,7].

Now let us consider another effect, i.e. the hysteresis of nonlinear dynamic susceptibility in the dc field for the case of the ac field being perpendicular to the dc one. We shall not give here the theory of this effect. The only thing to be taken into account in this situation is the difference between the fields within and out of the cylinder. It is easy to demonstrate that in the asymptotic case of the weak field $h_O \ll H$, the formulae for odd harmonics C_{2k+1} are as follows:

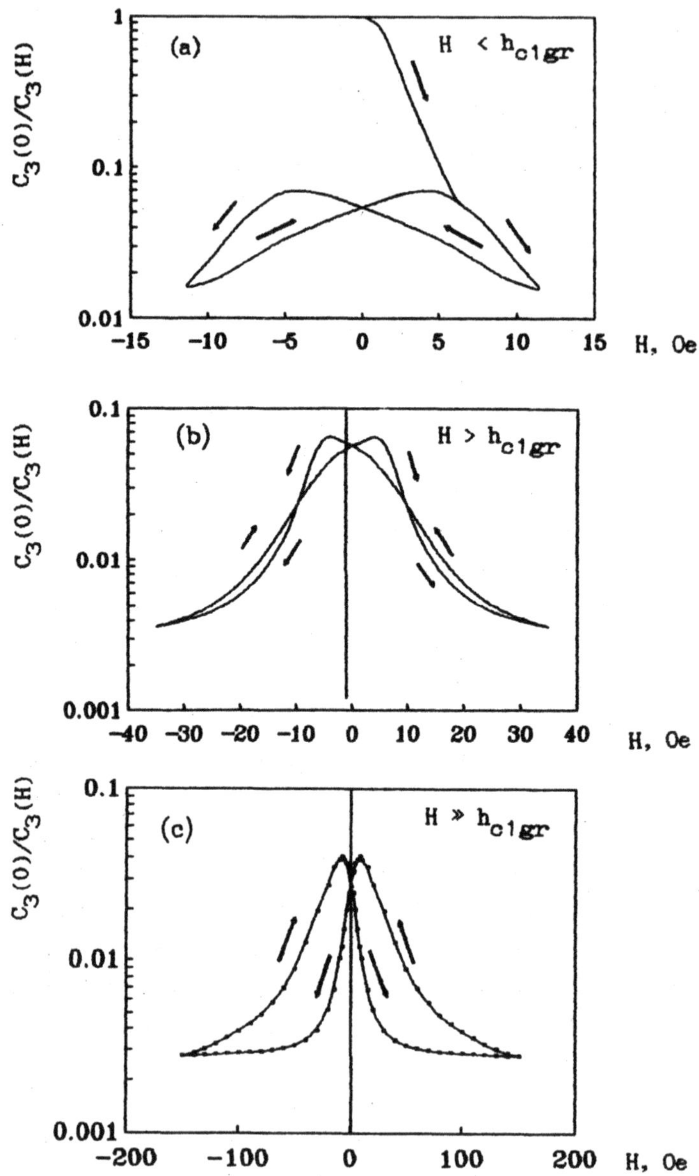

Fig.3. Hysteresis curve for inverse of third harmonic, C_3^{-1}, f = 20 kHz, h_0 = 0.2 Oe. Y-based ceramics.

$$C_{2k+1}(H) \sim \frac{1}{2} h_0^2 \{\frac{1}{j_c[h(R_1)]} + \frac{1}{j_c[h(R_2)]}\} \qquad (15)$$

Since $j_c(h)$ is a single-valued function of h, the hysteresis of $C_{2k+1}(H)$ may occur only due to the hysteresis of $h(R_1)$ and $h(R_2)$. Because of $h^{(0)}(R_2) = H$, the hysteresis of odd harmonics, as one can see from (10) and (11) is governed by the hysteresis of $h^{(0)}(R_1)$, $m^{(0)}$,$m^{(1)}$ and $M^{(0)}$. If the field doesn't penetrate into the grains, then, as it was noted above, terms of $m^{(0)}$,$m^{(1)}$, $M^{(0)}$ can be neglected and it is the hysteresis of $h^{(0)}(R_1)$ that remains. Since $h^{(0)}(R_1) < H$ in the ascending branch of the hysteresis loop and $h^{(0)}(R_1) > H$ in the descending one and $j_c(h)$ is a decreasing function of h, $C_{2k+1}(h_{asc}) < C_{2k+1}(h_{desc})$. Further, as one can see from (6) ,(7), $h(R_1)$ turns into naught not for H = 0, but for $H = -h_m < 0$ in the descending branch and for $H = h_m > 0$ in the ascending one. Therefore its clear that the minimums of $C_{2k+1}(H)$ have to be for $H = \pm h_m$ in the ascending and descending branches correspondingly. As one can see from the formula (15), $C_{2k+1}(H) \sim 1/j_c(H)$ in the asymptotic case of weak ac field; therefore, to understand the phenomena observed, it's more convenient to consider experimental hysteresis loops for the inverse value, C_{2k+1}^{-1}. Accordingly, minimums in the hysteresis loop for C_3^{-1} is clearly seen in Fig.3(a). It also follows from (15) that $C_{2k+1} \sim h_0^2$. Let us note that $C_3 - h_0$ quadratic dependence (for H = const) was observed in experiment up to dc field values of the order of 100 Oe.

Now let us consider the situation where the flux trapping into the grains occurs. In this case, h_{dip} appears as large enough. Let us assume that in the formula (10) $M^{(0)}$ plays a main role in the h(r); therefore, with allowance for (14), the expression (15) appears as follows:

$$C_{2k+1}(H) \sim h_0^2/\{j_c(H-4\pi nM^{(0)})\} \qquad (16)$$

In this case, the hysteresis $C_{2k+1}(H)$ is completely governed by hysteresis of $M^{(0)}$ and, due to negative sign ("-") before $nM^{(0)}$ in the (16), this gives the inverted C_3^{-1}- hysteresis loop with two maxima for $H = \pm 4\pi n|M^{(0)}|$ in its descending and ascending branches correspondingly. This kind of hysteresis loop for $C_3^{-1} \sim j_c(H)$ is shown in Fig.3b. It should be noted that in all experiments on the studying of transport current (e.g.see [3-5]) it is observed only the last (paramagnetic) kind of hysteresis loop, because of these works having dealt with simple-connected superconductors. Apparently, in the fields, when the competition between shielding and dipole fields is essential, in our case the inverse and the intersection of high harmonics hysteresis have to occur, which is observed indeed in experiment (see Fig.3b where hysteresis loop for C_3^{-1} is shown).

Let us discuss another interesting phenomenon observed in the experiment that was made as follows. At first, in the absence of dc magnetic field the value of third harmonic C_{30} was measured, then external field H_{max} was switched on. After switching-off that field the residual value of third harmonic $C_{3\ rem}$ was measured. Then, the sample was overheated up to $T > T_c$ and on returning its initial temperature the procedure of the measuring of $C_{3\ rem}$ was repeated for the new value of H_{max}. Fig.4 shows the ratio C_{30}/C_{3rem} versus H_{max}. The remarkable

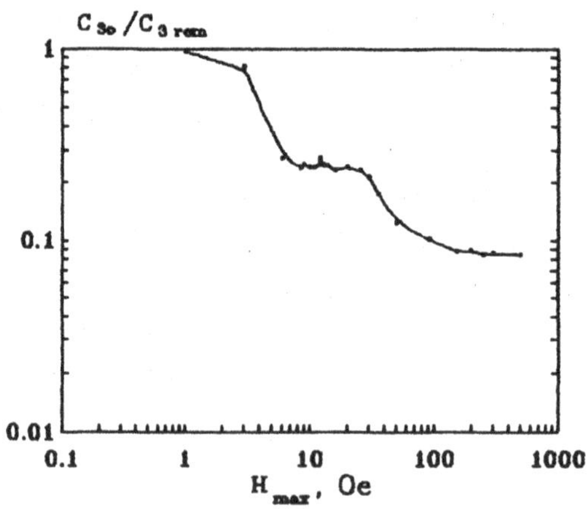

Fig.4. C_{30}/C_{3rem} plotted against H_{max} (f = 20 kHz, h_0 = 0.2 Oe). Bi-based ceramics.

feature of this dependence is an existence of two regions of H_{max} where C_{30}/C_{3rem} is not dependent on H_{max}. This is rather easy to be explain. In the geometry of hollow cylinder, the field applied parallel to the cylinder axis starts to penetrate the cylinder from its out site and, on certain value being obtained, passes throughout the cylinder wall. In this case, after the removing of H_{max}, the trapped field $h(R_1)$ should always be the same. Then, naturally, $C_{3\ rem}$ should be constant too. The second plateau is governed by the similar phenomena being related, however, to the field penetration into grains themselves. Let us note that in the range of $h_0 \leqslant 1$ Oe $C_{3\ rem}$ is always proportional to h_0^2, meaning the fulfilment of the ac field being small.

Thus the results of experimental and theoretical studies, given in the work, are in qualitative accordance with each other and show that intragranular irreversible magnetization strongly affects the transport and magnetic properties of ceramic high-T_c superconductors.

References

[1] J.R.Clem, Physica C, V. 153-155 (1988) 50.

[2] S.L.Ginzburg, V.P.Khavronin, G.Yu.Logvinova, I.D.Luzyanin, J.Herrmann, B.Lippold, H.Borner, H.Schmiedel, Physica C, V.174 (1991) 109-116.

[3] J.E.Evetts, B.A.Glowacki, Gryogenics, V.28 (1988) 641.

[4] K.Kwasnitza, Ch.Widmer, Gryogenics, V.29 (1988) 1035.

[5] K.Kwasnitza, Ch.Widmer, Physica C, V.171 (1990) 211.

[6] M.R.Cimberle, C.Ferdeghini, G.L.Nicchiotti, M.Putti, A.S.Siri, C.Rizzuto, C.A.Costa, M.Ferretti, C.L.Olcese, F.C.Matacotta, E.Olzi, Supercond.Sci.Technol., V.1 (1988) 30.

[7] V.Calzona, M.R.Cimberle, C.Ferdeghini, F.Pupella, M.Putti, C.Rizzuto, A.Siri, R.Vaccarone, Gryogenics, V.30 (1990) 569.

[8] Y.B.Kim, C.F.Hempstead, A.R.Strnad, Phys.Rev., V.131 (1963) 2486.

[9] A.Barone, G.Paterno Physics and Applications of the Josephson Effect, John Wiley & Sons, Inc., P.14

BEAN-LIVINGSTON SURFACE BARRIER IN HIGH TEMPERATURE SUPERCONDUCTORS

M. KONCZYKOWSKI
Laboratoire des Solides Irradiès, Ecole Polytechnique,
91128 Palaiseau, France

Experimental evidence for Bean-Livingston (BL) surface barrier contribution to magnetization of HTSC is presented. This is provided by Local Hall Probe Magnetometry (LHPM) measurements of BiSrCaCuO and untwinned $YBa_2Cu_3O_7$ crystals. Asymmetric form of hysteretic magnetization field curves indicates that close to T_c irreversible magnetization arises from BL barrier and not from bulk pinning. First magnetic penetration field, H_p is related to BL barrier and is higher than the lower critical field H_{c1}. Reinterpretation of other magnetic features of high temperature superconductors, implying active BL barrier, is suggested.

1. Introduction.

Currently, the analysis of magnetization of high-temperature superconductors (HTSC) is carried out in the framework of Bean critical state model [1]. This model implies strong bulk pinning and is appropriate for hard type II superconductors. In HTSC, giant flux creep [2] and wide region of reversible magnetization [3] is observed. Those features are related to very low pinning forces. One can therefore expect in HTSC some of magnetic features characteristic for reversible type II superconductors. Surface barrier was known to play an important role in magnetization and critical current of those materials [4]. In particular, BL surface barrier[5] is expected to be very pronounced in HTSC, because of very high Ginzburg-Landau parameter κ [6].

One of the very controversial magnetic properties of HTSC is the value and temperature dependence of H_{c1}. The results of various experiments agree only within factor of 4-5. Pronounced up-turn in $H_{c1}(T)$ curve is in general observed in low temperature region [7,8,9,10,11,12,13]. This up-turn of H_{c1} is hard to reconcile with temperature dependence of London penetration depth, λ, which exhibits saturation at low temperatures [14]. It was suggested that at low temperature, flux penetration is retarded by surface barrier [15].

Only recently, the improvement of quality of available crystals [16] and progress in LHPM experimental technique [17] allowed a closer insight into magnetic flux penetration. Some evidence for important contribution of BL surface barrier to magnetic properties was found in at least two representative HTSC compounds: $YBa_2Cu_3O_7$ [18] and $Bi_{2.2}Sr_{1.8}CaCu_2O_8$ [19]. This new information demands for careful analysis of a variety of magnetic measurements of HTSC, in particular the irreversible magnetization interpreted within Bean model, currently used for determination of critical current.

The purpose of this paper is a presentation of magnetization features of HTSC related to BL barrier. In section 2, I recall essential properties of BL surface barrier, underlining the differences in respect to Bean critical state model. In Section 3, I present experimental evidence for BL barrier in untwinned $YBa_2Cu_3O_7$ and $Bi_{2.2}Sr_{1.8}CaCu_2O_8$ crystals. Section 4 contains analysis of temperature dependence of first magnetic penetration field in those compounds. Some consequences of BL barrier on interpretation of other magnetic features are examined in section 5.

2. Bean-Livingston surface barrier vs. Bean critical state models.

The energy of the vortex line close to the surface was first analyzed by deGennes [20]. Boundary condition of vanishing vortex current component normal to the surface can be expressed in terms of interaction of a vortex line with its mirror image in free space. This results in attractive force and vortex energy varying as $e^{-2x/\lambda}$, where x denotes distance from the surface and λ, penetration depth. Bean and Livingston [5] and the Orsay group noticed independently that this interaction will produce an energy barrier impeding flux penetration to the volume of a superconductor. This barrier arises from the competition between Lorentz force produced by shielding current, repelling vortex line from the surface and contributing to vortex energy as $\approx e^{-x/\lambda}$ and attractive image force. Total energy of a vortex line contains negative term $\approx e^{-2x/\lambda}$ and positive term $\approx e^{-x/\lambda}$ and can exhibit a maximum. Bean and Livingston [5] pointed out that this surface barrier inhibits first flux penetration above thermodynamic lower critical field H_{c1}. First flux penetration through a smooth surface occurs only at external field approaching thermodynamic critical field H_c. Retardation of the flux penetration resulting from BL barrier should be very important in high κ materials as HTSC [6].

In their first paper [5], Bean and Livingston suggested that flux escape will be more inhibited by surface barrier that penetration. Later, it was found that on the contrary, flux escape is much less affected [21,22,23,24]. This asymmetry of BL barrier was one of features differentiating it from barrier arising from "surface sheath" superconductivity [25].

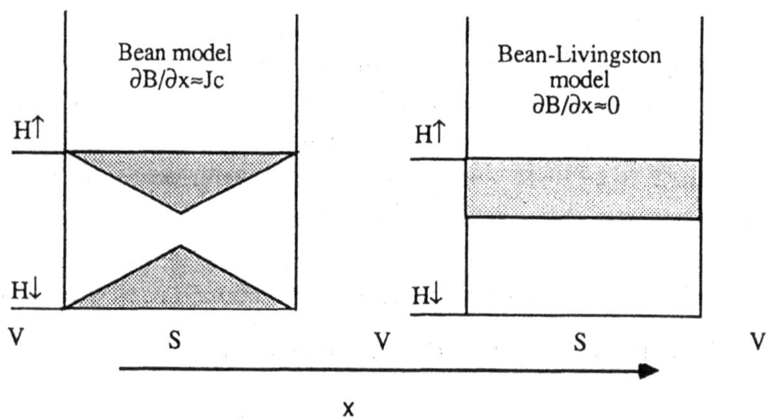

Figure 1. Schematic representation of magnetic field profiles and resulting magnetization (shaded areas) in Bean critical state model and Bean-Livingston surface barrier. S and V are the superconducting regions and vacuum respectively.

Detailed analysis of flux penetration and escape limited by BL barrier and of related irreversible magnetization is presented in ref. 4. On increasing field surface barrier leads magnetic induction in the volume of the sample lower then its equilibrium value. When external field start to decrease, magnetic induction in the sample remains constant because surface barrier inhibits flux escape until magnetization of the sample vanishes. At this point surface shielding current disappears and BL barrier collapses. Flux can escape freely until some barrier is restored. This feedback effect stabilizes the magnetization on some value close to zero on decreasing field branch of magnetic hysteresis loop.

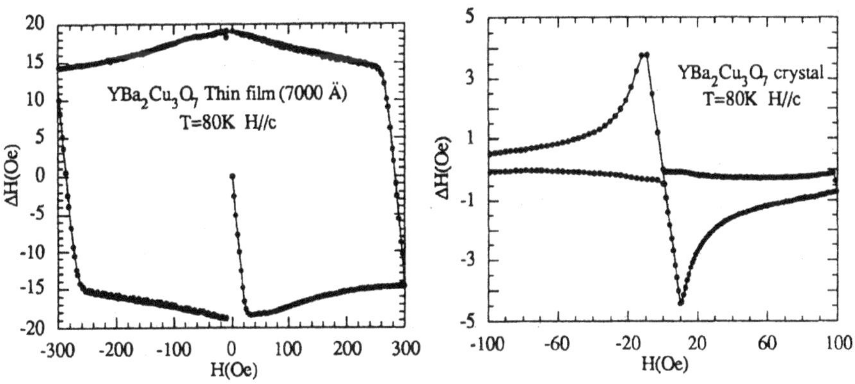

Figure 2. Magnetic hysteresis loops recorded on untwinned $YBa_2Cu_3O_7$ crystal at 80K on thin film obtained by laser ablation technique (left panel) and on untwinned crystal (right panel).

The asymmetry of magnetic hysteresis loop is a main fingerprint of irreversibility arising from surface barrier and not from bulk pinning [4]. In the Bean critical state model, neglecting field dependence of critical current, J_c, field profiles are symmetric and positive magnetization in decreasing field equals a negative one on increasing field. In contrast BL barrier produces only negative magnetization* . Figure 2 presents typical magnetic hysteresis loops corresponding to those two models: symmetric

* This holds only in absence of bulk pinning. In thin samples, self field can inverse the direction of magnetic field on the edge and restore BL barrier. This can produce positive magnetization at weak applied fields.

magnetization loop recorded on thin film corresponds to Bean critical state, while asymmetric loop recorded on untwinned $YBa_2Cu_3O_7$ crystals close to T_c seems to results from BL barrier.

Bean and Livingston [5] already noticed that surface irregularities and sample edges depress surface barrier. This problem was investigated in the late sixties [21,22,23,24] and overviewed by Campbell and Evetts in ref 4.

Attractive force between vortex and its mirror image arises on surface on the length scale comparable to vortex size, typically of the order of λ. Surface irregularities perturb current distribution and attenuate image force. Resulting degree of depression of surface barrier depends on relative size of those irregularities in respect to λ and ξ. This leads to possible temperature and surface roughness dependence of efficiency of surface barrier [26].

In real crystal of reversible type II superconductors, flux penetration and escape is likely to occur by "weak spots" on surface irregularities. The amount of irreversible magnetization or the value of transport critical current depends on the surface roughness and on the geometrical form of the sample. An example of such effect is the sensitivity of the critical current to polarity of the magnetic field and transport current in triangular [21] and semi-cylindrical [24] section samples. Much higher critical current was found in configuration corresponding to flux penetrating flat surface of the sample.

The effect of surface damage produced by ion implantation on transport critical current was investigated by Chang and Rose-Innes [27]. The depression of critical current for flux penetrating damaged surface was detected by rotating the specimen in external magnetic field.

The sensitivity of magnetic features to surface shape and roughness can be considered as a second fingerprint of BL surface barrier.

3. Evidence for BL barrier from magnetization curves of untwinned $YBa_2Cu_3O_7$ and $Bi_{2.2}Sr_{1.8}CaCu_2O_8$ crystals

Most crystals of the of $YBa_2Cu_3O_7$ exhibit dense twinning. Twin planes produce bulk pinning that probably dominates magnetic features of those crystals [28]. Only recently, untwinned single crystals of useful dimensions became available. Magnetic properties of those crystals were investigated by LHPM technique [17]. Data presented here were recorded on three untwinned samples of dimensions 1090x460x4 and 1000x460x4 μm^3 (these are two pieces of what was originally one crystal) and 1700x1400x23μm^3, the same as investigated in reference 18. We refer to those samples as U1a, U1b and U2, respectively.

3.1. Local Hall Probe Magnetometry Technique.

In the past, the field profiles of superconductors were investigated by Hall probes [29,30]. The key new technical point of LHPM is the use of a **miniaturized** Hall sensor with 80 x 100 μm^2 active area, of highly linear slope (typically 50 mΩ/G) that is almost temperature independent below 120K. Such a probe can be made of a narrow-gap semiconductor and put directly on the surface of the investigated sample. It should be noted that: (1) The current injected to the probe (typically 500 μA) produces negligible self field. (2) Even a common dc voltmeter allows a resolution of 4 mGs; this corresponds to a single flux line for the active area of the sensor. The calibration of the probe can be done as a function of the external field H_a either in the absence of the sample at any temperature or with the sample at temperature T well above the transition temperature T_c. Once the characteristics of the probe, namely $R_H(H_a)$ are established, the probe is ready for measurements of magnetization field as a function of temperature, field or time. An external field H_a is applied and the field H_m measured by the probe is detected. As a result of perturbation, by the sample, of the homogeneous applied field, we find $\Delta H = H_m - H_a$. This perturbation ΔH is related to the supercurrent induced in the sample, we refer to it as the magnetization field. This magnetization field depends on the Hall probe position in respect to the sample and on current distribution within the sample. Detection of characteristic points on magnetization curve, like first magnetic penetration or field of full penetration is easy by this technique. With few exceptions discussed later this is not affected by position of Hall probe. In contrast the analysis of the value of magnetization field requires model of current distribution. Two types of distributions and related magnetization fields are of particular interest; (1) shielding current (e.g. Meissner state), where supercurrent is confined to surface sheath [31] (2) Bean critical state [30, 32].

Magnetization field on the center of flat disk shaped sample in full critical state was found to be $a*J_c^b*t$, where a is numerical factor of the order of 1.5, J_c^b is bulk critical current and t sample thickness [32]. Determination of J_c^b from magnetization field is thus as straightforward as in classic magnetization measurement [1].

An advantage of LHPM is the possibility of monitoring magnetic field profiles. This can be realized either by displacement of the Hall probe or use of multiple Hall probes. The latter solution was used for investigation of YBCO crystals.

3.2 Magnetization field of untwinned $YBa_2Cu_3O_7$ crystals.

A typical example of magnetization field vs. applied field curve recorded by LHPM is presented in figure 3. On increasing field branch, magnetic field was turned off after each step and trapped field was recorded. Descending field branch was recorded by standard procedure.

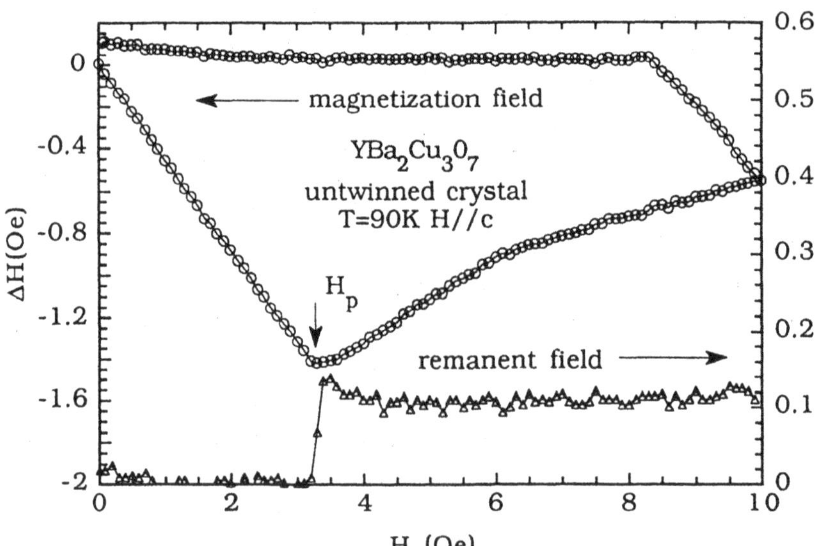

Figure 3. Magnetization field and trapped field recorded at 90K by Hall probe placed on the surface of untwinned crystal of $YBa_2Cu_3O_7$.

Initial linear decrease of magnetization field is related to Meissner state magnetization. Sharp kink associated with onset of residual trapped field is unambiguously identified as first magnetic penetration field, H_p*. Magnetization above H_p corresponds to equilibrium magnetization of Abrikosov lattice.

It should be noted that saturation of the remanent field indicating H_a equal twice H*, field of full penetration in Bean critical state model, occurs very close above H_p. This means that just above first penetration field, magnetic flux spreads in whole volume of the sample. Very low remanent field proves weak bulk pinning of this sample.

More information on magnetic field distribution was provided by simultaneous measurement of magnetization field in two locations; one on the sample surface, the other outside the sample close to its edge [33]. For this purpose, the Hall probe with two active areas was used.

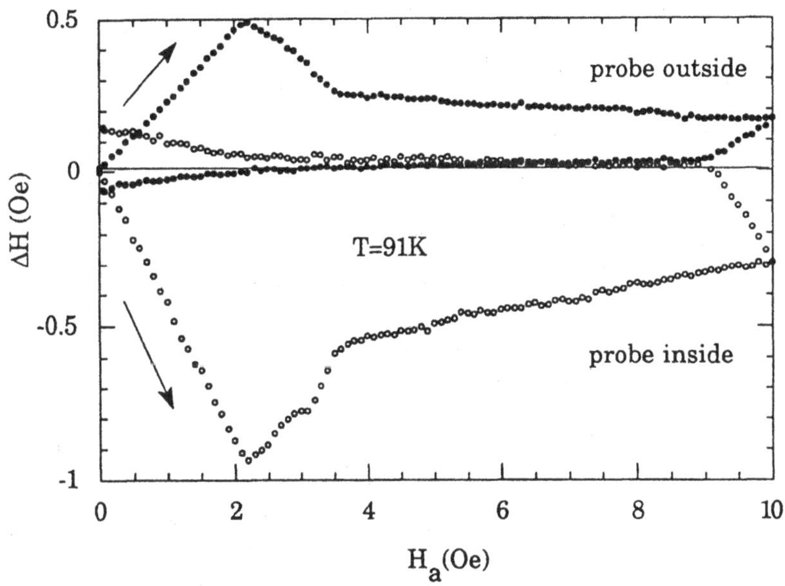

Figure 4. Magnetization field recorded by Hall probe on untwinned $YBa_2Cu_3O_7$ crystal U1b at two locations.

* The actual value of H_p has to be corrected by demagnetization factor.

Magnetization field loops recorded at 91K in this set-up are presented in figure 4. Initial linear magnetization is negative on the sample surface while it is positive for probe located outside the sample. Sharp kink indicating first magnetic penetration is observed exactly for the same applied field, independently on probe locations.

The most striking feature is that value of magnetization field recorded at both locations is almost zero on descending field branch of hysteresis loop. This is a main fingerprint of magnetic hysteresis resulting from BL barrier[4]. Field gradient on the sample surface is close to zero while deceasing magnetic field, this demonstrates very low bulk critical current.

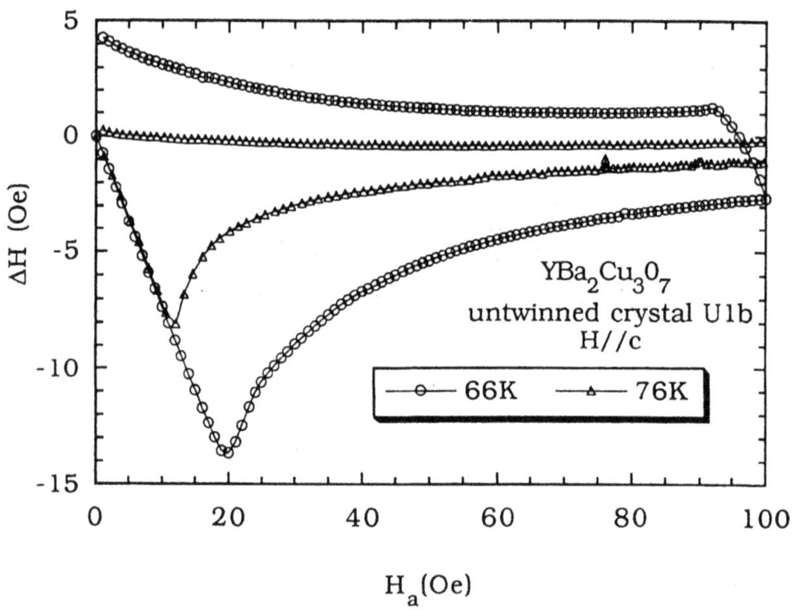

Figure 5. Hysteretic magnetization loops recorded on sample U1b at 66 and 76K

Magnetic hysteresis loops exhibiting almost zero magnetization field on decreasing field persist at temperatures down to 70K in most reversible samples. At lower temperatures positive magnetization starts to grow up on deceasing field and hysteresis loops becomes symmetric in all investigated samples.

In Bean critical state model in the absence of surface effect, any minor loop at $H_a>H_{c1}$ should produce irreversible magnetization. Flux starts to leave the sample on the beginning of decreasing field ramp. In BL barrier model, when external field starts to decrease flux expulsion is inhibited until total magnetization reaches zero. In minor loops, magnetic induction remains constant until an envelope hysteresis curve is reached. Therefore, low amplitude minor loops are reversible. The slope dM/dH in these loops is, in principe, the same as the initial Meissner slope.

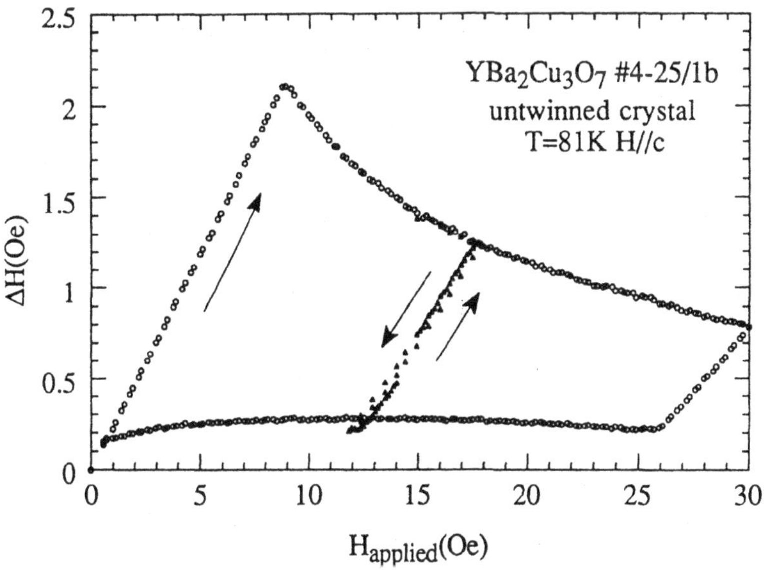

Figure 6. Full and low amplitude minor magnetic hysteresis loops, recorded on untwinned $YBa_2Cu_3O_7$ crystal U1b at 81K by Hall probe placed outside sample.

This reversible behavior of minor loop is well observed in untwinned $YBa_2Cu_3O_7$ crystals close to T_c. Figure 6 presents a typical example of minor loop recorded by Hall probe placed outside the sample. The same behavior was observed with a probe placed inside the sample. In this configuration of Hall probes, the reversibility is unambiguously related to BL barrier. It should be noted that apparently reversible minor loops of magnetization field can be recorded by Hall probes placed in fully screened position, e.g. inside a hollow cylinder or on the center of a large sample. In this case modification of external field is not reflected on Hall probe location. Slope $\partial\Delta H/\partial H$ is always -1 in this case.

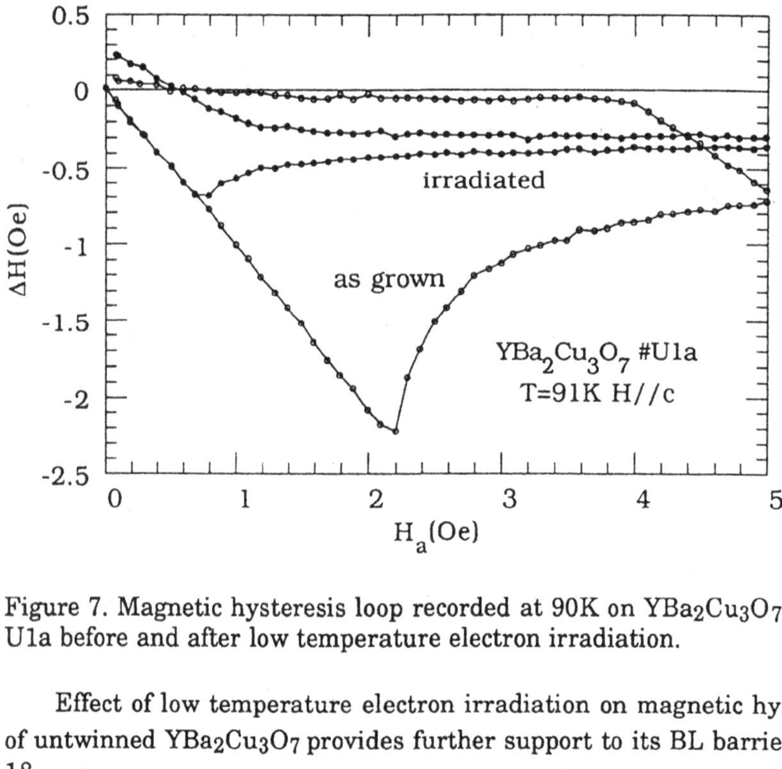

Figure 7. Magnetic hysteresis loop recorded at 90K on $YBa_2Cu_3O_7$ crystal U1a before and after low temperature electron irradiation.

Effect of low temperature electron irradiation on magnetic hysteresis of untwinned $YBa_2Cu_3O_7$ provides further support to its BL barrier origin [18]. After low dose irradiation (estimated damage of 0.6-1.2 10^{-4} d.p.a.), width of the loop was strongly depressed and field of first magnetic penetration decreased. This kind of irradiation (2.5MeV electrons impinging sample hold at 20K), produces uniform spread of Frenkel pair in whole volume of the sample. However, on warming the sample for transfer to magnetometer, migration of defects occurs [34]. Defects agglomerate in small cluster and sink to the surface. The latter process seems to affect BL barrier and depress magnetic hysteresis and field of first penetration. Magnetic hysteresis related to bulk pinning would only increase after introduction of additional pinning centers. From the other side, even small surface irregularities of 15Ä size, depress image force and the height of BL barrier [35]. Surface defects of this size are likely to be produced during annealing of irradiation defects.

3.3 Magnetization curves of $Bi_{2.2}Sr_{1.8}CaCu_2O_8$ crystals: evidence for BL barrier.

Most of BSCCO crystals used in early investigations exhibited mosaic structure and weak links. Flux penetration at very low fields complicates the interpretation of magnetic phenomena. Recent progress in crystal growing technique allows us to solve this problem. In particular crystals grown by floating zone technique exhibit very clean magnetization loops comparable to those obtained on untwinned crystals of $YBa_2Cu_3O_7$.

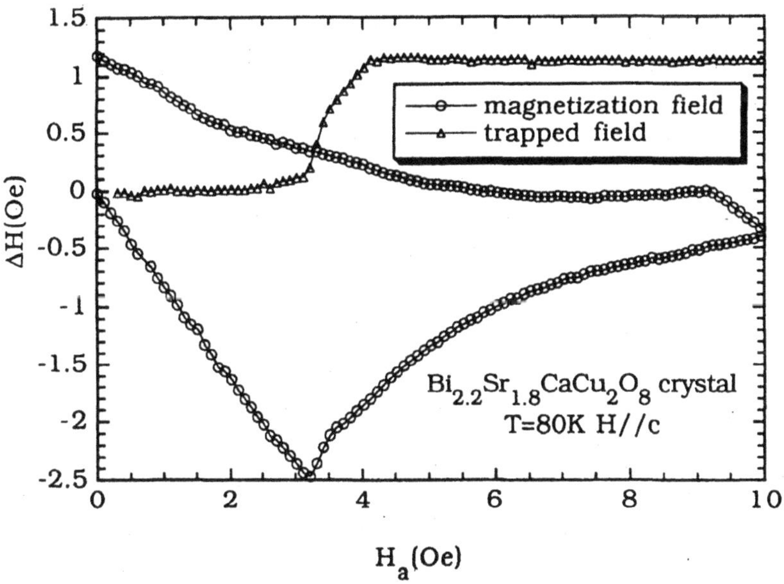

Figure 8. Magnetic hysteresis loop and trapped field vs. applied field variation recorded on $Bi_{2.2}Sr_{1.8}CaCu_2O_8$ crystal at 80K.

Typical magnetic hysteresis loop recorded by LHPM on $Bi_{2.2}Sr_{1.8}CaCu_2O_8$ crystal at 80K is presented in figure 8. Sharp kink at first magnetic penetration field, H_p and absence of trapped field below H_p, testify to the high quality of those crystals. This hysteresis loop presents characteristic features of BL barrier. Increasing field branch of the loop exhibits a rapid rise of magnetization above H_p incompatible with Bean critical state but close to Abrikosov lattice magnetization. Saturation of trapped field occurs at field only a few Oe above H_p which indicates very low bulk pinning. On the decreasing field branch of the loop magnetization field is close to zero except the applied field region close to origin.

4. Temperature dependence of first magnetic penetration field.

Temperature dependences of first magnetic penetration field measured in $YBa_2Cu_3O_7$ on as grown sample U2 and on sample U1a before and after electron irradiation are presented in figure 9. At low temperatures (below 60K) deviation from linear Meissner state magnetization field, $\partial\Delta H$ and corresponding rise of trapped field were slow. The commonly used technique of extracting H_p from intercept of square root of $\partial\Delta H$ vs. applied field was used [9].

At higher temperatures detection of H_p is unambiguous either from sharp kink on isothermal $\Delta H(H_a)$ curves, onset of trapped field (see figure 3) or break-down of Meissner state shielding on ΔH vs. temperature curves (see figure 10)[36]. All three methods lead to very similar values of H_p.

The long standing controversy about temperature dependence of H_{c1} seems to find its explanation in BL barrier limited flux penetration. Field of first flux penetration, observed in most experiments, always exceeds H_{c1} by temperature dependent factor, δ ($H_p=H_{c1}*(1+\delta)$). This factor is related to efficiency of BL barrier and depends on size of surface irregularities in respect to vortex scale lengths ξ_o and λ. The temperature dependence of ξ_o and λ can lead to variation of δ and to observed variation of H_p [18,35].

Three temperature regions can be distinguished in Hp vs. temperature variation presented in figure 9:

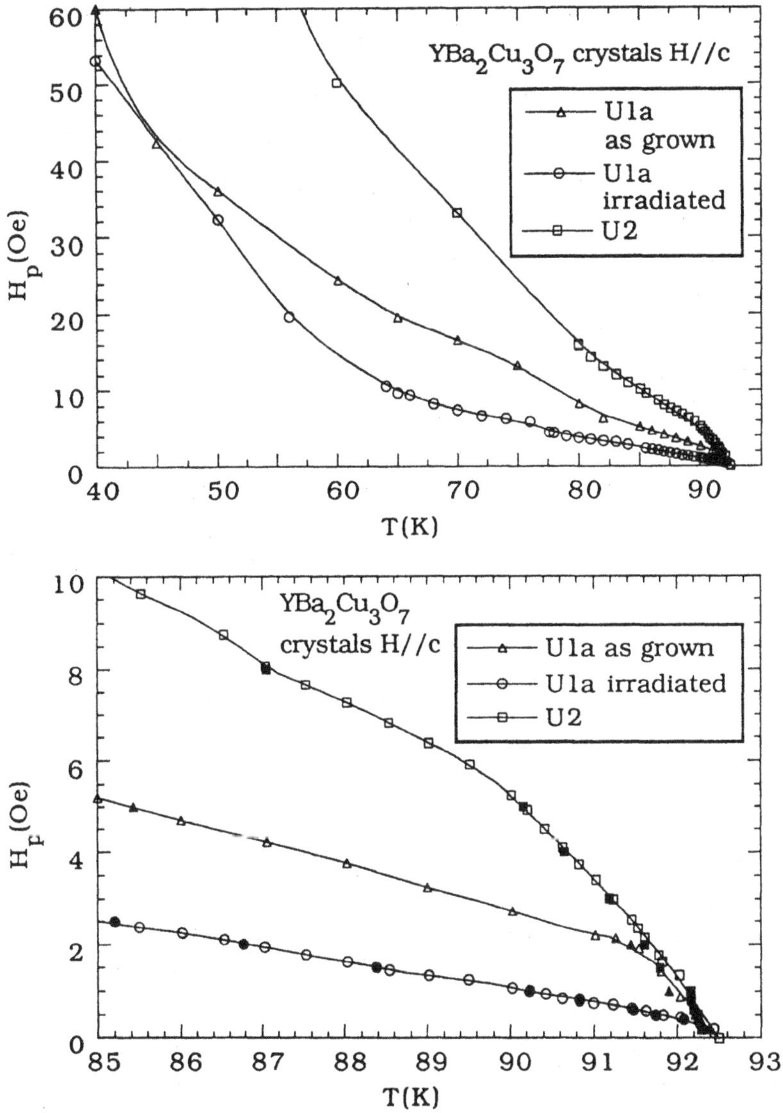

Figure 9. First magnetic penetration field recorded on two untwinned crystals of $YBa_2Cu_3O_7$, as grown sample U2 and on sample U1a before and after electron irradiation. On lower panel, open symbols represent H_p

values extracted from kink on isothermal $\Delta H(H_a)$ curves, full symbols H_p detected from onset on ZFC ΔH vs. temperature curve.

(a) Close to T_c existing surface regularities are small in comparison to ξ and λ and BL barrier is fully active. Penetration of magnetic fields occur when applied field reaches values close to H_c [5,20].

(b) Intermediate temperature region, where flux penetration occurs throughout the weak spots on surface irregularities but bulk critical current is low enough to allow spread of flux to the whole volume of the sample. Actual value of H_p in this temperature range corresponds to most depressed BL Position of the bump separating region from fully active BL barrier domain near T_c, is sample dependent. After electron irradiation this bump was pushed very close to T_c. In irradiated sample variation of H_p vs. temperature seems to be close to that of H_{c1}.

(c) Low temperatures where flux penetrating throughout spots is confined by bulk critical current to vicinity of the spot. The low temperature up-turn of H_p is related to the decreasing role of "weak spot" flux penetration when bulk critical current rises. It should be noted that linear variation $\div\partial\Delta H$ vs H_a is not an appropriate modeling of flux penetration in presence of "weak spots". Up-turn presented in figure 9 reflects only qualitative behaviour.

Temperature variation of H_p observed in $Bi_{2.2}Sr_{1.8}CaCu_2O_8$ crystals closely resembles that of $YBa_2Cu_3O_7$ presented in figure 9 [19]. Three regimes corresponding to low temperature up-turn, intermediate temperatures with linear H_p vs. T variation and down-turn close to T_c can be distinguished. It is appealing to interpret this variation in the framework of the same "weak spot" limited penetration throughout BL barrier. An alternative explanation of low temperature up-turn was given in ref. 6. BL barrier was assumed to limit flux penetration, but jumps of fragmented vortex line was suggested to depress barrier height and give rise to exponential increase of H_p on decreasing temperature. In this model, linear part of H_p at intermediate temperatures represents H_{c1}. This model fails to explain down-turn of $H_p(T)$ close to T_c and depression of H_p after electron irradiation of $YBa_2Cu_3O_7$.

5. Conclusions.

The importance of BL barrier is well proved by the experiments presented above. Subsequent questions to be answered are: (1) in what circumstances BL barrier dominates carrying current capacities of HTSC superconductors, (2) to what extent interpretation of magnetic features of HTSC should be revised.

A partial answer to the first question can be found in the magnetization curve presented in Fig. 3. Upper limit on magnetization field of bulk critical current $J_c^{b}*t$, can be set by value of remanent field. When this field becomes small, another source of irreversible magnetization, surface barrier starts to dominate. This crossover from bulk pinning to surface pinning seems to occur when magnetization field of bulk critical current is comparable to the height of surface barrier. Low $J_c^{b}*t$ can results from low pinning force and from small thickness of the sample. This means that in thin samples, surface barrier effect is likely to dominate hysteretic magnetization. This explains also why clear evidence for surface barrier where found in LHPM technique and not in conventional magnetometers. The sensitivity of conventional techniques depends on sample volume while LHPM requires mainly that the sample surface be bigger than dimension of the Hall probe. Small volume, thin samples of weight <20μgram, can be easily investigated by LHPM.

There are many implications of BL surface barrier on current carrying capacities of HTSC. First, BL barrier enhances critical current limit in fully reversible no-critical-state-penetration mode. This regime can be of potential interest for low-loss AC operation of current transport lines.

BL barrier is a known source of strong pinning by normal state inclusions in superconducting matrix [37]. This kind of pinning is likely to be present in quench-melt-grown processed high J_c HTSC ceramics.

Anomalous hysteretic enhancement of transport critical current in magnetic fields was observed in HTSC ceramics [38]. This effect seems to be related to BL barrier limited intra-grain magnetization [39].

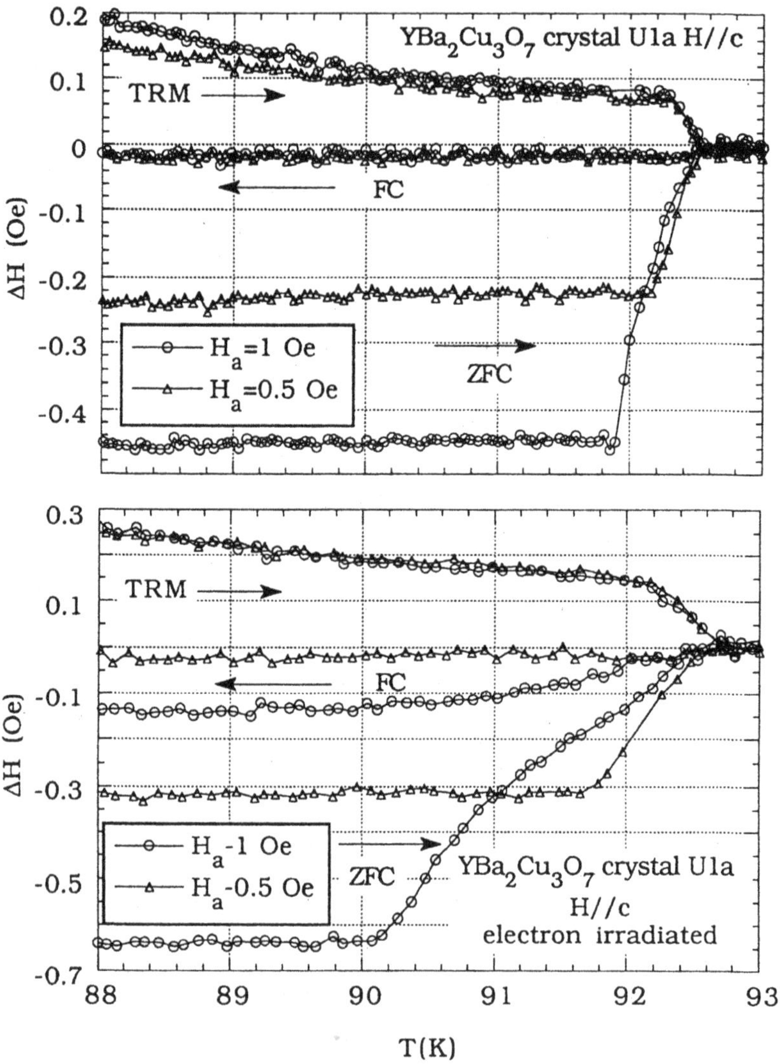

Figure 10. ZFC-FC-TRM cycle recorded on $YBa_2Cu_3O_7$ crystal U1a at fields of 1 and 0.5 Oe before (upper panel) and after (lower panel) electron irradiation. The actual values of ΔH depend on Hall probe position.

An answer to the second question is beyond the scope of this paper. In particular, dynamic of flux transfer by BL barrier, and related AC magnetic field response need to be investigated.

Here, I analyze the meaning of Meissner fraction (MF), defined as ratio of FC and ZFC magnetizations. MF close to 1 in low magnetic field is currently considered as a test of sample quality and superconducting state in whole sample volume. As can be seen in figure 10, MF is close to 0 in untwinned $YBa_2Cu_3O_7$ crystal in fields as low as 0.5 and 1 Oe. This absence of flux expulsion on cooling below T_c is directly related to BL barrier. The symmetry of ZFC - FC- thermoremanence (TRM) cycle is not fulfilled here. The difference between FC and ZFC magnetization field is much higher then TRM magnetization. It should be noted that, in Bean model M_{FC}-M_{ZFC}=M_{TRM} [40].

After electron irradiation and depression of BL barrier, some FC Meissner effect starts to be observed. MF becomes non-zero at 1 Oe field, but decreases at lower fields.

6. Acknowledgments.

It is a pleasure to thank my colleagues, Y. Yeshurun, L. Burlachkov, F. Holtzberg and N. Chikumoto for collaboration in works covered by this review. I am greatly indebted to J. Gilchrist for guidelines in earlier works on BL barrier.

7. References.

1. C. P. Bean, *Rev. Mod. Phys.* **36**,(1964) 31
2. Y. Yeshurun and A. P. Malozemoff, *Phys. Rev. Lett.* **60**,(1988) 2202
3. K. A. Müller, M. Takashige and J. G. Bednorz, *Phys. Rev. Lett.* **58**,(1987) 1143
4. A. M. Campbell and J. E. Evetts, ***Critical Currents in Superconductors*** (Taylor and Francis Ltd., London, 1972).
5. C. P. Bean and J. D. Livingston, *Phys. Rev. Lett.* **12**,(1964) 14
6. V. N. Kopylov, A. E. Koshelev, I. F. Schegolev and T. G. Togonidze, *Physica* **C 170**,(1990) 291

7. R. B. Goldfarb, A. F. Clark, A. I. Braginski and A. J. Panson, *Cryogenics* **27**,(1987) 475
8. A. Shaulov and D. Dorman, *Appl. Phys. Lett.* **53**,(1988) 2680
9. A. Umezawa, et al., *Phys. Rev.* **B 38**,(1988) 2843
10. Y. Yeshurun, A. P. Malozemoff, F. Holtzberg and T. R. Dinger, *Phys. Rev.* **B 38**,(1988) 11828
11. E.-W. Scheidt, et al., *Solid State Comm.* **71**,(1989) 505
12. C. J. Liu, et al., *Physica* **C 162-164**,(1989) 1609
13. H. Adrian, W. Assmus, A. Höhr, J. Kowalewski, H. Spille and F. Steglich, *Physica* **C 162-164**,(1989) 329
14. L. Krusin-Elbaum, R. L. Greene, F. Holtzberg, A. P. Malozemoff and Y. Yeshurun, *Phys. Rev. Lett.* **62**,(1989) 217
15. M. W. McElfresh, Y. Yeshurun, A. P. Malozemoff and F. Holtzberg, *Physica* **A 168**,(1990) 308
16. F. Holtzberg and C. Field, *Eur. J. Sol. State Inorg. Chem.* **27**,(1990) 107
17. M. Konczykowski, F. Holtzberg and P. Lejay, *Superconductivity Science and Tech.* **4**,(1991) S331
18. M. Konczykowski, L. Burlachkov, Y. Yeshurun and F. Holtzberg, *Phys. Rev.* **B 43**,(1991) 13707
19. N. Chikumoto, M. Konczykowski, N. Motohira, K. Kishio and K. Kitazawa, *Physica C, Proceedings M^2HTSC, Kanazawa Japon* ,(1991)
20. P. G. deGennes, *Superconductivity of Metals and Alloys* (W.A. Benjamin Inc., New York, 1966).
21. P. S. Swartz and H. R. Hart, *Phys. Rev.* **137**,(1965) 818
22. P. S. Swartz and H. R. Hart, *Phys. Rev.* **156**,(1967) 412
23. H. R. Hart and P. S. Swartz, *Phys. Rev.* **156**,(1967) 403
24. J. Lowell, *J. Phys. C (Solid St. Phys.)* **2**,(1969) 372
25. D. Saint-James and P. G. deGennes, *Phys. Lett.* **7**,(1963) 306
26. A. M. Campbell, J. E. Evetts and D. Dew-Hughes, *Phil. Mag.* **18**,(1968) 313
27. C. C. Chang and A. C. Rose-Innes, E. Kandas, Eds., Low temperature Physics, LT-12 (Academic Press of Japan, 1971), pp. 381.

28. L. J. Swartendruber, A. Roitburd, D. L. Kaiser, F. W. Gayle and L. H. Bennet, *Phys. Rev. Lett.* **64**,(1990) 483

29. H. T. Coffey, *Cryogenics* ,**7**,(1967) 73

30. D. J. Frankel, *J. Appl. Phys.* **50**,(1979) 5402

31. S. Reich and V. M. Nabutovsky, *J. Appl. Phys.* **68**,(1990) 668

32. M. Daümling and D. C. Larbalestier, *Phys. Rev.* **B 40**,(1989) 9350

33. M. Konczykowski, L. Burlachkov, Y. Yeshurun and F. Holtzberg, *Physica C, Procedings M^2 HTSC, Kanazawa, Japon* ,(1991)

34. H. Vichery, F. Rullier-Albenque, H. Pascard, M. Konczykowski, R. Kormann and D. Favrot, *Physica* **159**,(1989) 689

35. L. Burlachkov, Y. Yeshurun, M. Konczykowski and F. Holtzberg, (*Proceedings of MMM-Intermag* Pittsburgh June 18-21, 1991),

36. L. Krusin-Elbaum, A. P. Malozemoff, Y. Yeshurun and F. Holtzberg, *Phys. Rev.* **B 39**,(1989) 2936

37. J. E. Evetts, *Phys. Rev.* **B 2**,(1970) 95

38. M. E. McHenry, M. P. Maley and J. O. Willis, *Phys. Rev.* **B 40**,(1989) 2666

39. J. Gilchrist, *Solid State Communs.* **75**,(1990) 225

40. A. P. Malozemoff, in *Physical Properties of High Temperature Supercondutors* D. M. Ginsberg, Eds. (Word Scientific Publishing, 1989), vol. 1, pp. 73.

NMR AND NQR STUDIES OF HIGH-T_c SUPERCONDUCTORS

Takao Kohara, Yoh Kohori and Koh-ichi Ueda

Department of Material Science, Himeji Institute of Technology, Akoh-gun, Hyogo 678-12, Japan

ABSTRACT

NMR and NQR results of some heavy fermion- and high-T_c oxide-superconductors are reported. From the temperature dependences of the nuclear relaxation and the Knight shift, the property of Cooper pairing is discussed in both systems. The T_1^{-1} behaviors on high T_c-oxides below T_c are similar, irrespective of these systems. In the heavy fermion-superconductors, the nuclear relaxation indicates an anisotropic energy gap having nodes at lines on the Fermi surface (p- or d-wave superconductivity). In the high-T_c cuprates, the d-wave like behavior observed in Cu relaxation may be explained by a rapid reduction of antiferromagnetic spin fluctuation localized on Cu sites below T_c, which enhanced T_1^{-1} of Cu in the normal state. Superconductivity and antiferromagnetism compete with each other in the high-T_c oxides, while they microscopically coexist or compete in the heavy fermion-compounds.

1. INTRODUCTION

NMR and NQR techniques have proved extremely powerful for the investigation of classical superconductors[1,2] and also offered an attractive way to study microscopically the magnetic and superconducting properties, since the response of inequivalent sites can be resolved in the nuclear resonance spectra and the dynamical properties on them can also be studied by the nuclear magnetic relaxation at respective sites. The cornerstones of success in this area include the value of the energy gap, the Knight shift collapse due to Cooper pairing,[2] the field distribution in the vortex lattice,[3] the competition between magnetism and superconductivity,[4] to name just a few. Particularly, it is so important for understanding the low-lying elementary excitations of quasiparticle to make a systematic measurement of the nuclear spin-lattice relaxation time, T_1.

In the electronic strong-correlation systems where superconductivity and magnetism coexist or compete with each other, heavy fermion (electron) superconductors [HFS] and high-T_c oxide superconductors [HTS], in particular, have attracted a great deal of attention, because the superconductivity on them is unexpected and the magnetic properties on these superconductors are also very unusual.[5] In both systems, superconductivity and/or antiferromagnetic ordering appear in their region by doping some elements or varying temperature and/or magnetic field and so on. This has led to the speculation that these materials may exhibit an unconventional type of superconductivity involving a new type of pairing and a pairing-mechanism. Though any direct evidence for the mechanism of unconventional superconductivity is not proved yet, there is a possibility of non-BCS mechanism or an attractive interaction through spin fluctuations, being responsible for the occurrence of superconductivity. Especially, considerable amounts of experimental evidences have suggested that the energy gap in HFS or HTS is vanishing at points or lines on the Fermi surface, in analogy to superfluid ^{3}He-A phase where the gap zeros of points are associated with the anisotropic triplet Cooper pairing.

Among variety of our NMR/NQR studies, we restrict ourselves to report and to discuss the following two kinds of materials on the magnetism and superconductivity associated with highly correlated electrons in this paper. The first is on uranium based HFS. The second is on cuprate HTS doped by oxygen or divalent Ca ions.

2. NMR MEASUREMENTS IN THE SUPERCONDUCTING STATE

2.1. T_1 IN THE SUPERCONDUCTING STATE

As is well known, in the framework of BCS theory, the T_1 ratio in the normal and superconducting states is expressed as,[6]

$$\frac{T_{1N}}{T_{1S}} = \frac{2}{k_B T} \int [N_S^2(E)+M_S^2(E)]f(E)[1-f(E)]dE$$

where $N_S(E)=E/(E^2-\Delta^2)^{1/2}$, which is the normal density of states [DOS] of the quasiparticles, $M_S(E)=\Delta/(E^2-\Delta^2)^{1/2}$, which is the anomalous DOS arising from the coherence factor characteristic for singlet Cooper pairing. $f(E)$ and 2Δ are the Fermi distribution-function and the energy gap, respectively. Owing to a uniform energy gap in usual BCS superconductors, which give rise to a divergence of both $N_S(E)$ and $M_S(E)$ at the gap edge, T_1^{-1} shows an enhancement just below T_C and then exponentially decreases with decreasing temperature. The absence of the enhancement just below T_C and the non-exponential behavior, nearly T^{α}-dependence of T_1^{-1}, provide a valid evidence of the anisotropic nature of superconductivity. In the anisotropic triplet superconductor like superfluid ^{3}He-A phase, the anomalous DOS arising from the coherence factor, $M_S(E)=0$, and the gap vanishes at points with $2\Delta(\theta)=\Delta_0\sin\theta$ (axial-model) or lines with $2\Delta(\theta)=\Delta_0|\cos\theta|$ (polar-model) on the Fermi surface.[7] Thus the system has an anisotropic gap, suggesting the p- or d-wave superconductivity being realized owing to a strong on-site repulsion. The anisotropy of the energy gap is also reflected in the T-dependence of T_1^{-1}. In the axial- and the polar-models, $N_S(E)$ diverges more weakly compared with BCS states (varies linearly with E^2 as E tends to 0) and remain finite (varies linearly with E) at the gap edge, respectively. Thus T_1^{-1} is expressed to decrease as T^5 (axial-model) or T^3(polar-model) at low temperatures without an enhancement just below T_C.

2.2. THE KNIGHT SHIFT IN THE SUPERCONDUCTING STATE

As the spin susceptibility, χ_S, reflects on the Knight shift, we obtain the information of χ_S from the Knight shift. The χ_S below T_C is modified depending on the nature of Cooper pairing. In BCS model, χ_S varies as the so-called 'Yosida function' and tends to zero at T=0, if spin-orbit scattering is neglected.[2] Other even-parity mechanism also holds for collapse of the Knight shift. In the anisotropic triplet model, χ_S below T_C is the same as that in the normal state, $\chi_S(T)=\chi_N$. In the isotropic model (BW model) like ^{3}He-B phase, χ_S

decreases to 2/3 of χ_N, as T tends to zero.[8]

3. HEAVY FERMION SUPERCONDUCTORS [HFS]

A variety of ground states have been found in the heavy fermion systems of lanthanide- or actinide-compounds; paramagnetic-normal conducting, paramagnetic-superconducting, antiferromagnetic-normal conducting and coexisting antiferromagnetic-superconducting. As is well known, this system is characteristic of a strongly interacting Fermi liquid with heavy effective mass. At low temperatures, Pauli susceptibility and an anomalously huge specific heat coefficient, γ, 1J/mole·K^2, which has two or three orders of magnitude larger than that of an ordinary metal, show the f-local moments strongly coupled with conduction electrons to be like a clod of heavy itinerant-electrons. On the contrary, a Curie-Weiss behavior is observed in the susceptibility at high temperatures, showing the system of a weakly interacting-assembly of local moments and conduction electrons. This HFS system is located nearby the antiferromagnetically ordered state, which often appears by doping a small amount of transition-metal impurities. These compounds, have many anomalous properties in comparison with conventional superconductors; for instance, the transport and thermodynamical properties with T^{α}-dependence at low temperatures. These results have generated a great deal of excitements since they may indicate an unusual state in which the gap vanishes at points or lines on the Fermi surface.

3.1. UPt_3

UPt_3 of the hexagonal $MgCd_3$ type has very anisotropic resistivity and susceptibility. The typical values of T_c and γ are 0.5 K and 450 mJ/mole U·K^2, respectively. Neutron diffraction experiment has firstly found an antiferromagnetic ordering at 5 K with a small magnetic moment ($\sim 0.02\mu_B$).[9] As mentioned before, the antiferromagnetic ordering appears by the substitution of small amount of Th or Pd. The maximum T_N is the same temperature of 6 K at which Th or Pd doping induce a much larger magnetic moment ($\sim 0.7\mu_B$).[10]

Fig.1 shows the T-dependence of T_1^{-1} of ^{195}Pt. In the normal state, T_1^{-1} varies in proportion to T, which means that the system is in the Fermi liquid state, subsequently deviating slightly above 8

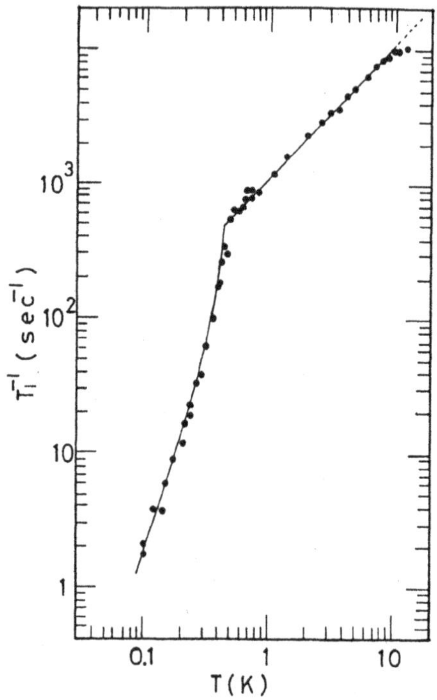

Fig.1. T_1^{-1} of ^{195}Pt vs. temperature in UPt_3. The resonance frequencies (fields) are 4.9 MHz (~5.2 kOe) and 1.9 MHz (~2.0 kOe) above and below 0.5 K, respectively.

K.[11] T_1 measurements show that any indication of magnetic ordering (the critical slowing down at T_N or the rapid decrease of T_1^{-1} associated with the gap-opening of SDW) is not observed at 5 K, which will be mentioned hereafter by the results of the NMR spectrum. T_1^{-1} decreases drastically just below T_c, subsequently in proportion to T^3 down to the lowest temperature of 0.1K. When we try to analyze the T_1-results on the framework of the simple triplet polar model (M_s=0), we can get the calculated curve fitted by taking $2\Delta(\theta)=10.5k_BT_c|\cos\theta|$ with an assumption of $\Delta(\theta)=\Delta_{BCS}(T)|\cos\theta|$, as shown in Fig.1. This value is about two times larger than the weak coupling value. If we assume the anisotropic gaps for $CeCu_2Si_2$[12] and UBe_{13},[13] the experiment can be reproduced when $2\Delta(\theta)=5k_BT_c|\cos\theta|$ and $2\Delta(\theta)=7k_BT_c|\cos\theta|$ are adopted, respectively. The Knight shifts on both compounds de-

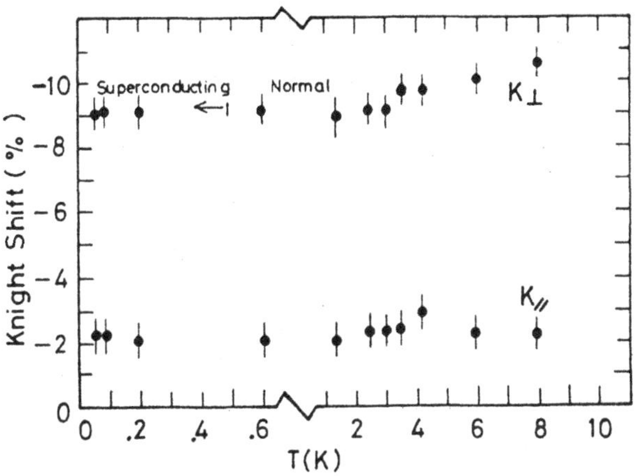

Fig.2. T-dependence of the Knight shift. The resonance frequency (field) is about 4.4 MHz (~4.6 kOe).

crease below T_c with decreasing temperature indicating the even parity pairing.[14,15] Next, NMR spectrum of ^{195}Pt in UPt_3 is also quite anisotropic with a large negative shift. From the edge and the peak of the spectrum, the parallel and perpendicular components of the Knight shift to the hexagonal c-axis, $K_\perp$ and $K_{//}$, are obtained, respectively. Shown in Fig.2 is the T-dependence of $K_\perp$ and $K_{//}$. In the normal state, $K_\perp$ decreases with decreasing temperature, whereas $K_{//}$ is nearly T-independent. Despite of the existence of the long range antiferromagnetic ordering below 5 K reported by the neutron diffraction,[9] no extra broadening of the spectrum associated with the magnetic ordering is observed around 5 K.[16] The transferred hyperfine field is about 85 kOe/μ_B. Since $K_\perp$ and $K_{//}$ are on the same line plotted as K-χ diagram, the hyperfine field is considered to be isotropic. So K is induced mainly by the core-polarized interaction of the 4d conduction electrons which is spin-polarized by the uranium 5f electrons. The detailed analysis of Pt Knight shift was discussed in the previous papers.[11] Both $K_\perp$ and $K_{//}$ do not change even in the superconducting state, which is consistent with the absence of Pauli limit in upper critical field.[17] This result implies that the superconductivity in UPt_3 is favorable for the odd parity pairing. However we cannot exclude the possibility of the even parity pairing with strong impurity- or surface-scattering, which keeps the spin susceptibility unchanged even in the superconducting state. On the contrary, the low residual resistivity in UPt_3 seems to be unfavorable for the impurity scattering mechanism. At present a triplet polar superconductivity on UPt_3 is concluded to be more probable, although not decisively. The preliminary NMR results on the antiferromagnetic $U(Pt\text{-}Pd)_3$ have also been reported.[16]

3.2. URu_2Si_2

URu_2Si_2 of the tetragonal $ThCr_2Si_2$ type is a moderately HFS with a large γ-value of 65.5mJ/mol U·K^2 [18] or 75mJ/mol U·K^2.[19] This system shows an antiferromagnetically ordered state below 17.5 K and a superconducting state below 1.3 K.[18,19] The magnetic moment is reported to be $(0.04\pm0.02)\mu_B$ along the tetragonal c-axis.[20] The ^{29}Si spin echo spectra at 1.4 K (normal state) and 0.5 K (superconducting state) are shown in Fig. 3.[21] The spectra have an anisotropic nature presumably due to the anisotropy of the susceptibility and/or the hyperfine coupling constant. As seen in this figure, no appreciable change between the normal and superconducting states is observed, showing that the spin-susceptibility does not change through T_c. This suggests that the system is a triplet superconductor or a singlet superconductor with large spin-orbit scattering by impurities.

Shown in Fig. 4 is the T-dependence of T_1^{-1} of ^{29}Si.[21] T_1^{-1} is T-independent above the Kondo temperature (T_K) of about 60 K. Below 60K, T_1^{-1} is proportional to T. T_1^{-1} rapidly decreases below 17 K, which is the ordering temperature (T_N) of SDW state, and attain the relation of 'T_1T=const.' again. The 'T_1T=const.' obtained below 10 K is associated with the ungapped electrons in the SDW state. The decrease of T_1^{-1} is attributed to the reduction of the DOS in the SDW state. From the T-dependence below 17 K, we can estimate the energy gap of the SDW state to be about 100 K, which is also comparable in magnitude to those obtained from the specific heat[18] and the inelastic neutron scattering[22] measurements. Below T_c, T_1^{-1} rapidly decreases again associated with the appearance of the superconductivity. The continuous decrease of T_1^{-1} from the normal to the superconducting state shows that the ungapped electrons in SDW state enter the superconducting condensed-state indicating the coexistence of the SDW and the superconducting states. Otherwise if the SDW state were destroyed by the superconductivity, T_1^{-1} at T_c would recover discontinuously the large value be-

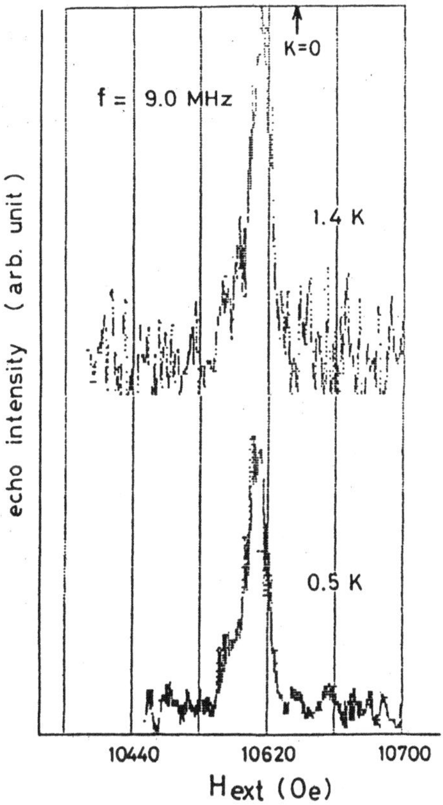

Fig.3. ^{29}Si spin-echo spectra at 1.4 K (normal state) and 0.5 K (superconducting state).

cause of the disappearance of energy gap associated with the SDW state. Again we have observed no enhancement just below T_c. The change of T_1^{-1} is close to the T^3 dependence with a slight deviation below 0.65K. This result suggests that superconductivity has an anisotropic nature having gap zero at lines on the Fermi surface. If we tentatively try to fit the result with the calculated value as in the case of UPt_3, we can get an agreement assuming $2\Delta(\theta)=3.5k_BT_c|\cos\theta|$. In this compound, however, the absolute value of T_1 is so large that any effect of magnetic impurities on nuclear relaxation may give rise to the deviation from the T^3 dependence at low temperatures. Furthermore, the large residual resistivity in this system may suggest a possibility of the impurity scattering which produces finite spin

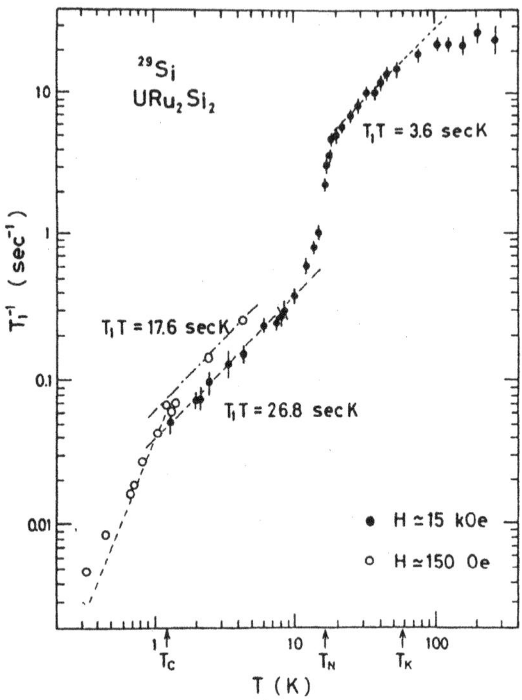

Fig.4. T_1^{-1} of ^{29}Si vs. temperature in URu_2Si_2.

susceptibility even in the even parity. At present we are now measuring ^{29}Si NMR to study the electronic states of antiferromagnetic $U(Ru\text{-}Rh)_2Si_2$[23] and ferromagnetic $U(Ru\text{-}Re)_2Si_2$ in detail.

4. HIGH T_c CUPRATES [HTS]

Oxide superconductors have many common features such as the two-dimensional structure and the presence of CuO_2 layers, which are considered to play a key role in giving high T_c superconductivity The one striking feature is that antiferromagnetism is commonly situated near the superconducting phase. By the study of the spin dynamics on the many kinds of oxide superconductors, the antiferromagnetic spin correlations is considered to remain even in the superconducting state and to be responsible for the unusual properties in the normal state. Both of the identification of the symmetry on the order parameter (s-wave or d-wave) in oxide superconductors and whether the normal state may be described by a Fermi-liquid picture or not are controversial.

4.1. $La_2CuO_{4+\delta}$

La_2CuO_4 is a Mott insulator in which Cu moments undergo three dimensional antiferromagnetic ordering at $T_N\sim325$ K. Hole doped $La_2CuO_{4+\delta}$ by adding excess oxygen have both the reduced T_N and the induced superconductivity below 40 K. This material has been studied extensively because it embodies much of the physics and the chemistry generic to the high-temperature conductors and the magnetic superconductors. Jorgensen et al. have recently reported that a reversible phase-separation into two crystallographically distinct structures occurs near 320 K in powdered $La_2CuO_{4+\delta}$.[24] Diffraction studies on single crystal of $La_2CuO_{4+\delta}$ have also suggested that the phase-separation temperature, T_P, is about 280 K.[25] Below T_P, the oxygen loaded samples are reported to be composed of both a non-superconducting, oxygen-poor phase (Bmab; presumably with high resistance) and a superconducting, oxygen-rich phase (Fmmm; presumably with less resistance) with a closely related orthorhombic structure. The detailed process for the sample preparation is described in the literature.[26] Shown in Fig.5 is Cu spin echo spectrum of $La_2CuO_{4+\delta}$ annealed under oxygen gas of 1200 bar. Since the frequency- and the integrated intensity-ratios of the two lines correspond to the ratios of the electric quadrupole moment and natural abundances of ^{63}Cu and ^{65}Cu, respectively, these peaks are assigned to be Cu NQR signals with no internal field. Comparing with that of $(La\text{-}Sr)_2CuO_4$,[27] the spectrum has a rather wider width and a tail observed in higher frequency, which may be caused by a random distribution of electric field gradient [EFG] due to excess oxygen. As the NQR signal is approximately proportional to the oxygen pressure in annealing, this paramagnetic signal comes from the oxygen-rich (Fmmm) phase. On the contrary, the antiferromagnetic Cu NMR signal in $La_2CuO_{4+\delta}$ was also observed under zero external field in the same frequency-range (75~115 MHz) as that in undoped La_2CuO_4,[28] that means the antiferromagnetic NMR signal comes from the oxygen-poor (Bmab) phase. These results show that there coexist paramagnetic (Fmmm) and antiferromagnetic(Bmab) phases in the oxygen-loaded sample.

Fig.6 shows the relative nuclear relaxation rate, $(T_1^{-1})/(T_1^{-1})_{T=Tc}$ of ^{63}Cu in the CuO_2 planes of several compounds plotted against T/T_c.

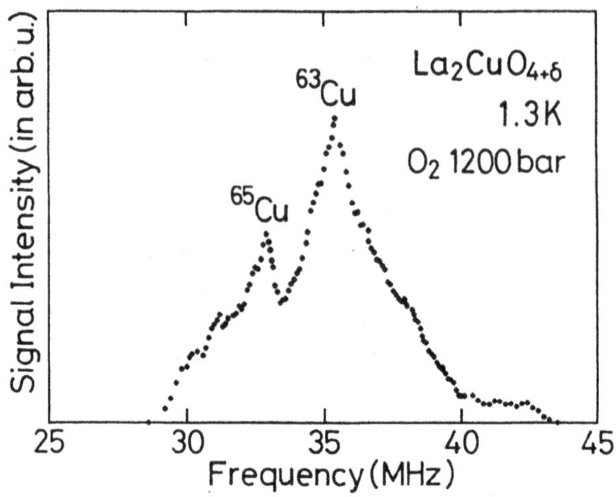

Fig.5. Cu NQR spectrum in $La_2CuO_{4+\delta}$ at 1.3 K.

As seen in this figure, the T-dependence of T_1^{-1} in the oxygen-doped La system below T_c is similar to those of other (LaSr)-,[27] Y-,[29] Bi-[30] and Pb-systems.[31] Neither the enhancement just below T_c nor the exponential decrease in T_1^{-1} is observed in the superconducting state, both of which are characteristic of usual BCS superconductors. Although there are discrepancies above T_c, the normalized T_1 values below T_c fall approximately on a single line, suggesting that the superconducting nature (for example, the properties on energy gap) is common for these cuprates. It is also reported that the T-dependence of T_1^{-1} below T_c is explained by a d-wave model with $2\Delta(\theta)=10k_BT_c|\cos\theta|$.[32] The T-dependences of the penetration depth and the Knight shift of Cu in Y and Tl compounds, however, seem to support a s-wave pairing.[32,33] This discrepancy still remains for future problems. As seen in this figure, T_1^{-1} above T_c has weaker T-dependence than T-linear. The T-dependence of T_1 in the normal state is as important as that in the superconducting state to study the T-dependence of χ_Q, which is expected to follow a Curie-Weiss law at high temperatures becoming constant below a characteristic temperature with further decreasing temperature. So we prepare the $La_2CuO_{4+\delta}$ with more loaded oxygen to measure the Cu NQR signal in CuO_2 plane with higher accuracy in the wider temperature range of the normal state.

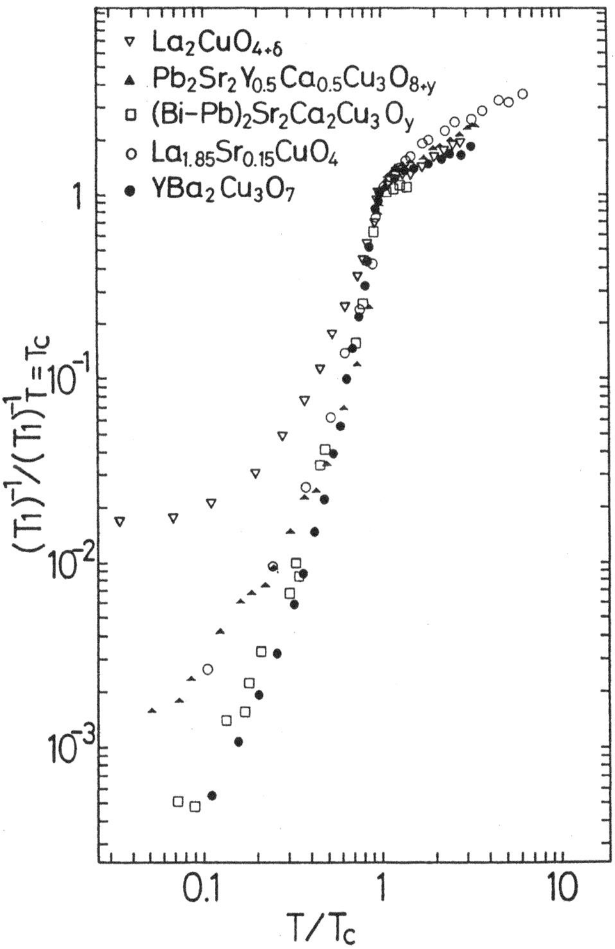

Fig.6. T_1^{-1}-values of ^{63}Cu in the CuO_2 planes normalized by their values at T_c are plotted against T/T_c for $La_{1.85}Sr_{0.15}CuO_4$, $YBa_2Cu_3O_7$, $(Bi\text{-}Pb)_2Sr_2Ca_2Cu_3O_y$ and $Pb_2Sr_2Y_{0.5}Ca_{0.5}Cu_3O_y$.

In Fig.7 we show the La NMR spectrum taken at 195K in the magnetic field of approximately 7.8 T applied to the c-axis. The spectrum is composed of a nearly unshifted line (K=0) and a broad line having a substantial shift, which may originate from oxygen-rich and from oxygen-poor regions, respectively. The broad line may be caused by the incomplete alignment of c-axis oriented parallel to the magnetic

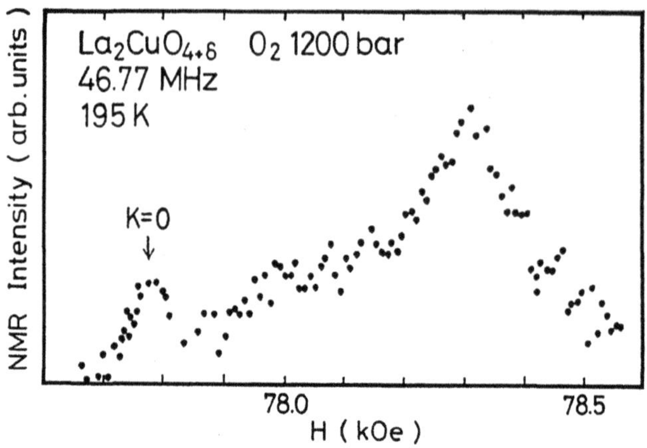

Fig.7. NMR spectrum of ^{139}La.

field. The ratio of the integrated intensity in a nearly unshifted line to that in a broad line (i.e. the ratio of the volume in both regions) was found to be about 1:9 at 195 K. Unshifted NMR signal, gradually increasing the intensity above 280 K, was still observed up to room temperature. On the contrary, the intensity of the broad line drastically decreased accompanied by decreasing the shift with increasing temperature, and the signal vanished upon warming above 270K. The abrupt disappearance of the broad line may partly be due to the warming through T_N of 255K. The precise intensity measurements on both La NMR signals are in progress to clarify microscopically the magnetic behaviors in poor-oxygen region and a phase-separation around 300 K.

4.2. $Pb_2Sr_2Y_{0.5}Ca_{0.5}Cu_3O_{8+y}$

A 70 K class superconductor, $Pb_2Sr_2Y_{0.5}Ca_{0.5}Cu_3O_{8+y}$, contains two CuO_2 planes (Cu2 site) and oxygen-free Cu1 plane (Cu1 site), which has monovalent Cu locating between two PbO layers.[34] A band calculation study predicts that Cu2 sites have main contribution to the DOS at E_F, while Pb and Cu1 sites have little one to DOS.[35] It is expected that the CuO_2 plane plays an important role on both the normal and superconducting behaviors like other HTS.

By an ordinary four-terminal resistivity measurement, the onset- and zero-resistance temperatures were 80 K and 76 K, respectively. A clear Meissner effect of about 80 % diamagnetization was also con-

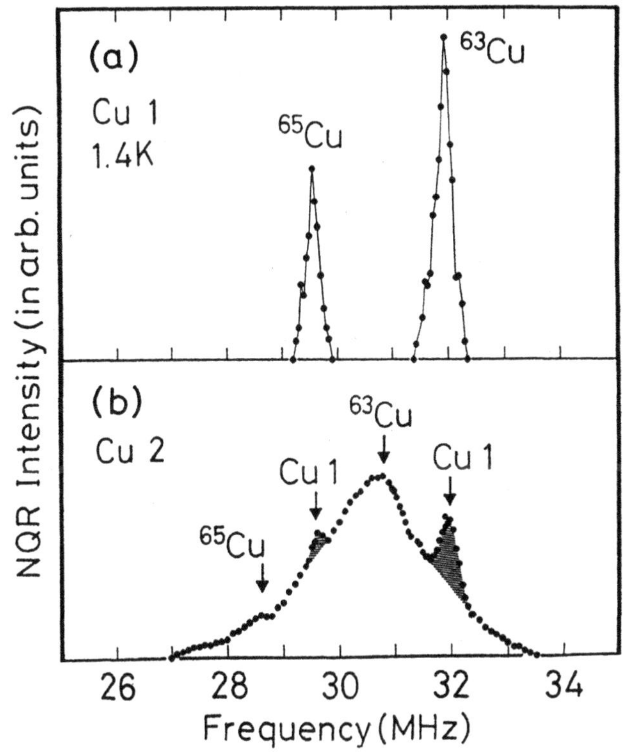

Fig.8(a). NQR spectrum of Cu1.
Fig.8(b). NQR spectrum of Cu2.

firmed by ac susceptibility measurement. In Fig.8(a) and (b) we show the spin echo signal obtained between 26~34 MHz. No other signal was found in the frequency range of 4~90 MHz. The observed spectrum was decomposed into a couple of sharp peaks (Fig.8a) and a broad line (Fig.8b) by use of the differences of T_1 and spin-spin relaxation time, T_2, between two sorts of signals. Considering the frequency ratio of two sharp peaks and the area-ratio of broad line to that of sharp signal, the sharp and broad lines can be interpreted as the signals which may originate from Cu1 and Cu2 sites, respectively.[31] The broadening of Cu2 spectrum may be caused by a substitution of Ca for Y.

The T-dependence of T_1 is shown in Fig.9. T_1^{-1} of Pb in the normal state is nearly proportional to T while that of Cu2 in the normal state has weaker T-dependence than T-linear. In the normal state, the

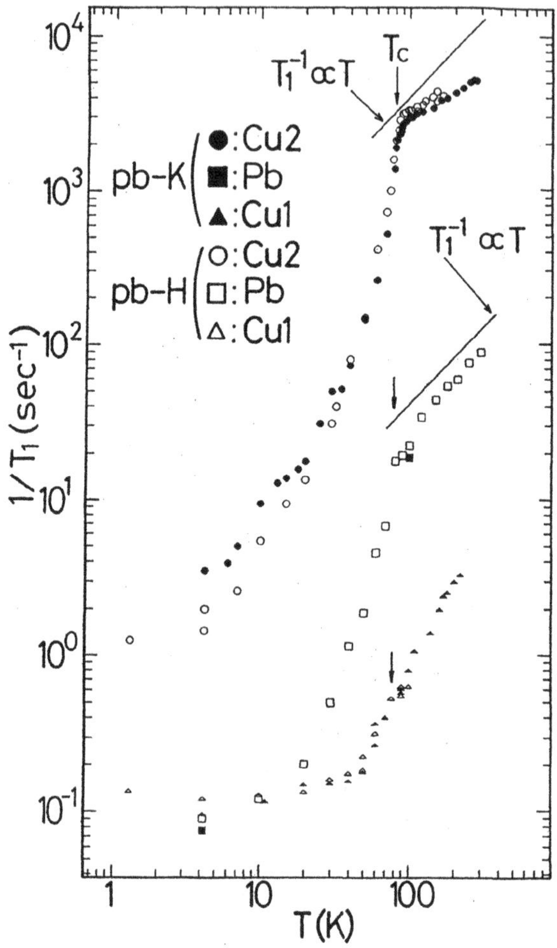

Fig.9. T_1^{-1} of Cu1, Cu2 and Pb vs. temperature in Pb-system. The samples labeled as pb-K and pb-H are independently prepared by almost the same manner.

T_1-value of Cu1 has three orders of magnitude larger than that of Cu2. In the superconducting state, T_1^{-1} of Cu and Pb, which are similar to those of other high T_c oxides, decrease with decreasing temperature with no enhancement just below T_c. The T-dependence of T_1 of Cu in CuO_2 plane is expressed by a universal function below T_c, which is mentioned in the former part. On the contrary, any anomaly of T_1 on

Cu1 is also not detected at T_c. Moreover, the measurement of the isotope effect on T_1 showed that the relaxation for Cu1 is not due to magnetic origin but to quadrupolar origin. These results suggest that DOS of Cu1 is absent even in metallic region and the effects of spin fluctuations from Cu2 moments are considered to be canceled at oxygen free Cu1 site, which are confirmed qualitatively by a band calculation.[35]

Pb NMR was observed under the external field of 7.3 T. Shown in Fig.10 is the T-dependence of the Knight shift, K, of Pb. By using Pb metal as a reference (0.111% at room temperature),[36] the T-independent Knight shift in the normal state is evaluated to be 0.28%, indicating the spin-susceptibility to be T-independent. On the contrary, broadening and diamagnetization shift of spectrum are expected, if inhomogeneous distribution of magnetic field exists in the supercon-ducting sample. In the present case, no appreciable difference of the

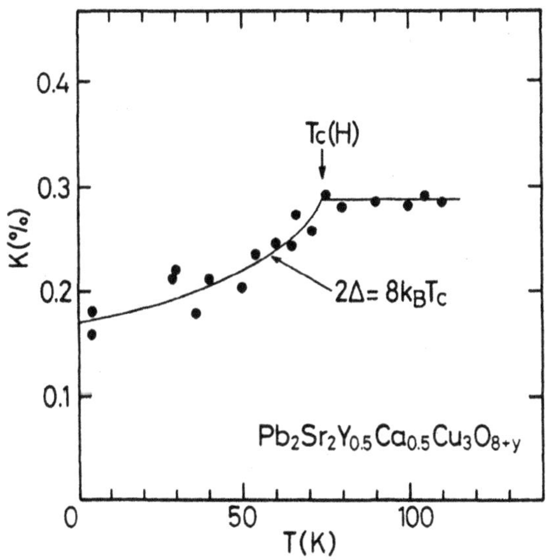

Fig.10. T-dependence of the Knight shift. The solid and dotted lines are calculated values based on a d-wave model (with a energy gap of $2\Delta(\theta)=8k_BT_c|\cos\theta|$ and gap-zero line on the Fermi surface) and a s-wave model (with a energy gap of $2\Delta=4.5k_BT_c$), respectively.

line width was observed between the normal and superconducting states. By subtracting the T-independent term (orbital or chemical shift) from the observed value, the T-dependence of the spin susceptibility, which decreases with decreasing temperature below T_c, is obtained. As seen in the figure, it is hard to adopt a s-wave model (with a larger energy gap than BCS gap) or a d-wave model (with a larger energy gap, for example $2\Delta(\theta)=8\sim10k_BT_c|\cos\theta|$) for the nature of a pairing. However the T-dependence of the Knight shift on this system may have a similar behavior to those of TlBaCaCuO-,[32] and YBaCuO-systems,[33] which may support a s-wave pairing. As mentioned in the former part, this discrepancy between s-wave and d-wave pairings remains a question for further investigation.

5. CONCLUDING REMARKS

Superconductivity and antiferromagnetism compete with each other in high-T_c oxides, while they microscopically coexist or compete in the heavy fermion-compounds. In the superconducting state, T_1^{-1} of respective nucleus decreases with decreasing temperature without any enhancement just below T_c and varies as T^3 at low temperatures in all the heavy fermion- and high T_c-compounds studied here. This indicates the existence of an anisotropic superconducting energy gap which vanishes at lines on the Fermi surface. The calculated magnitudes of the gap parameter Δ_0 in $2\Delta(\theta)=\Delta_0|\cos\theta|$ in UPt_3, $CeCu_2Si_2$, UBe_{13}, URu_2Si_2 and high T_c cuprates are 10.5, 5, 7, 3.5 and 10 in unit of k_BT_c, respectively. The results of nuclear relaxation together with the Knight shift support d-wave and p (or d)-wave pairing models in cuprates and heavy fermion-superconductors, respectively.

However, to determine the parity of the pairing and the anisotropy of the energy gap for the superconductivity on HTS and HFS decisively, there remain many outstanding questions of the consistency with other experimental results (for example, the penetration depth and the impurity scattering) to be solved in future.

ACKNOWLEDGEMENTS

The authors wish to express their thanks to Professors K.Asayama and Y.Oda for stimulating discussions. The work concerning La-system has been done in collaboration with Drs. M.Motoyama and Kashiwai.

Thanks are also due to Professors Y.Saito, Y.Koike and Dr. T.Noji for supplying one part of Pb cuprates.

References

1. for example, L.G.Hebel and C.P.Slichter, Phys.Rev.,113,1504(1957).
2. K.Yosida, Phys.Rev.,110, 769 (1967).
3. for example, A.G.Redfield, Phys.Rev., 162, 367 (1967).
4. for example, Y.Kitaoka, S.Hiramatsu, K.Ishida, T.Kohara and K.Asayama, J.Phys.Soc.Jpn., 56, 3024 (1987).
5. Review for HTS: for example, H.R.Ott, Physica, B148, 1 (1987); Z.Fisk, D.W.Hess, C.J.Pethick, D.Pines, J.L.Smith, J.D.Thompson and J.O.Willis, Science, 239, 33 (1988).
6. D.E.MacLaughlin, Solid State Phys.,(Academic Press) 31, 1 (1976).
7. P.W.Anderson and P.Morel, Phys.Rev., 123, 1911 (1961).
8. R.Balian and N.R.Werthamer, Phys.Rev., 131, 1553 (1963).
9. G.Aeppli, E.Bucher, C.Broholm, J.K.Kjems, J.Baumann and J.Hufnagl, Phys.Rev.Lett., 60, 615 (1988).
10. A.I.Goldman, G.Shirane, G.Aeppli, B.Batlogg and E.Bucher, Phys. Rev., B34, 6561 (1986); P.H.Frings, B.Renker and C.Vettier, J.Mag. Mag.Mater., 63-64, 202 (1987).
11. Y.Kohori, T.Kohara, H.Shibai, Y.Oda, Y.Kitaoka and K.Asayama, J.Phys.Soc.Jpn.,56, 867 (1987); ibid., 57, 395 (1988); Jpn.J.Appl. Phys., 26, 1239 (1987).
12. Y.Kitaoka, K.Ueda, T.Kohara, Y.Kohori and K.Asayama, Theoretical and Experimental Aspects of Valence Fluctuations and Heavy Fermions, (Plenum) pp.297, (1987).
13. D.E.MacLaughlin, C.Tien, W.G.Clark, M.D.Lan, Z.Fisk, J.L.Smith and H.R.Ott, Phys.Rev.Lett., 53, 1833 (1984).
14. K.Ueda, Y.Kitaoka, H.Yamada, Y.Kohori, T.Kohara and K.Asayama, J.Phys.Soc.Jpn., 56, 867 (1987).
15. R.H.Heffner, D.W.Cooke, Z.Fisk, R.L.Hutson, M.E.Schillaci, J.L.Smith, J.O.Willis, D.E.MacLaughlin, C.Boekema, R.L.Lichti, A.B.Denison and J.Oostens, Phys.Rev.Lett., 57, 1255 (1986).
16. Y.Kohori, M.Kyogaku, T.Kohara, K.Asayama, H.Amitsuka and Y.Miyako, J.Mag.Mag.Mater., 90-91, 510 (1990).
17. F.Steglich, Theory of Heavy Fermions and Valence Fluctuations, (Springer-Verlag) pp.23, (1985).

18. M.B.Maple, J.W.Chen, Y.Dalichaouch, T.Kohara, C.Rossel, M.S.Torikachvili, M.W.McElfresh and J.D.Thompson, Phys.Rev.Lett., 56, 185 (1986).
19. W.Schlabitz, J.Baumann, B.Pollit, U.Rauchschwalbe, H.M.Mayer, U.Ahlheim and C.D.Bredl, Z.Phys., B62, 171 (1986); T.T.M.Palstra, A.A.Menovsky, J.van den Berg, A.J.Dirkmaat, P.H.Kes, G.J. Nieuwenhuys and J.A.Mydosh, Phys.Rev.Lett., 55, 2727 (1985).
20. C.Broholm, J.K.Kjems, W.J.L.Buyers, P.Matthews, T.T.M.Palstra, A.A.Menovsky and J.A.Mydosh, Phys.Rev.Lett., 58, 1467 (1987); C.Broholm, H.Lin, P.T.Mattews, T.E.Mason, W.J.L.Buyers, M.F.Collins, A.A.Menovsky, J.A.Mydosh and J.K.Kjems, Phys.Rev., B43, 12809 (1991).
21. T.Kohara, Y.Kohori, K.Asayama, Y.Kitaoka, M.B.Maple and M.S. Torikachvili, Solid State Commun., 59, 603, (1986); Theoretical and Experimental Aspects of Valence Fluctuations and Heavy Fermions, (Plenum) pp.337, (1987); Jpn.J.Appl.Phys., 26, 1239 (1987).
22. U.Walter, C.-K.Loong, M.Loewenhaupt and W.Schlabitz, Phys.Rev., B33, 7875 (1986).
23. Y.Kohori, Y.Noguchi, T.Kohara, K.Asayama, H.Amitsuka and Y.Miyako, in preparation.
24. J.D.Jorgensen, B.Dabrowski, S.Pei, D.G.Hinks, L.Solderholm, H.Morosin, J.E.Schirber, E.L.Venturini and D.S.Ginley, Phys.Rev., B38, 11337 (1988).
25. C.Chaillout, J.Chenavas, S.W.Cheong, Z.Fisk, M.S.Lehmann, M.Marezio, B.Morosin and J.E.Schirber, Physica, C162-164, 57 (1989).
26. K.Ueda, T.Sugata, Y.Kohori, T.Kohara, Y.Oda, M.Yamada, S.Kashiwai and M.Motoyama, Solid State Commun., 73, 49 (1990).
27. K.Ishida, T.Kondo, Y.Kitaoka and K.Asayama, J.Phys.Soc.Jpn., 58, 2638 (1989).
28. T.Tsuda, T.Shimizu, H.Yasuoka, K.Kishio and K.Kitazawa, J.Phys.Soc.Jpn., 57, 2908 (1988).
29. for example, Y.Kitaoka, S.Hiramatsu, T.Kondo and K.Asayama, J.Phys.Soc.Jpn., 57, 30 (1988); T.Imai, T.Shimizu, H.Yasuoka, Y.Ueda and K.Kosuge, J.Phys.Soc.Jpn., 57, 2280 (1988); P.C.Hammel, M.Takigawa, R.H.Heffner, Z.Fisk and K.C.Ott, Phys.Rev.Lett., 63, 1992 (1989).
30. K.Fujiwara, Y.Kitaoka, K.Asayama, H.Sasakura, S.Minamigawa, K.Nakahigashi, M.Kogachi, N.Fukuoka and A.Yanase, J.Phys.Soc.Jpn., 58, 380 (1989).

31. T.Kohara, K.Ueda, H.Takenaka, Y.Kohori and Y.Oda, Physica, B165-166, 1307 (1990).
32. K.Fujiwara, Y.Kitaoka, K.Asayama, Y.Shimakawa, T.Manako and Y.Kubo, J.Phys.Soc.Jpn., 59, 3459 (1990).
33. for example, S.E.Barret, D.J.Durand, C.H.Pennington, C.P.Slichter, T.A.Friedmann, J.P.Rice and D.M.Ginsberg, Phys.Rev., B41, 6283 (1990).
34. R.J.Cava, B.Batlogg, J.J.Krajewski, L.W.Rupp, L.F.Schneemeyer, T.Siegrist, R.B.van Dover, P.Marsh, W.F.Peck,Jr, P.K.Gallagher, S.H.Glarum, J.H.Marshall, R.C.Farrow, J.V.Waszczak, R.Hull and P.Trevor, Nature, 336, 211 (1988).
35. L.F.Mattheis and D.R.Hamann, Phys.Rev., B39, 4780 (1989).
36. R.K.Harris, J.D.Kennedy and W.McFarlane, NMR and the Periodic Table, (Academic Press) pp.368, (1978).

PINNING CENTERS IN C-AXIS ORIENTED Bi-Sr-Ca-Cu-O THIN FILMS

Grzegorz Jung
Instytut Fizyki PAN, Al.Lotników 32, PL 02-668 Warszawa, Poland, and Dipartimento di Fisica, Universita' di Salerno, I 84-081 Baronissi (SA) Italy

Bonaventura Savo, Luigi Maritato, Serghei Prishepa*, Carmine Attanasio, and Antonio Vecchione
Dipartimento di Fisica, Universita' di Salerno, I 84-081 Baronissi (SA) Italy

ABSTRACT

Pinning centers in c-axis oriented Bi-Sr-Ca-Cu-O thin films have been characterized by flux creep and low frequency noise measurements. Two types of pinning centers have been found. Short range centers with low activation energy and high attempt rates manifest themselves in the flux creep controlled transport properties at low temperatures, while long-range, low attempt rate and high activation energy centers constitute the source of two-level fluctuators at temperatures close to sample critical temperature.

1. Introduction

High T_c superconductors (HTSC) are characterized by critical temperatures orders of magnitude higher, and vortex pinning energies orders of magnitude weaker in comparison to low T_c superconductors. This leads to the pronounced manifestations of various thermally activated flux phenomena, such as giant flux creep[1], activated resistivity[2], vortex phase transitions[3], etc. Vortices dynamics lays behind the principal phenomena determining dissipation in oxide superconductors. Vortex dissipation, both in *dc* and *ac* driving fields, limits critical current and deteriorates the performance of practical HTSC devices[4].

The dissipation depends on the type and properties of pinning centers that are spontaneously created during the HTSC crystalization, or are deliberately introduced to the material by controlled particle irradiation. Due to very short HTSC coherence length many structural defects, including oxygen atom vacancies, may act as weak pinning sites. From the experimental standpoint, pinning energies can be determined by various techniques such as transport measurements, *dc* and *ac* magnetization measurements, decoration techniques, neutron diffraction, microwave absorption, harmonic generation, and noise techniques[5].

In this paper we concentrate on evaluation of the pinning properties of c-axis oriented thin BiSrCaCuO (BSCCO) thin films. In particular, we confront the activation energies, attempt rates, and flux hopping ranges as evaluated from the shape of I-V curves recorded in the flux creep regime at low temperatures with the results provided by a novel technique based on investigations of random telegraph signal (RTS) voltage noise in HTSC samples.

The paper is organized as follows: section 2 contains information on the fabrication procedure and basic crystallographic and superconducting properties of the film. In section 3 we discuss theoretical aspects of flux creep controlled I-V characteristics, section 3.1, and experimental evaluation of the pinning properties of BSCCO films, section 3.2. Noise techniques and experimental results based on the investigation of random telegraph voltage signals are presented in section 4. The co-existence of two types of pinning centers, confirmed by 1/f noise and thermally activated resistivity measurements, is discussed in sections 5 and 6. Final conclusions are presented in section 7.

2. Sample fabrication and basic properties

Thin film BSCCO samples were fabricated by simple e-beam deposition technique relying on completely evaporating a weighted amount of a sintered BSCCO pellet and subsequent ex-situ annealing. The deposition was performed in oxygen-free vacuum onto unheated (100) MgO substrates. Typically, 1200 mG of a starting pellet provided 1μm thick film in about 45 minutes. After the deposition, residuals of the starting pellet (not more than 5% in weight) were present in the e-gun crucible. Taking this into account, by correcting the stoichiometry of the starting pellet to 1.6:1.8:1.05:2.4 (Bi:Sr:Ca:Cu) atomic ratio, we were able to produce reproducibly superconducting films with an atomic ratio 2:2:1:2. After the deposition, an annealing process was held using a muffle furnace pre-heated to 880° C. Samples were rapidly introduced into the furnace, sintered in air for 10 hours and cooled down to 500° C in 2-3 hours.

The structural properties of the sintered samples have been studied by means of EDS microprobe and X-Ray diffraction analysis. Large uniform areas with plate-like structures, separated by 1 μm wide fractures with strongly reduced film thickness, were shown by scanning electron micrographs. EDS analysis confirmed the TEM data indicating a large homogeneity of the film. The X-Ray spectra showed a strong preferential orientation with the c-axis perpendicular to the substrate. Peaks related to the 2:2:0:2 and 2:2:2:3 phases were initially present in the samples but their intensity decreased with increasing annealing time. This was accompanied by a change in the $T_c(R=0)$ of the film. After 10 hours annealing only the 2:2:1:2 phase showed out in the X-Ray spectra.

Films were patterned into 50 μm wide strips, in such a manner that the strips were contained within a single uniform area and were not crossed by any fracture. Patterning was performed using a standard photolithography and wet etching in 4% nitric acid. The patterning process did not deteriorate the superconducting properties of the films. Evaporation of the silver contact pads on both ends of the strip completed the sample fabrication process.

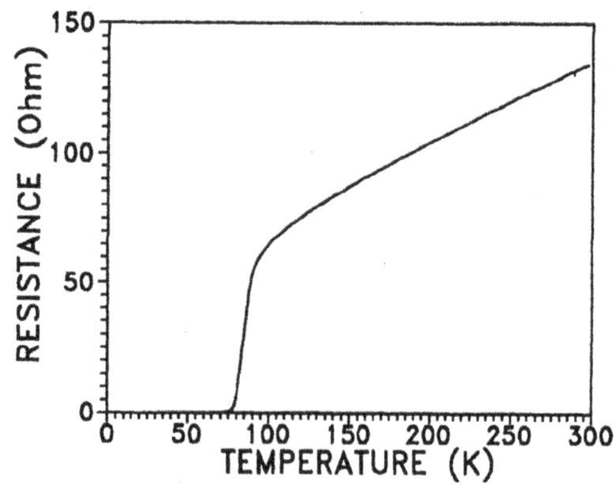

Fig.1. Resistive transition of the BSCCO thin film strip.

Figure 1 demonstrates R(T) transition curve of a patterned film, recorded by passing a small *d.c.* current trough the sample and recording the corresponding voltage. The critical current density was measured with 5 μV/cm field criterion. We have found J_c values of the order of 5×10^4 A/cm^2 at 4.2 K. Although the J_c is not very high, it was reduced only to 1×10^4 A/cm^2 in an external magnetic field of 2 T applied along the c-axis.

3. Flux creep

3.1. Theoretical considerations

If a quantized flux vortex sits at the bottom of a potential well of depth U it can, by thermal activation, hop out of the well at the rate[6]

$$r = \Omega_o \exp\left(-\frac{U}{kT}\right). \tag{1}$$

In the particular case of pinning potential especially at high temperatures close to T_c, one has to take into account the temperature dependence of the activation energy.

$$U = U(T) = U_o\beta(T). \tag{2}$$

The functional form of $\beta(T)$ will be discussed in detail in section 4. In the absence of any driving force, the hopping probability is isotropic in space, but transport current or applied magnetic fields impose a stress through the Lorentz force action which favors hopping in one direction, recognized as the forward direction. Therefore, the net forward flux creep is determined by the difference between the forward $r_F = \Omega_o \exp[-(U - \Delta W)/kT]$ and backward $r_B = \Omega_o \exp[-(U + \Delta W)/kT]$ hopping rates[6-9], i.e.

$$r_{net} = 2\Omega_o \exp\left(-\frac{U}{kT}\right) \sinh\left(\frac{\Delta W}{kT}\right). \tag{3}$$

At low reduced temperatures $t = T/T_c \ll 1$, or at high stress levels $\Delta W \gg kT$, the sinh term can be approximated by $\sinh(x) = \exp(x/2)$. Moreover, for $t \ll 1$ we neglect the temperature dependence of the pinning potential, $\beta(T) = 1$, $U(T) = U_o$, and equation (3) can be simplified as follows:

$$r_{net} = 2\Omega_o \exp\frac{\Delta W - U_o}{kT}. \tag{4}$$

The functional form for $\Delta W(I)$ depends on the assumed model. In the classical flux-creep model[6-8], this dependence is linear. However, for the HTSC case a nonlinear equation was proposed[10-14]:

$$\Delta W(I) = gU\left(I/I_{co}\right)^{-\frac{1}{\mu}}, \tag{5}$$

where I_{co} is a phenomenological parameter having the sense of the critical current that would be in the absence of any flux creep and g is a parameter close to unity. Nevertheless, in low temperatures and at low stress levels μ =1, and ΔW is again a linear function of a transport current[15-17].

Moving flux lines generate electric field:

$$E = r_{net}lB, \tag{6}$$

with B being the magnetic field and l the distance at which flux jumps. Voltage developed across the sample with thermally activated flux creep may be expressed as:

$$V = V_o \exp\left(-\frac{U_{eff}}{kT}\right), \tag{7}$$

where at low temperatures and in weak magnetic fields the effective activation energy,

$$U_{eff} = -k\frac{d(\ln V)}{d(1/T)} = U_o\left(1 - g\frac{I}{I_{co}}\right). \tag{8}$$

Weak field approximation, assumes that vortices do not interact between themselves. This requires that an average distance between the flux lines $a_o \gg \lambda$. The minimum separation between vortices a_o^{min} at the maximum applied magnetic field used in the experiments ($B_{max} \approx 10$ Gauss), assuming a triangular vortex lattice, is

$$a_o^{min} = \left(\frac{2\Phi_o}{\sqrt{3}B_{max}} \right)^{1/2} = 1.54\mu m \gg \lambda. \tag{9}$$

Therefore, the weak field approximation will be used throughout the entire paper.

3.2. Flux creep regime, experimental results

Experimental procedure of determination of pinning energies relied on measuring the I-V characteristics of our BSCCO strips in the flux creep regime at low temperatures and in weak magnetic field range. We have measured I-V characteristics in the

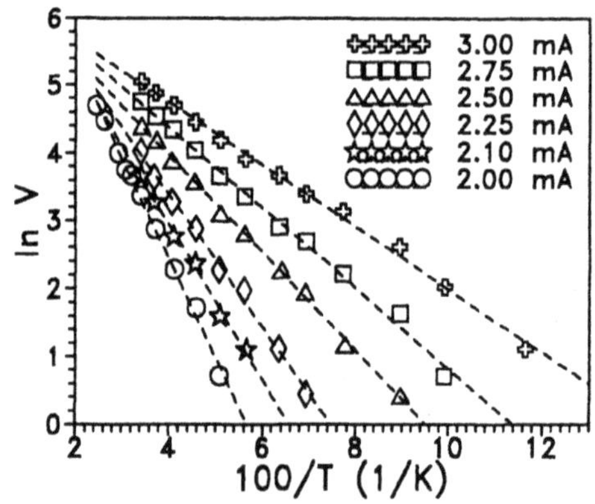

Fig.2. Flux creep voltage at various bias current levels as a function of temperature. Dashed lines are the best fits to Eq.(7).

temperature range between 5.7 K and 30 K. Using this data we have plotted $\ln(V)$ versus $100/T$ curves, see Fig.2, for different values of the current flow (from 2 to 3 mA), just above the strip critical current. From the best fits to the experimental data, using equation (7) we have obtained the U_{eff} as a function of a bias current. This is plotted in Fig.3. As follows from equation (8), the linear extrapolation to $I = 0$ of these data gives the value of zero temperature activation energy, which in our case turned out to be 25 meV. Observe that this result is independent from the actual J_{co} value and is in a good agreement with the data already reported in the literature for BSCCO superconductor[2,18,19].

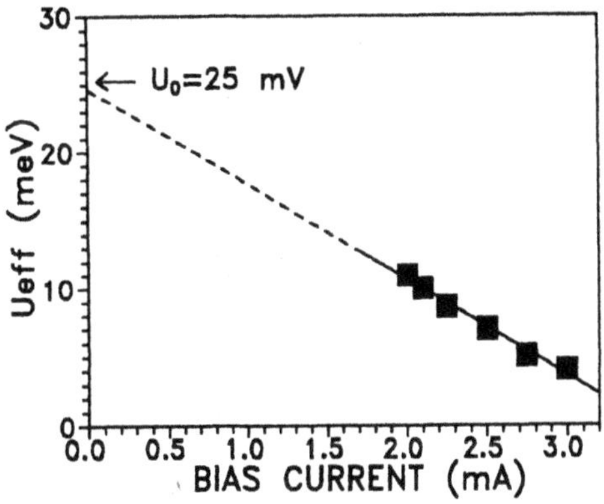

Fig.3. Effective activation energy as a function of the bias current.

Proceeding along the lines proposed by Hylton and Beasley in their model of flux pinning in HTSC films[20], we estimate the distance between pinning centers along c-axis, as:

$$d_{p-p} = 0.5 \frac{U_o}{\varepsilon_c} \frac{L}{L_p} \tag{10}$$

where, ε_c is the condensation energy in the core region per unit length of a vortex, and L/L_p is the ratio between the total length of a vortex to the length of the fraction effectively pinned. Taking for Bi-Sr-Ca-Cu-O[20]: $\Gamma = m_{ab}/m_c = 1000$, λ_{ab} =150 nm, ξ_{ab} =1.5 nm, $L_p/L = 0.25$ one finds that pinning energy of $U_o = 25$ meV provides $d_{p-p} = 23\mathring{A}$, i.e. the value close to the distance between CuO planes in 2:2:1:2 BSCCO phase. This result indicates that BSCCO pinning centers at low temperatures are of a 2-D nature.

For a complete characterization of pinning centers one should evaluate the attempt frequency Ω_o, i.e. the frequency of self-oscillation of a pinned flux line in the pinning site energy well, as well as an average distance at which activated flux lines jump. In order to evaluate the hopping range we have to consider, first of all, the bias current distribution along the film width. We have found that an approximate solution for the case of an insulated film of thickness d and width w, with $wd \gg \lambda^2$ and d of the order of λ (as in our experimental situation), provides almost uniform current distribution inside the film[21]. Current distribution close to the film center, if the x-axis origin is in the center of the film, can be expressed as:

$$J(x) = J(0)\sqrt{1 - (\frac{2x}{w})^2}. \tag{11}$$

Only very close to the film edges the distribution becomes exponential,

$$J(x) = J\left(\frac{w}{2}\right)\exp\left(-\frac{w/2 - |x|d}{a\lambda^2}\right), \tag{12}$$

where a is a constant close to unity. Solutions (11) and (12) intercept at the points

$$x_{1,2} = \pm\left(\frac{w}{2} - \frac{a\lambda^2}{2d}\right). \tag{13}$$

Confronting (11) and (12) at the intercept points we find the relation between current density in the center of the film and at the edge:

$$J\left(\frac{w}{2}\right) = J(0)\frac{1.165}{\lambda}\sqrt{\frac{wd}{a}}. \tag{14}$$

It follows from (14) that current density at the edge is 55 times larger with respect to that in the film center. However, the exponential distribution is restricted to regions only 0.012 μm wide, i.e. it occupies only about 0.1% of the total (50 μm) film width. Moreover, already at a distance of only 1.5 μm from the film edge $J(x) = 1.03\ J(0)$. Therefore, we are allowed to approximate current density distribution in our sample by an effective uniform current density:

$$J_{eff} = \int_{-w/2}^{w/2} J(x)dx = \frac{I}{wd}. \tag{15}$$

Assuming that $J(x)$ for $x \in [x_1, x_2]$ is given by Eq. (11) and outside of this interval by Eq.(12), by integrating (15) we have

$$J_{eff} = 1.56\ J(0). \tag{16}$$

The exponential contribution to the effective current density is only 0.02 $J(0)$.

Within the effective uniform current distribution approximation we can easily estimate an average hopping distance in the sample. Lorentz force acting on a vortex, if we approximate the vortex length by the film thickness d, is:

$$F_L = J\Phi_o d = \frac{I\Phi_o}{w}. \tag{17}$$

Therefore, the stress $U(I)$ imposed by the Lorentz force on a pinning site

$$U(I) = F_L l = \frac{I\Phi_o l}{w}. \tag{18}$$

The unknown hopping distance l can be calculated as:

$$l = \frac{\partial U}{\partial I}\frac{w}{\Phi_o}. \tag{19}$$

Since a vortex may hop at a certain angle δ with respect to the direction of the Lorentz force we should modify equation (19) to:

$$l = \frac{\partial U}{\partial I}\frac{w}{\Phi_o}\cos\delta. \tag{20}$$

The derivative $\partial U/\partial I$ can be obtained from the slope of the U_{eff} vs. bias current dependence from Fig.3. We find in this manner

$$\frac{\partial U}{\partial I} = 1.12 \times 10^{-18}\ [J/A]. \tag{21}$$

Assuming $\cos\delta = 1$, we estimate the hopping distance $l = 270\mathring{A}$. Observe that l is much larger then the separation between pinning sites along the c-axis, confirming the 2-D nature of the pinning. The evaluated hopping range is close to the distance between extended defects observed frequently in the transmission electron micrographs[22].

Now, we can proceed to evaluate the attempt frequency of a pinned vortex. To this purpose we will use the temperature dependence of the critical current.

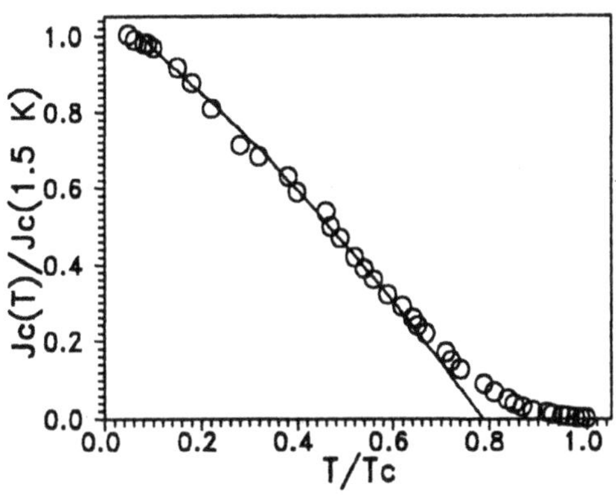

Fig.4. Temperature dependence of the critical current. Solid line is drawn according to Eq.(22) with $\alpha = 1.1$ and $\beta = 0.41$.

The classical Anderson-Kim flux creep theory predicts that at low temperatures, $t \ll 1$, the critical current changes according to the following rule[7,23]:

$$J_c(B,t) = J_c(B,0)[1 - \alpha t - \beta t^2] \tag{22}$$

where β is a parameter of the order of 1, and α is given by:

$$\alpha = \frac{kT_c}{U(B,0)}\ln\left(\frac{v_o}{v_{min}}\right), \tag{23}$$

where v_o is the phenomenological creep velocity in the absence of energy barriers and v_{min} is the minimum velocity giving detectable creep effects. In experimentally measurable quantities α can be expressed as:

$$\alpha = \frac{kT_c}{U(B,0)} \ln\left(\frac{lB\Omega_o}{E_{min}}\right), \tag{24}$$

where l is the hopping distance, B magnetic field, Ω_o attempt frequency, and E_{min} is the field criterion used for the critical current determination. Figure 4 shows the temperature behavior of the normalized critical current. Solid line represents the the best fit to the equation (22) obtained with $\alpha = 1.1$ and $\beta = 0.41$ for $t < 0.7$. From (24) we calculate Ω_o as:

$$\Omega_o = \frac{E_{min}}{lB} \exp\left(\frac{\alpha U_o}{kT_c}\right) = 1.4 \times 10^{-10} \text{ s}. \tag{25}$$

To conclude this section let us repeat that pinning sites evaluated from flux creep measurements turn out to be 2-D defects located in the CuO planes. Flux lines oscillate inside the sites with frequencies of order of 7 GHz, and are thermally activated over the barrier of 25 meV, hopping at a short range of about 27 nm. The hopping range consists only about 1/50 of the average spacing between flux lines, indicating that the experimental conditions are faraway from those providing collective flux effects[11-12,24-25].

4. Random telegraph signal noise

BSCCO thin film strips 50 μm wide exhibited a pronounced random telegraph (RTS) noise in a very narrow temperature range close to the critical temperature. RTS was visible in the voltage drop across a strip current biased and placed in weak magnetic field perpendicular to the film surface. The effect occurred only within a limited range of carefully tuned current flow and applied magnetic field. The narrow bias range at which the RTS persists will be referred to as "noisy window". The duty cycle of the RTS, defined as the ratio of the average life time in the high voltage state ("up" state) to the sum of average life times in the low voltage state ("down" state) and high voltage state of the RTS, was strongly dependent on bias conditions and temperature. Since this type of noise, also observed in YBCO granular films at liquid helium temperatures[26,27], originates from activated flux hopping between two distinct pinning sites[28] we may employ it for independent characterization of pinning centers in BSCCO thin films.

4.1. Pinning energy estimation

Let us start with the pinning energy. Accepting the model in which RTS is generated by an elementary two-level fluctuator[29] constituted by a system of two pinning centers and a thermally activated flux line hopping between them, the up and down decay rates can be expressed as[26]:

$$\tau_{dn} = \tau_o \exp\left(\frac{U + \Delta W_{dn}}{kT}\right), \tag{26}$$

$$\tau_{up} = \tau_o \exp\left(\frac{U - \Delta W_{up}}{kT}\right), \tag{27}$$

where $\Delta W_{up,dn}$ are stresses imposed on the relevant pinning sites by transport current and magnetic field. As already mentioned, the RTS duty cycle strongly depends on bias conditions and temperature. Therefore, in order to avoid any ambiguities in collecting data for Arhenius plot, we have used only symmetric RTS waveforms ($\tau_{dn} = \tau_{up}$) to determine the activation energy of an active TLF. For this purpose we have symmetrizied the RTS signal at each temperature by proper adjustment of current flow and applied magnetic field.

The symmetric bias points were experimentally found as crossing points of up and down RTS decay rates, determined from statistical histograms of pulse durations measured as functions of bias current and applied field. An example of such procedure, performed at 85.1 K is shown in Figs. 5 and 6. In the symmetric conditions we write for the decay rates:

$$\tau_{dn} = \tau_{up} = \tau = \tau_o \exp\left(\frac{U(T)}{kT}\right). \tag{28}$$

The activation energy can be now determined simply from the slope of the $\ln \tau$ versus $1/T$ plot

$$U = k\frac{d(\ln \tau)}{d(1/T)}. \tag{29}$$

The relevant Arhenius plot is shown in Fig. 7. It follows from the figure that hopping process is thermally activated, but a direct fit of equation (29) to the data provides clearly unphysical values for the activation energy and attempt rate, U = 1.47 eV and $\tau_o = 1.8 \times 10^{-90}$ s, respectively. The reason is, that in the temperature range close to T_c , where the RTS was observed, one cannot neglect the temperature dependence of the activation energy. The appropriate functional form of the $\beta(T)$ factor, Eq. (2), remains controversial[15,30–32]. In this paper we accept the point of view in which $U(T)$ is proportional to the square of the critical field B_c , Ginzburg-Landau coherence length ξ, and temperature independent pinning length L determining the dimensionality of pinning effects.

$$U(T) \propto \frac{B_c^2}{2\mu_o}\xi^n(T)L^{3-n}. \tag{30}$$

Assuming for the penetration depth and critical fields the following temperature dependencies[8]:

$$\xi(T);\lambda(T) = \xi(0);\lambda(0)\sqrt{(1-t^4)} \tag{31}$$

$$B_c(T) = B_c(0)(1-t^2) \tag{32}$$

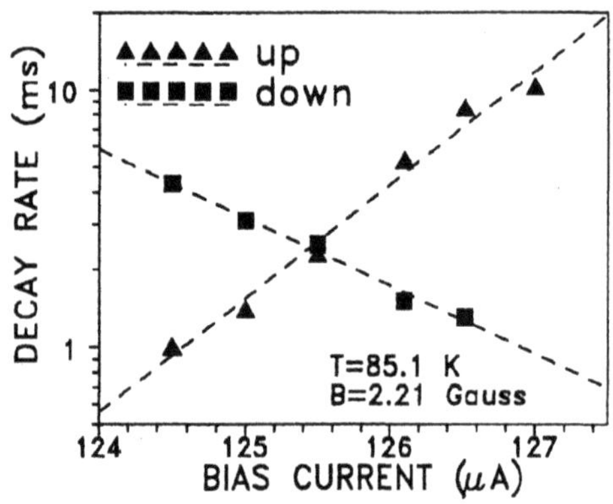

Fig.5. Decay rates dependence on bias current. Dashed lines represent the exponential fits.

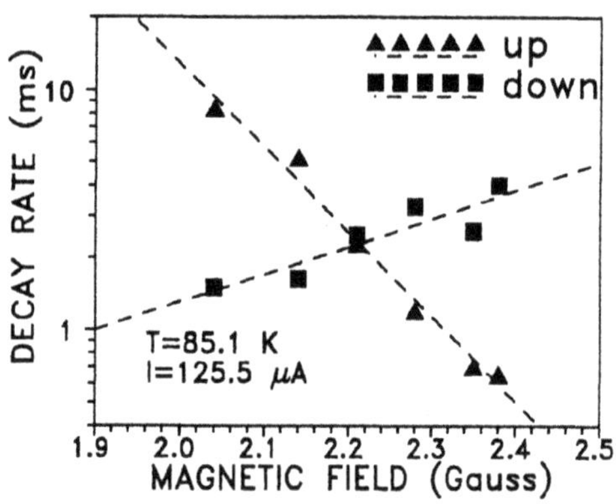

Fig.6. Random telegraph signal decay rate dependence on magnetic field at 85.1 K. Dashed lines represent the exponential fit to the data. Note the deviations from the exponential dependence, see sec.4.3 for details.

and assuming successively n=3, 2, and 1, we obtain

$$\beta(T) = (1 + t^2)(1 - t^4)^{-1/2}\,; \qquad n = 3 \tag{33}$$
$$\beta(T) = (1 - t^4)\,; \qquad n = 2 \tag{34}$$
$$\beta(T) = (1 - t^2)(1 - t^4)^{1/2}\,; \qquad n = 1. \tag{35}$$

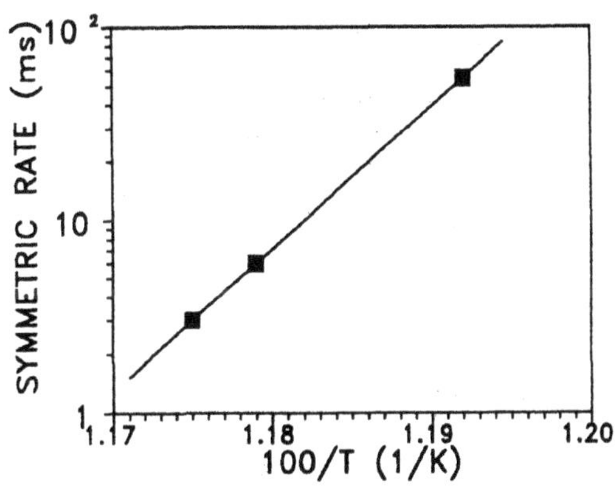

Fig.7. Temperature dependence of the symmetric RTS rates.

The model with n=1 was developed[1,31] for strongly interacting vortices, i.e. it applies for strong fields $B > 0.2B_c$, and we will not consider it here. However, present theoretical situation provides no basis to choose between the 2-D and 3-D case. Consequently, we have fitted equation (29) to the experimental data using both forms for $\beta(T)$, Eq.(33) and Eq.(34). Eq.(29) contains yet another unknown parameter, the attempt rate τ_o. Instead of accepting an arbitrary τ_o value we have performed the nonlinear fitting for all physically plausible τ_o ranging from 10^{-5} to 10^{-12} s. We have also used the critical temperature T_c as a fitting parameter, since the exact critical temperature in the region where TLF is located may be different from the one determined on the basis of percolative resistive transition shown in Fig.1. The results of the fitting procedures are presented in Table 1. Observe, that different zero temperature activation energies result in the same activation energy values at high temperatures when the relevant model is applied to calculate $U(T)$. Generally, 3-D model gives lower activation energies and lower critical temperatures in comparison to 2-D case. The activation energies determined from the RTS noise are an order of magnitude larger respect to those evaluated from the flux creep in low temperatures, but are in a surprisingly good agreement with the data obtained by Berkeley group in their ingenious SQUID-based experiment[32] dedicated to the flux noise in single crystal BSCCO flakes. Unfortunately, neither the literature, nor the above data do

TABLE I. Activation energies and critical temperatures as obtained from nonlinear fitting procedures

$\tau_o[s]$	$T_c \pm \Delta T_c$ [K]	$U_o \pm \Delta U_o$ [meV]
	2-D case ⟶ n = 2	
1×10^{-12}	93.67 ± 0.19	498.4 ± 7.74
1×10^{-11}	92.79 ± 0.17	485.4 ± 7.74
1×10^{-10}	91.90 ± 0.15	473.4 ± 7.74
1×10^{-9}	91.01 ± 0.13	460.9 ± 7.74
1×10^{-8}	90.11 ± 0.12	448.4 ± 7.74
1×10^{-7}	89.21 ± 0.10	435.9 ± 7.74
1×10^{-6}	88.30 ± 0.08	423.5 ± 7.74
1×10^{-5}	87.39 ± 0.06	411.0 ± 7.74
	3-D case ⟶ n = 3	
1×10^{-12}	88.94 ± 0.09	206.0 ± 1.6
1×10^{-11}	88.53 ± 0.08	193.0 ± 1.6
1×10^{-10}	88.11 ± 0.07	179.7 ± 1.5
1×10^{-9}	87.69 ± 0.06	166.0 ± 1.4
1×10^{-8}	87.26 ± 0.05	151.7 ± 1.3
1×10^{-7}	86.83 ± 0.05	136.8 ± 1.3
1×10^{-6}	86.40 ± 0.04	121.0 ± 1.2
1×10^{-5}	85.97 ± 0.03	109.7 ± 1.0

not enable to choose between various U_o since all associated critical temperature values are realistic.

4.2. Electromagnetic interactions.

In order to evaluate the attempt frequency and flux jumping distance from the RTS noise measurements we have concentrated on the magnetic field influence upon the RTS. In general, we were able to separate two types of magnetic field interactions with a pinning site. The first one, hereafter named the core interaction, is the well known effect of the direct influence of magnetic field on the condensation energy in the core region[1,31–34], while the second one, hereafter named the electromagnetic interaction, relies on canceling the stress due to bias current by means of magnetically induced circulating screening current. This effect is used in our experiment to provide the symmetric waveform RTS. Since the mechanism of converting flux jumps to the voltage RTS is associated with the resistivity fluctuations converted to voltage signal by means of current flow, one needs to polarize the film with a significant bias current in order to have a measurable voltage waveform. This current imposes a stress on the TLF that has to be balanced by magnetic field induced screening current flowing in the opposite direction.

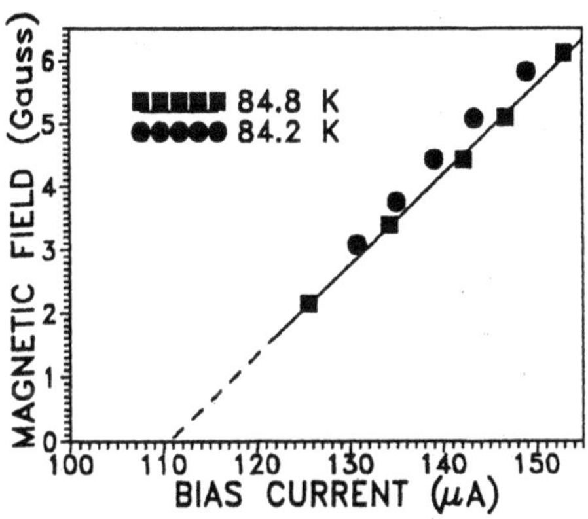

Fig.8. Symmetrizing current vs. symmetrizing magnetic field, see text.

The electromagnetic interactions were examined by finding noisy windows at different current flows by means of readjusting the applied magnetic field. Fig.8 presents the dependence of the magnetic field required to get a symmetric RTS on the current flow in the sample, as measured for two different temperatures. The relations are linear and we are entitled to define a field-current coupling constant σ as the ratio of the current and the magnetic field that cause a symmetric signal. Those bias values are positioned at the very centers of noisy windows. On the other hand, σ must be equal to the ratio of the slopes of the relevant activation energy derivatives.

$$\sigma \equiv \frac{I_{sym}}{B_{sym}} = \left(\frac{\partial U}{\partial B}\right)\left(\frac{\partial U}{\partial I}\right)^{-1} \tag{36}$$

It follows from the examination of Fig.8, that in order to fulfill Eq.(36) we have to accept the presence of an offset field B_{off}.

$$I_{sym} = \sigma\left(B_{sym} + B_{off}\right) = 6.2 \times 10^{-2}\left(B_{sym} + 1.7 \times 10^{-3}\,[T]\right). \tag{37}$$

However, there are no physical reasons for such enormous 17 Gauss offset. In fact, the offset several times exceeds any plausible field that might be due to trapped flux, environment magnetic field or magnetic shield imperfections. Therefore, we have applied an alternative approach in which the offset is regarded as due to the current

$$(I_{sym} - I_{off}) = \sigma B_{sym} = \left(I_{sym} - 1.1 \times 10^{-4}\ [A]\right). \tag{38}$$

The existence of 110 μA offset current means that the Lorentz force is exercised on pinning sites only above a certain threshold current flow determined by I_{off}. In other words, it means that the current can flow in the vicinity of the pinning site only above I_{off} . Such physical situation can be created in the presence of a superconducting by-pass possessing critical current I_{off}. This hypothesis is strongly supported by the experimentally measured dependence of the RTS pulse height on bias current, shown in Fig.9.

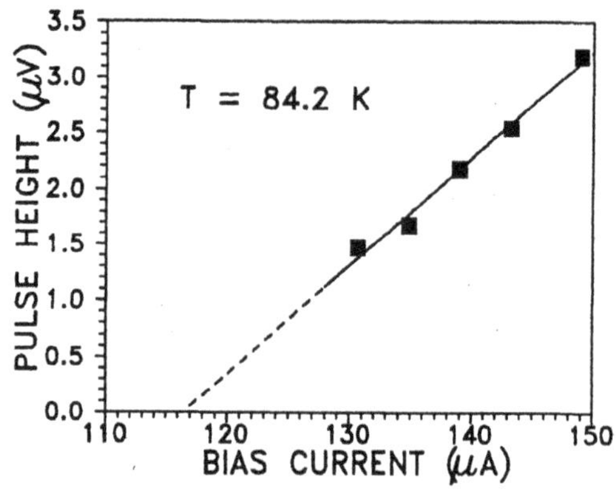

Fig.9. Amplitude of the RTS waveform vs. bias current

Resistivity fluctuations can not be probed by the current flow below a critical current. Pulse height vs. bias current characteristics clearly demonstrates the threshold current, that should be identified with a critical current. Within the experimental error the threshold current determined from Fig.9 equals I_{off}. Although the I-V curves recorded in the RTS temperature range are linear and demonstrate no traces of any critical currents, we conclude that inside the percolation path there exists a region at which non-zero supercurrent flows. This region is inaccessible by a standard transport measurements, because it is masked by series resistances of the other parts of the sample. Its existence can be revealed only by local measurements, like noise measurements that probe just the fluctuator area. We would like to stress that the existence of the critical current of the order of that measured in our experiment is required to provide an effective Josephson coupling that exceeds thermal fluctuations in the temperature range of the experiment.

4.3. Core interactions

Symmetric rate of the RTS decreases when the noisy window is pushed towards higher fields. The experimentally measured symmetric rate dependence on applied magnetic field is shown in Fig.10. Rate decrease should be ascribed to the decrease

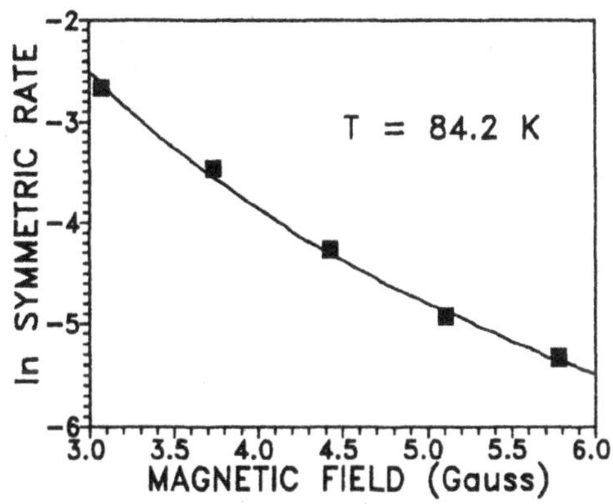

Fig.10. Symmetric rate dependence on the applied magnetic field.

of the symmetric TLF energy barrier, i.e. to the decrease of the core energy with increasing magnetic field. Therefore, we have fitted the experimental data to the following equation.

$$\ln \tau(B,T) = \ln(\tau_o) + \frac{U_o B^m \beta(T)}{kT} = \ln(\tau_o) + AB^m, \tag{39}$$

where m equals 1, 2, or 3 and A is temperature dependent constant. The choice of m is related to the problem of the dimensionality of pinning centers[35], as in the case of the temperature dependence of pinning energy. For the case of $n = 1$ we have $m = -1$, but this is valid[1,31,36] only for strong fields, $B > 0.2B_c$ and does not apply to our experimental situation. In the 2-D case, $n = 2$, we have to assume[2,35] $m = -1/2$, while no general predictions are available[25] for $n = 3$. We have obtained the best fit to the equation (39) for $m = -0.49$, solid line in Fig.10. This result strongly indicates the 2-D nature of BSCCO pinning centers. The fit was performed with τ_o and A taken as fitting parameters. The attempt frequency determined from the fit is 3×10^{-6} s, and the energy factor A =0.176 [SI units].

The core interactions provide thus an excellent tool for uniquely choosing only one value of the activation energy and critical temperature from those listed in Table 1. The activation energy can be obtained by the best fit to the Eq.(29) taking for

$\beta(T)$ the 2-D formula, Eq.(34), and assuming $\tau_o = 3 \times 10^{-6}$ s. In this way we have obtained the activation energy $U_o = 421$ meV and the critical temperature T_c =87.9 K. As a cross-check of the applied procedures we have calculated the theoretical value of the A factor. Putting into Eq. (39) the above values of U_o and T_c we obtain $A =$ 0.160 [SI units], very close to the value obtained by the fitting procedures.

The flux hopping distance has been evaluated with a procedure analogous to that employed in section 3. First, we have checked that the effective current density approximation can be applied at the RTS temperature range. Taking into account the temperature dependence of the penetration depth in integral (15) we obtain $J_{eff} = 1.54\ J(0)$, where the exponential distribution contributes $0.052\ J(0)$. Thus, we conclude that the approximation is still valid. The derivative $\partial U/\partial I$ can be obtained from the rate versus bias current plots by fitting them, as indicated by the solid lines in Fig.5, to the equation:

$$\tau = \tau_o \exp\left(\frac{1}{kT}\frac{\partial U}{\partial I}I\right). \tag{40}$$

The hopping distance calculated from Eq.(20), assuming $\cos\delta = 1$, is very high, of the order of the sample width. Observe, that the flux hopping distance of the same range was reported for BSCCO single crystals from direct SQUID measurements of RTS flux noise[32]. However, in a marked difference to those results, we find that the range at which flux jumps depends on temperature and increases with decreasing temperature. The temperature dependence of the hopping range is presented in Fig.11. The physical reason for such dependence becomes obvious if one concerns temperature dependence of the activation energy in Eq. (19). In fact, the experimental data from Fig.11 can be reasonably fitted to the temperature dependence described by Eq. (34), taking T_c a fitting parameter. The best fit, solid line in Fig.11, was obtained for T_c=86.6 K, i.e. very close to the T_c value obtained from the core interactions. Detailed discussion of the temperature dependence of the hopping range, $\partial U/\partial I$, and $\partial U/\partial B$ derivatives is beyond the scope of this paper and will be published separately[28].

Let us observe that noise measurements revealed a different type of pinning centers, active in the temperature range close to T_c . Their activation energy is almost 20 times larger than the activation energy estimated from flux creep measurements. Flux lines oscillate inside the sites with low frequencies, of the order of 0.3 MHz, and hop at very large distances, comparable to the strip width. This suggests that deep pinning centers are likely not magnetic in origin. Magnetic pinning is the result of changes in the field and current distribution in the area of the radius λ, close to the vortex core bound by a pinning site. Hopping distances $l \gg \lambda$ make us to associate pinning sites of the active TLF with film borders, known to constitute natural pinning sites. Another plausible candidates are growth defects such as extended grain boundaries or stalk faults. The pinning across grain boundary, in the 2-D Josephson coupled grains HTSC model, is proportional to the maximum Josephson supercurrent flowing across the defect acting as a pinning site[37]. Josephson current is periodically modulated in applied magnetic field with the period of flux quantum Φ_o. For this reason the intergranular pinning potential is periodic in magnetic field[38]. The period is determined by the value of the field producing the flux quantum in the junction area, A_j, in the plane perpendicular to the applied field direction. In the experiment the field was applied along the c-axis and

$$A_j(T) = 2\lambda_{ab}(T)\, l_j, \tag{41}$$

where l_j is the junction length and λ_{ab} is London penetration depth in the a-b plane.

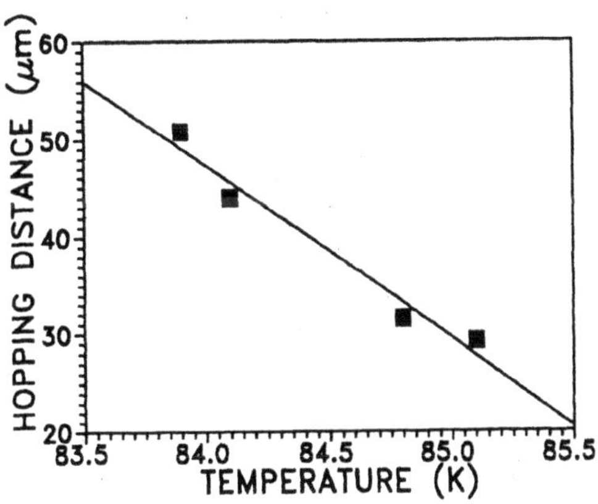

Fig.11. Temperature dependence of the vortex hopping range evaluated from the RTS rate dependence on bias conditions. Solid line is the best fit to the 2-D temperature dependence of the activation energy, Eq.(34).

The periodic behavior of the pinning potential can be detected in the magnetic field dependence of the decay rates. Remember, that the slope of $\ln \tau$ vs magnetic field dependence is proportional to the pinning energy. One can notice the periodic behavior in Fig.6, but it became even more pronounced at lower temperatures, as demonstrated in Fig.12 showing the low voltage state ("down") rate dependence on the applied magnetic field at 84.8 K. Indeed, the departures of experimental points from the exponential fit, dashed line, are not due to the measurement errors, but result from the modulation of the pinning potential in the applied magnetic field. Moreover, the period increases with decreasing temperature, consistently with the temperature dependence of the penetration depth λ. The magnetic field causing the flux quantum is of the order of 2×10^{-5} T, corresponding to the area of about μm. Consequently, Eq. (41) gives the defect length $l_j \approx 12.5$ μm. Let us point out, that even rejecting the Josephson network model at temperatures close to T_c, the periodic modulation of the pinning energy can be always explained by the heuristic statement that the pinning potential can not depend on the number of flux quanta contained in a pinned flux line[39].

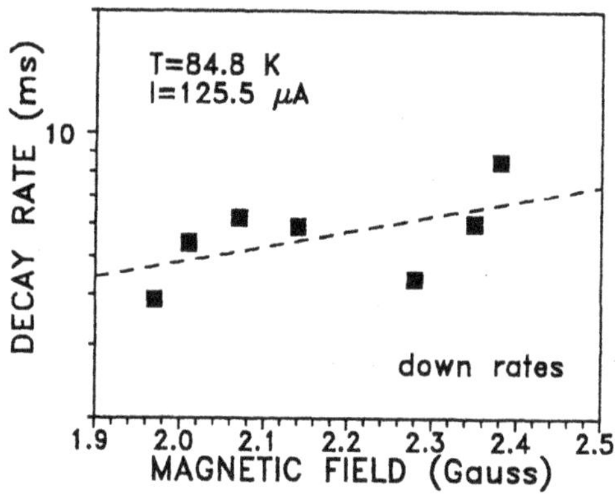

Fig.12. Decay rate of the low voltage RTS state at 84.8 K as a function of the applied magnetic field. Note the regular deviations from the exponential dependence plotted with the dashed line.

Magnetic interactions indicate the 2-D nature of deep pinning centers even at temperatures where one may expect a crossover to 3-D behavior in BSCCO due to the increase of the coherence length and the Josephson coupling between the CuO planes. In general, 2-D behavior in HTSC should be restricted to the high field range[40], $B \gg B_{2-D}$, above the cross-over field

$$B_{2-D} \approx \frac{\Phi_o}{\Gamma d^2}, \tag{42}$$

where Γ is the anisotropy factor and d is the spacing between CuO planes. Below the cross-over field, for BSCCO $\Gamma \approx 1000$ and $B_{2-D} \approx 0.3$ T, magnetic coupling between 2-D vortices[41] located in the adjacent CuO layers should cause them to exhibit a 3-D character[13,42–43]. Our findings do not confirm the theoretical predictions and indicate a pronounced 2-D behavior even in weak magnetic fields, well below B_{2-D}. Note, that a similar "surprising" 2-D BSCCO vortices behavior in weak fields was already observed experimentally[19,40].

5. Pinning energy from 1/f noise spectra.

Another powerful tool for the investigations of pinning energy distribution is constituted by measurements of power spectra of low frequency noise. It is well known that an elementary two-level fluctuator producing a RTS waveform has a Lorentzian spectrum[44]

$$S_x^i(\omega) \propto \frac{\tau}{1+\omega^2\tau^2}, \tag{43}$$

where the effective time τ determines the cut-off frequency ω_c.

$$\frac{2\pi}{\tau} = 2\pi \frac{\tau_{up}\tau_{dn}}{\tau_{up}+\tau_{dn}} = \omega_c. \tag{44}$$

According to the widely accepted Dutta-Horn-Dimon model[45], 1/f noise originates from incoherent superposition of many TLF with an appropriate distribution of cut-off frequencies.

$$S_x(\omega) \propto \int \frac{\tau}{1+\omega^2\tau^2} D(\tau)d\tau, \tag{45}$$

where $D(\tau)d\tau$ is the number of processes with cut-off frequencies between $1/\tau$ and $1/(\tau + d\tau)$ In particular, if $D(\tau) \propto \tau^{-1}$ for $\tau_1 \leq \tau \leq \tau_2$, then the spectrum is of the 1/f type.

$$S_x(\omega) \;\propto\; \omega^{-1} \text{ for } \tau_2^{-1} \;\ll\; \omega \;\ll\; \tau_1^{-1}. \tag{46}$$

For thermally activated processes, $\tau = \tau_o \exp(U/kT)$, and the activation energy distribution required to provide 1/f spectrum is[45,46]:

$$D(U) = const. \text{ for } \quad kT\ln(\tau_1/\tau_o) \leq U \leq kT\ln(\tau_2/\tau_o). \tag{47}$$

Therefore, there is a peak in the distribution of the activation energies $D(U)$ whenever the exponent m of the 1/f-like power spectrum, measured as a function of temperature, deviates from unity. The corresponding activation energy can be calculated as

$$U_o(\Delta f, T) = -kT\frac{\ln(\Delta f \tau_o)}{\beta(T)}, \tag{48}$$

where Δf is the frequency bandwidth in which the 1/f power spectra were recorded, and τ_o is the attempt rate.

In the frequency domain measurements we have always observed a Lorentzian spectrum of the voltage RTS imposed on the background of a strong 1/f noise. The origin of 1/f noise in HTSC remains still an open question[47]. Nevertheless, the pronounced noise close to the superconducting transition region is generally believed to be related to the motion of flux vortices[18,32,48–50]. Flux noise is converted to voltage fluctuations by means of directly induced voltage, intrinsic SQUID loops[26], or flux induced resistivity fluctuations[51]. Therefore, we ascribe the background 1/f noise to the contributions from low energy, short range fluctuators, similar to those manifesting themselves in flux-creep measurements. However, a direct evidence that two distinct types of pinning centers, with low and high pinning energies, exists and are due to BSCCO intrinsic properties and not to the particular fabrication procedures comes from the independent measurements of the low frequency flux noise in single crystalline BSCCO bulk samples containing small inclusions of 110 K phase[18]. Fig.13 presents the temperature dependence of the exponent measured as a function of temperature in the bandwidth of 10 Hz (data obtained from Fig.2 of the ref.[18]). The

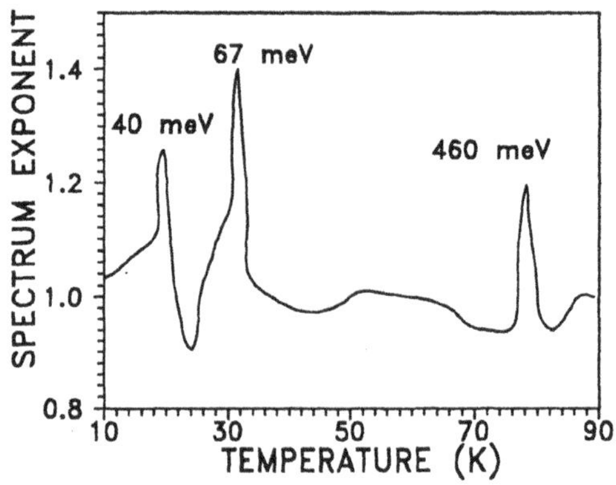

Fig.13. Temperature dependence of 1/f noise power spectrum exponent from Ref.[18]. The spectra were recorded in 1 Hz bandwidth and attempt rate $\tau_0 = 10^{-11}$ s was assumed.

exponent deviates from unity at three temperatures indicating three peaks in the pinning energy distribution. The 40 meV and 67 meV peaks correspond to low energy pinning sites of the low-T_c and high-T_c phase, respectively while a clear peak at 78 K indicates the pinning energy of 460 meV.

6. Thermally activated resistivity

If low energy pinning sites are not completely smeared out at high temperatures by large thermal fluctuations, they should be responsible for the resistance seen in transport measurements. One may expect that in the flux creep regime the resistance should be thermally activated[9]

$$R(T) = R_o \exp \frac{U\beta(T)}{kT}. \tag{49}$$

We have fitted the shape of the sample resistive transition, from Fig.1, to Eq.(49) in the temperature range where RTS was observed. For fitting procedures we have assumed 2-D functional form of $\beta(T)$, and T_c =87.9 K, taking R_o and U_o as fitting parameters. The best fit was obtained for R_o = 54Ω, and U_o = 30.8 meV. The physical sense of R_o becomes clear if one calculates the sample resistance at T=T_c using Eq. (48):

$$R(T_c) = R_o \exp\left(U_o \frac{1-(T/T_c)^4}{kT}\right) = R_o. \tag{1}$$

Observe that the measured resistance of the sample at $T = 87.9$ K is 47Ω, in a reasonable agreement with the value obtained by the fitting procedures. We believe that sample resistance in the temperature range where the RTS noise shows out is due to thermally activated resistivity caused by the flux creep, or thermally activated flux flow over the shallow pinning sites that are responsible for the flux creep at low temperatures.

7. Conclusions

Random telegraph voltage noise has proved to be an efficient tool in characterization of the pinning properties of HTSC thin films. Independently BSCCO films were characterized through the measurements of the I-V curves in the flux creep regime. The detailed results are listed at the end of relevant sections of the paper. Generally, there are two types of pinning sites in BSCCO films. Both types have proved, even at high temperatures to be of a two-dimensional nature. Low energy sites, $U \approx 25$ meV, can be associated with point defects in the CuO planes, probably oxygen vacancies, while deep centers of $U \approx 420$ meV may be attributed to film edges or pronounced grain boundaries. Strong pinning sites can not be detected at low temperatures because the energy of thermal fluctuations is too low to enable an activated hopping from such deep wells. On the other hand, as follows from 1/f noise and thermally activated resistivity measurements, low-energy sites are still active in high temperatures close to T_c . They are responsible for the background 1/f noise and for sample resistance. In this picture the "buried" critical current revealed by the RTS measurements is due to the action of deep centers that locally prevent any dissipation through the flux-creep or flux-flow phenomena. In conclusion we suggest a modification to the BSCCO pinning sites model[3] by introducing, similarly to the YBCO case, two distinct types of pinning centers.

8.Acknowledgments

The authors thank S.J.Lewandowski for his useful advises and help in preparing the manuscript. This research was performed in the framework of the research project 2 0479 91 01

9. References

* on leave from the Institute of Radioelectronics, Minsk, USSR

1. Y.Yeshurun and A.P.Malozemoff, *Phys.Rev.Lett.* **60** 2202, (1988)
2. T.T.M.Palstra, B.Batlog, L.F.Schneemayer, and J.V.Waszczak, *Phys.Rev.Lett.* **61**, 1662, (1988)
3. N.C.Yeh, *Phys.Rev.* **B40**, 4566, (1989)
4. J.R.Clem in: *Progr.Low Temp.Phys.*, R.Nicolski, ed. (World Scientific, Singapore (1990)) vol.**25** p.64
5. H.Ullmaier, *Irreversible Properties of Type II Superconductors*, Springer-Verlag, Berlin 1975
6. P.W.Anderson, *Phys.Rev.Lett.* **9**, 309, (1962)
7. P.W.Anderson and Y.B.Kim, *Rev.Mod.Phys.* **36**, 36, 39, (1964)
8. M.Tinkham, *Introduction to Superconductivity,* McGraw-Hill (1980)
9. D.Dew-Huges, *Cryogenics*, **28**, 674, (1988)
10. B.M.Lairson, J.Z.Sun, T.H.Geballe, M.R.Beasley, and J.C.Bravman, *Phys.Rev.* **B43**, 10405, (1991)
11. M.V.Feigel'man, V.B.Geshkenbeim, A.I.Larkin, and V.M.Vinokur, *Phys.Rev.Lett.* **63**, 2303, (1989)
12. V.M.Vinokur, M.V.Feigel'man, V.B.Geshkenbeim, and A.I.Larkin, *Phys.Rev.Lett.* **65**, 259, (1990)
13. D.S.Fisher, M.P.A.Fisher, and D.A.Huse, *Phys.Rev.* **B43**, 130, (1991)
14. E.Zeldov, N.M.Amer, G.Coren, A.Gupta, M.W.McElfresh, and R.J.Gambino, *Appl.Phys.Lett.* **56**, 680, (1990)
15. R.C.Budhani, D.O.Welch, M.Suenaga, R.L.Sabatini, *Phys.Rev.Lett.* **64** 1666, (1990)
16. J.Hettinger at al.*Phys.Rev.Lett.* **62** 2044, (1989)
17. K.Enpuku, R.Saho, T.Kisu, and K.Yoshida, *Jap.J.Appl.Phys.* **29**, L 1069, (1990)
18. M.J.Ferrari, M.Johnson, F.C.Wellstood, J.Clarke, D.Mitzi, P.A.Rosenthal, C.B.Eom, T.H.Geballe, D.Kapitulnik, and M.R.Beasley *Phys.Rev.Lett.* **64**, 72, (1990)
19. C.J.van der Beek and P.H.40, *Phys.Rev.* **B43**, 13032, (1991)
20. T.L.Hylton and M.R.Beasley, *Phys.Rev.* **B41**, 11669, (1990)
21. T.Van Duzer,*Principles of Superconducting Devices and Circuits*, Elsevier 1981
22. A.F.Marshall, R.W.Barton, K.Char, A.Kapitulnik, B.Oh, R.H.Hammond, and S.S.Laderman at al.*Phys.Rev.* **B37**, 9353, (1988)

23. M.Tinkham,*Helv.Phys.Acta* **61**, 443, (1988)

24. K.H.Fisher and T.Natterman, *Phys.Rev.* **B43**, 10372, (1991)

25. Y.Y.Xue, Z.J.Huang, P.H.Hov, and C.W.Chu. *Phys.Rev.* **B43**, 13598, (1991)

26. G.Jung, M.Bonaldi, S.Vitale, and J.Konopka, *J.Appl.Phys*, in print

27. G.Jung, M.Bonaldi, J.Konopka, and S.Vitale, *Physica* **B 165&166**, 1375, (1990)

28. G.Jung, B.Savo, and A.Vecchione , to be published

29. M.J.Kirton and M.J.Uren, *Adv.Phys.* **38**, 367, (1989)

30. O.Dobbert, K.P.Dinse, W.Goldacker, and T.Wolf, *Physica* **C 177**, 17, (1991)

31. M.Tinkham,*Phys.Rev.Lett.* **61**, 1658, (1988)

32. M.Johnson, M.J.Ferrari, F.C.Wellstood, J.Clarke, M.R.Beasley, A.Inam, X.D.Wu, L.Nazar, and T.Vanketesan, *Phys.Rev.* **B42**, 10792, (1990)

33. E.Zeldov, N.M.Amer, G.Coren, A.Gupta, R.J.Gambino, and M.W.McElfresh, *Phys.Rev.Lett.* **62**, 3093, (1989)

34. R.Griessen, *Phys.Rev.Lett.* **64**, 1674, (1990)

35. C.W.Hagen and R.Grissen, *Phys.Rev.Lett.* **62**, 2857, (1989)

36. E.Zeldov, *Phys.Rev.Lett.* **62**, 2044,(1989)

37. C.J.Lobb, W.Abraham, and M.Tinkham, *Phys.Rev.* **B27**, 150, (1983)

38. K.H.Muller, M.Nikolo, and R.Driver, *Phys.Rev.* **B43**, 7976, (1991)

39. Th.M.Nieuwenhuizen and J.Pankert, *Europhys.Lett.* **11**, 457, (1990)

40. P.H.Kes, J.Aarts, V.M.Vinokur, and C.J. van der Beek, *Phys.Rev.Lett.* **64**, 1063, (1990)

41. J.R.Clem,*Phys.Rev.* **B43**, 7837, (1991)

42. M.V.Feigelman, V.B.Geshkenbeim, and A.Larkin, *Physica* **167**, 117, (1990)

43. V.M.Vinokur, P.H.Kes, and K.V.Koshelev, *Physica* **C 168**, 29, (1990)

44. S.Machlup, *J.Appl.Phys.* **25**, 341, (1954)

45. P.Dutta, P.Dimon, and P.M.Horn *Phys.Rev.Lett.* **43**, 646, (1979)

46. M.B.Weissman, *Rev.Mod.Phys.* **60**, 537, (1988)

47. G.Jung, M.Bonaldi, S.Vitale, and J.Konopka, *Physica* **C 180**, 92, (1991)

48. M.J.Ferrari, M.Johnson, F.C.Wellstood, J.Clarke, P.A.Rosenthal, R.H.Hammond, and M.R.Beasley *Appl.Phys.Lett.* **53**, 695, (1988)

49. Y.Song, A.Misra, P.P.Crooker, and J.R.Gaines, *Phys.Rev.Lett.* **66**, 825, (1991)

50. R.Gross, P.Chaudari, M.Kawasaki, M.B.Ketchen, and A.Gupta, *Physica* **C 170**, 315, (1990)

51. L.Wang, Y.Zhu, H.L.Zhao, and S.Feng, *Phys.Rev.Lett.* **64**, 3094, (1990)

The preparation methods leading to practical Y-Ba-Cu-O and Bi-Pb-Sr-Ca-Cu-O wires and tapes

B.A.Glowacki
Interdisciplinary Research Centre in Superconductivity, University of Cambridge, Madingley Road, Cambridge CB3 OHE, United Kingdom.

ABSTRACT

The paper describes the fabrication of superconducting YBCO and BPSCO wires and tapes by use of almost all of the conventional methods available for the ceramic and metallurgical processes such as low pressure plasma spraying, zone melting, extrusion, solid-liquid diffusion, magnetic alignment, cold pressing, cold swaging and cold rolling leading to the manufacture of practical superconducting conductors. Attention is focussed on the powder-in-tube technique process where a Ag-$(Bi,Pb)_2Sr_2Ca_2Cu_3O_y$ composite is formed into a tape by the conventional wire forming process. The limitation of the critical current in the conductors manufactured is discussed with respect to the microstructure, grain alignment and chemical composition of the superconductors as well as to the magnetic flux creep effect. The influence of the preparation techniques on the anisotropy of the critical current in a magnetic field, which provides data about the alignment of the local current paths as well as morphology of the pinning centers in the textured tapes, has been discussed in detail.

1. Introduction

Although increasing the transport critical current value, J_c, in bulk samples of superconductors is an extremely important task, the preparation of practical forms of the above materials capable of carrying a high J_c is an equally important area for research. The potential for large scale application of the high T_c superconductors such as $YBa_2Cu_3O_7$ (T_c=91.5K), $Bi_2Sr_2CaCu_2O_x$ (T_c=85K) and $(Bi,Pb)_2Sr_2Ca_2Cu_3O_y$ (T_c=110K), depends on the high critical current capability of the polycrystalline materials which can be fabricated in a reproducible way in the form of long conductors. Although there is a thallium based ceramic which is characterised by T_c=120K, because of health and safety reasons its applicability is strongly limited at the present stage of development. Both the Y-based "123" and Bi-based "2223" phases can be mechanically aligned along the a-b plane in silver clad tapes.[1,2,3] The "2223" phase however is much more promising for tape form conductor manufacture because of its high critical temperature value and its ease of alignment during mechanical cold deformation, a common technique in materials processing. Magnetic flux creep may however limit its high magnetic field applications at liquid nitrogen temperatures [4,5].

2. $YBa_2Cu_3O_7$ wires and tapes

A remarkable number of different techniques for the manufacture of the bulk $YBa_2Cu_3O_7$ superconductor have been developed or adopted from ceramic processing; however, the present development status of conductors is that a J_c of $3x10^3$ Acm^{-2} at 77.4K is barely achievable. The whole range of preparation techniques can be grouped in to two sections: the alignment of the superconducting particles and the melting processes.

2.1 The alignment of the superconducting particles

The magnetic alignment [6,7] as well as the mechanical alignment [8] of the superconducting plate like particles provides a strong texture, in volume, of the bulk superconductor required for the high current capability in "123" anisotropic material. However the above techniques can not distinguish between the a and b directions in c plane, resulting in high-angle grain boundaries after sintering, which are not capable of conducting a high current, and the typically achieved J_c value of 1000 Acm^{-2}. For these reasons, the techniques have not been applicable for conductor manufacture.
Strong development of the powder in the tube technique for wire and tape manufacture has been carried out from very early after the discovery of "123" compound. Most of the investigations have been dedicated to basic silver clad wire [9]; however, some of the work has been directed towards reinforced conductors [10,11] especially for internal diffusion of oxygen [10], Fig.1. Silver clad wires are very weak after annealing at temperatures above 900°C and show irreversible degradation under load, in comparison the composites clad in steel are able to withstand high loads without

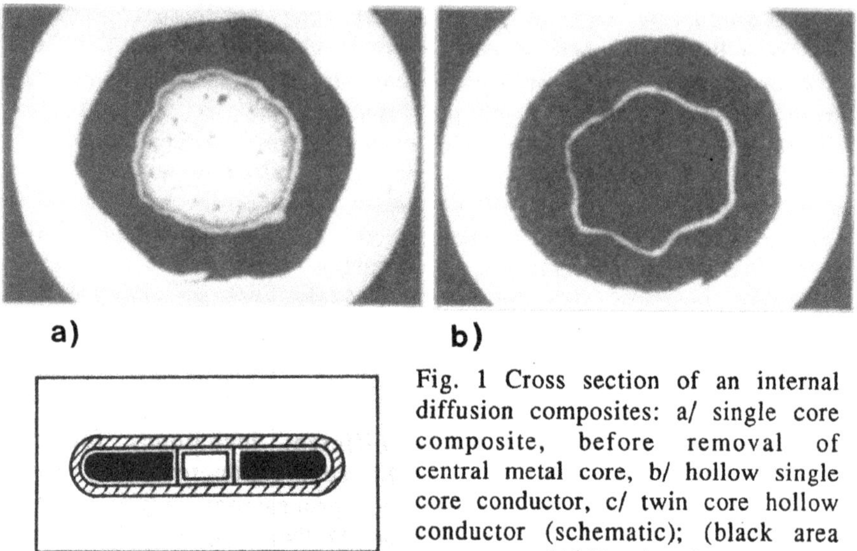

a) b) c)

Fig. 1 Cross section of an internal diffusion composites: a/ single core composite, before removal of central metal core, b/ hollow single core conductor, c/ twin core hollow conductor (schematic); (black area represents "123" phase).

degradation of I_c. The improvement in properties arises firstly because the additional metallic reinforcement does not soften at the sintering temperature, and secondly because the thermo-expansion coefficient of the applied reinforcement can be better matched to the thermo-expansion coefficient of the ceramic core than that of silver, placing the superconductor under compression. The difference in thermo-expansion coefficients between silver and the "123" core in basic Ag-"123" wire causes cracking and significant degradation of J_c. Because the "123" phase surface undergoes a texturing effect during pressing and rolling [12,13] the fabrication of the tape form conductors appeared the most promising direction in conductor manufacture by powder alignment [12,14] reaching J_c values up to 3500 Acm^{-2}; however, the reinforcement is still necessary with this technique.

2.2 Melting processes

A wide range of the different processing techniques for increasing J_c has been presented in the literature, some of with include directional solidification [15], melt texture growth [16], quench melt growth [17], liquid phase sintering [18], Joule heat zone melting [19].

The above techniques have been used successfully to prepare (in crucible) bars and bulk pieces of well-aligned $YBa_2Cu_3O_7$ characterized by a high J_c of about 10^4 Acm^{-2} at 77.4K in zero applied magnetic field. While the J_c results for these techniques are very encouraging and promising, almost all of the above processes involve a temperature gradient along the sample, which results in chemical segregation. Chemical demixing, caused by the migration of $BaCuO_2$ and CuO [20,21] or by the transport of a liquid phase down the temperature gradient and along the length of the sample [20], makes the formation of a continuous piece of uniform "123" phase longer than a few centimetres impossible.

Another area of research activity has been dedicated to the continuous densification of $YBa_2Cu_3O_7$ wires and fibres by zone melting. The laser zone melting process used for the melt growth of "123" fibres showed limited promise because of the very high evaporation rate of the constituents, which resulted in a loss of the overall stoichiometric "123" composition after zone melting [22]. A stoichiometric composition is necessary for the re-formation of pure $YBa_2Cu_3O_7$ superconducting phase.The high power laser and electron beam techniques are more suited to the evaporation of targets for thin film preparation and patterning than for the zone melting of "123" fibres and wires [23]. The application of a vertical resistance furnace for the melting of an extruded 1mm diameter "123" wire was totally unsuccessful [24] because of the necking of the wire and the consequent difficulties in obtaining pieces longer than 10 mm . However, a novel process for continuous densification of "123" wires by vacuum infrared zone melting can be applied [25]. The maximum transport critical current density obtained for such wires at 77.4K in zero magnetic field was 2.6×10^3 Acm^{-2}. Normalized critical current values as a function of magnetic field indicated that the quality of the intergrain connections for vacuum infrared zone melt wires is significantly better than for conventionally processed wires and tapes, Fig.2.

There are other techniques which involve melting processes like solid-liquid diffusion and low pressure plasma spraying. Solid -liquid diffusion involves infiltration of the Y_2BaCuO_5 tape form substrate by low melting point compound which is a mixture of barium copper oxide and copper oxide containing Ba:Cu in nominal atomic ratio 3:5 respectively [28,29]. This process seems to be quite promising but still requires future investigations of the diffusion processes involved, in respect to the texturing and J_c value of the final "123" diffusion layer.

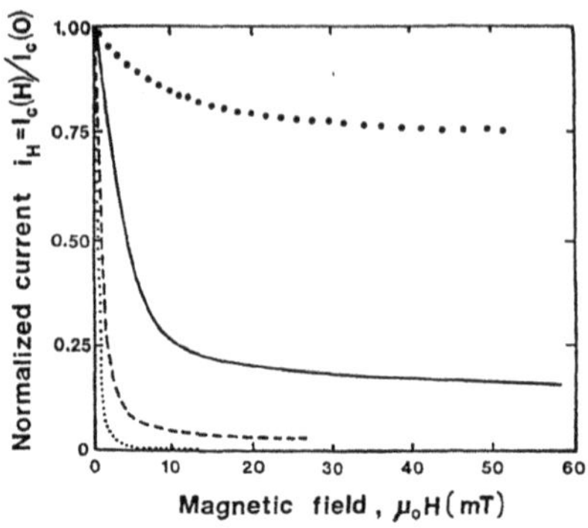

Fig.2 Normalized critical current i_h vs external applied magnetic field $\mu_0 H$ for "123" wires: ____ vacuum infrared zone melted wire [25] , - - - silver clad swaged wire [10], ... semi-wet extruded wire [26] and ● ● ● liquid phase sintering [27].

The low pressure plasma spray is a technique which can be used for large scale application [30], but the effect of the crystal orientation on the J_c is a particular problem when considering plasma spraying. Sprayed layers exhibit an amorphous/micro-crystalline structure and the post-spray anneal introduces only weak preferential orientation. This problem needs much more research work to establish the relation between spray conditions, substrate material and the texture of the final $YBa_2Cu_3O_7$ coating.

3. Ag-$(Bi,Pb)_2Sr_2Ca_2Cu_3O_y$ tapes

At the present time, considerable progress has been made in the development of different techniques for manufacture of the practical "2223" conductors capable of carrying high critical currents. Essentially two major techniques appear to be the most promising : the coating and powder in tube techniques.

3.1 The coating technique

The coating techniques such as doctor-blade [31,32] or spray coating [33] where the superconducting powder was mixed with binder, solvent, dispersant and plasticizer, and deposited on the surface of a silver ribbon, appeared very simple and promising.

After the deposition process, the coated tapes were heated up to burn off the organic components in the system and then cold-rolled to improve the density and texture of the coating. The cold-rolled ribbons were then subjected to melt processing (860-880°C). The partial melting process is necessary for this technique but unfortunately causes the final superconducting layer to be of the "2212" phase rather than "2223" phase. There are a few reasons why the formation of the pure "2223" phase did not take place. Firstly, during the partial melting at temperatures above 860°C in air, evaporation loss of volatile components of the initial material from the surface of the ribbon takes place and secondly, at such a high temperature the formation of an insulating layer of $(Sr,Ca)_2CuO$ at the interface with silver takes place [34]. For the above reasons, the coating technique is more appropriate for the formation of the "2212" ribbons where melting does not cause a significant degradation of the "2212" phase [35].

3.2 The powder in tube technique

Essentially there are two major routes for the preparation of Ag-"2223" tapes by cold rolling. The first involves packing a precursor powder of nominal composition "2223" into a silver tube, and subsequently swaging, drawing and rolling. The precursor powder has not been converted to the high temperature "2223" phase and in principle consist of a mixture of "2212", Ca_2PbO_4 and CuO phases [36,37]. During sintering of the precursor powder it appeared that formation of the "2223" phase from multi-phase composition involved a liquid phase reaction which improved connections between plate-like grains. The second procedure involves packing a pure "2223" phase into a silver tube [38,39]. Differences in the initial powder composition were found to affect the cold rolling and sintering procedures adopted. In both cases it is necessary to repeat the cold rolling and sintering procedures in order to minimise the effect of the retrograde densification; this takes place because of the growth of plate-like particles which push the structure apart, causing reduction of the density of the superconducting core [40]. Because retrograde densification considerably reduces the critical current carrying capabilities of the superconducting tapes, multi-stage cold rolling and sintering procedures are necessary [37,41,42].

Critical current values in the range of 10^4 Acm^{-2} have been presented in the literature data for silver clad wires manufactured by the cold pressing technique ie multi-stage uniaxial pressing and sintering of the Ag-"2223" tape form conductors [37,43,44]. Uniaxial pressing of a wire only results in radial spreading of the material, which helps to minimise the effect of retrograde densification but pressing of the long wires can not be viewed as an applicable technique. Additionally, uniaxial pressing involves a time and pressure dependent densification process improving connectivity between plate-like grains (as an effect of creep) which is very different from the dynamic compaction and densification which occurs during the cold rolling process. For these reasons, most of the research activities with regard to metal-superconductor ceramic composite tapes are concentrated at present on the cold rolling technique.

Recently it was shown that silver powder doped into "2223" phase during tape manufacture enhanced texturing and connectivity of the "2223" grains in the superconducting core, simultaneously improving J_c vs H properties of final tapes without degradation of the T_c [45]. Such a procedure used to complement for the powder in tube technique seems to be very promising.

Although the preparation of tape form silver clad ceramic wires by the cold rolling technique is a relatively simple task, the specification of the processing procedure and parameters for the manufacture of a long superconducting tape by this method , capable of carrying a high current of at least 10^4 Acm^{-2} at 77K, appears to be very complex [39,41,42,46,57,48]. For the future design, optimization and manufacture of superconducting tape, an understanding of the deformation process during the first cold-rolling treatment is essential.

Although the foundations of the cold rolling procedure seem to be clear, the handling of such thin tapes is not practical at all. One of the solutions can be the use of a multifilamentary tape form conductor. However, even in such a case the reinforcement of the multifilamentary wires will be unavoidable.

Additionally, an interesting correlation between the optimum critical current value versus the sintering temperature and non-isothermal DTA measurements, which indicates the optimum conditions for formation of the "2223" ribbons [34] is presented in Fig.3. Critical current measurements for Ag-"2223" tape after a single sintering treatment, over a temperature range of 842°C to 860°C under an atmosphere of composition $N_2:O_2=3.7$ by volume, showed that the highest J_c value was obtained for a

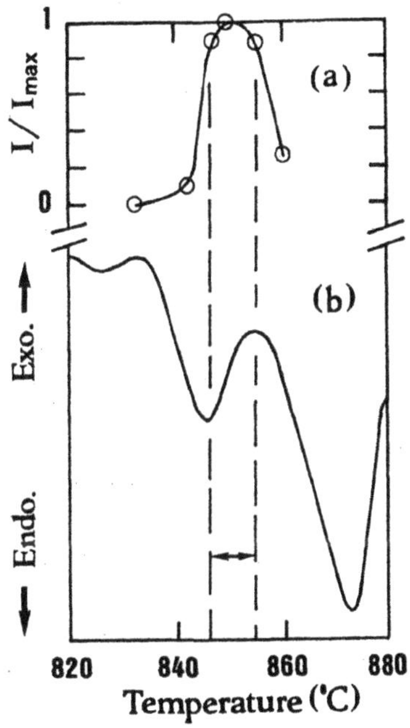

Fig.3 The comparison of the normalized critical current ($I_c/I_{c,\ max}$) measurements versus the sintering temperature of the tape with the non-isothermal DTA measurements, obtained under an atmosphere of composition $N_2:O_2$=3.7. The sintering time for the tapes was 10 hours. The sample used for DTA measurements was almost pure "2223" phase.

sintering temperature of 850 °C , however sintering between 847 °C and 855 °C is acceptable, Fig.3a. The above behaviour may be explained on the basis of the DTA results obtained for a sample of almost pure "2223" phase in the region of partial melting,Fi.3b. These results show that at 842 °C the "2223" phase is not formed at a significant rate. The rate of formation is slow due to the low mobility of the atoms at this temperature. As the temperature is increased to about 848°C the mobility of the atoms, during solid state diffusion, is greater and the optimum formation and sintering conditions are achieved resulting in a maximum I_c value for tapes sintered at this temperature. Increasing the temperature above 855°C causes partial melting of the "2223" phase [49] followed by some reaction with silver [34,50].

3.2.1 Critical current distribution in Ag-"2223" tapes

The critical current measurements of the Ag-"2223" tapes indicate that for a given magnetic field two different angular rotation sequences (in the first sequence the tape was rotated from $I_c(ab)$ perpendicular to $\mu_oH(c)$, towards $I_c(ab)$ parallel to $\mu_oH(ab)$; in the second sequence the tape was rotated from $I_c(ab)$ perpendicular to $\mu_oH(c)$, towards $I_c(ab)$ perpendicular to $\mu_oH(ab)$) both gave identical values for $I_c(\mu_oH)$ versus the angle of rotation [51]. This result can be explained by the spatial distribution of the local two dimensional paths for the supercurrent flowing through the boundaries between the individual plate-like grains; so that the Lorentz force free configuration for H parallel to the tape surface is impossible to detect. In addition to the dispersion of the current paths the local distortion of the flux lines in polycrystalline layer structure materials may contribute to the above effects [52]. The significant difference in morphology between the surface of the ceramic layer (thickness of a few μm) peeled off with the silver, Fig.4a and the opposing surface of the remaining ceramic core, Fig.4b, (after sintering) can be attributed to the surface texturing of the part of the ceramic core which is in immediate contact with the silver. The ceramic grains which are attached to the silver appeared to be very highly textured (c-axis) and almost parallel to the theoretical rolling plane and matching mutually with very small misalignment so well that the identification of the grain boundaries along the ab planes was not easy. The surface texturing by mechanical deformation, is accelerated during the subsequent sintering by the presence of silver, which lowers the melting point of the "2223" ceramic [34].

The ceramic core under the layer described above is of a different morphology, Fig.4b. The grains are still textured but misaligned from the theoretical rolling plane by up to 20 degrees, and grain boundaries can be very easily identified, suggesting an inefficient connectivity along the CuO_2 planes between the grains . The participation of the central part of the ceramic core in the conduction of the supercurrent is likely to be far less effective than the surface layer because critical current preferentially will flow in the conducting planes and not across them. Concluding, the cold rolling procedure improves the crystallographic alignment in the ceramic core which is related to the texturing of the silver-ceramic interface up to a depth of few micrometers into the ceramic core interior. The above effect of surface texturing requires reduction of

a)

b)

Fig.4 An SEM picture of the ceramic core close to the Ag-$(Bi,Pb)_2Sr_2Ca_2Cu_3O_y$ interface after the longitudinal fracture by peeling off the silver cover; a/ the internal surface of the ceramic layer (thickness 4 μm) stuck to the silver tape, b/ the internal surface of the ceramic core under the layer presented in Fig.3a.

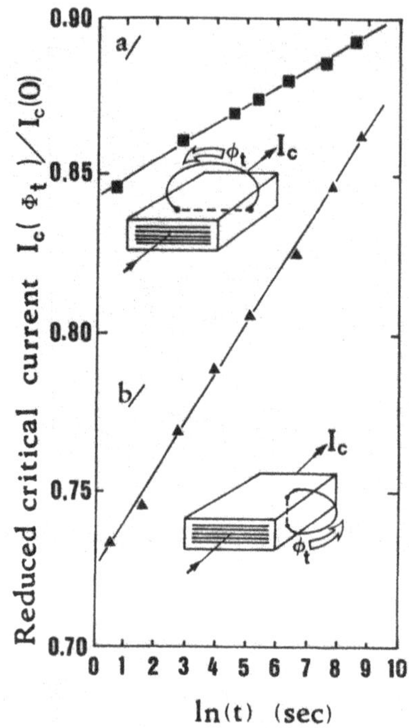

Fig.5 The semi-logarithmic dependence of the reduced transport critical current $I_c(\phi_t)/I_c(\phi)$ of the tape versus the time of relaxation of the trapped magnetic flux lines; a/ the trapped magnetic flux parallel to surface of the tape, b/ the trapped magnetic flux perpendicular to surface of the tape.

the ceramic core thickness up to tens of micrometers in order to achieve the highest critical current density in nearly fully textured Ag-"2223" tapes.

3.2.2 I_c anisotropy during relaxation of the trapped magnetic flux

In this section the effect of the decay rate of trapped magnetic flux in the volume of strongly textured Ag-"2223" tape on the transport I_c value is discussed. Because the decay rate for the two configurations of the trapped magnetic flux (ϕ_t) parallel and perpendicular to surface of the tape, for highly textured "2223" silver clad tapes is different [51], Fig.5, one may expect that the microscopic process of the flux relaxation from the pinning centers which affect the overall critical current value is different as well.

For magnetic field parallel to the surface of the tape, Fig.5a, the trapped flux lines have a core size such that they may easily fit into non-superconducting layers. Such flux lines in typical polycrystalline tape are possibly tilted by the misaligned grains as well as passing through the ab planes of some of the grains at a very low angle, less than 20 degrees. Because the pinned vortices, after the reduction of the external magnetic field to zero, form closed loops outside the superconductor, the "annihilation" of the trapped flux lines across the superconducting ab planes is difficult, and a lower decay rate is observed. For magnetic field perpendicular to the surface of the tape, Fig.5b, the trapped vortex punctures the ab planes in such a way that it breaks up into "pancake vortices" [53]. Assuming that the pinned vortices form closed loops outside the superconducting core , the "annihilation" of the trapped flux lines perpendicular to the CuO_2 planes is easier than in the parallel configuration and there is an increase of the decay rate. These results show that for highly textured material there is a strong anisotropy in the logarithmic decay rate of the remanent flux. This can be related to the anisotropy in the activation energy for flux depinning in layer structure superconductors.

4. Summarizing remarks

The wide range of the preparation methods for the fabrication of superconducting materials such as $YBa_2Cu_3O_7$ and $(Bi,Pb)_2Sr_2Ca_2Cu_3O_y$ in the form of wires and tapes has been reviewed and discussed. At the present stage the"2223" phase seems to be more promising for long tape form conductor manufacture than the "123" phase.

Although the foundations for the cold rolling procedure of the Ag-"2223" tapes seem to be more clear, the handling of such thin (less than 100 μm) tapes containing, in big proportion, silver which become very plastic after annealing, at temperatures around 850°C for prolonged time, is an extremely difficult task. One of the solutions can be multifilamentary tape form conductors; however, even in this case the reinforcement of the multifilamentary wires is unavoidable.

5. Acknowledgements

Although most of the presented work reflects my own involvment in IRC in Superconductivity in Cambridge, I am especially grateful to my collegues P.A. Freeman,Wai Lo, N. Ozkan, E.A. Robinson, P.D. Hunneyball and A.P. Baker for great engagement in research which I was able only to indicate in this paper; as well I would like to thank J.E. Evetts and E.C.G. Kirk for valuable discussion.

6. References

1. S. Masuda, M. Okada, T. Morimoto, T. Matsumoto and K. Akihara, *Mat. Res. Soc. Symp. Proc.* **99**(1988)695.
2. M. Kawashima, M. Nagata, Y. Hozada, S. Tanakano, N. Shibuta, H. Mukai and T. Hikata, *IEEE Trans. Mag.* **25**(1989)2168.
3. H. Kumakura, K. Togano and H. Maeda, *J. Appl. Phys.* **67**(1990)3443.
4. A. Oota, A. Yaata, K. Hayashi and K. Ohba, *J. Appl. Phys.* **69**(1991)3182.
5. Y. Yamada, S. Murase, K. Yamamoto and Y. Kamisada, *Cryogenics* **30**(1990)581.
6. J. D. Livingston, H. R. Hart Jr. and W. P. Wolf, *J. Appl. Phys.* **64**(1988)5806.
7. J. M. Ferreia, M. B. Maple, H. Zhou, R. R. Hake, B. W. Lee, C. L. Seaman, M. V. Kuric and R. P. Guertin, *Appl. Pyhs. A* **47**(1988)105.
8. X. Zheng, H. Kuriyaki and K. Hirakawa, *Jpn. J. Appl. Phys.* **28**(1989)L52.
9. S. Jin, R. C. Sherwood, R. B. vanDover, T. H. Tiefel and D. W. Johnson Jr., *Appl.Phys.Lett.* **51**(1987)203.
10. B. A. Glowacki and J. E. Evetts, *Mat. Res. Soc. Proc.* **99**(1988)741.
11. R. Flukiger, T. Muller, T. Wolf, I. Apfelstedt, E. Seibt, H. Kupfer and W. Schauer, *Physica C* **153-155**(1988)1574.
12. M. Okada, A. Okayama, T. Matsumoto, K. Aihara, S. Matsuda, K. Ozawa, Y. Morii and S. Funahashi, *Jpn. J. Appl. Phys.* **27**(1988)L1715.
13. B. A. Glowacki and J. E. Evetts, Mat. Res. Soc. Proc. 99(1988)419.
14. K. Osamura, T. Takayama and S. Ochiai, *Cryogenics* **29**(1990)430.
15. L. Fengsheng, M. Huilin, W. Keguang, Z. Tingjie, W. Xiaozu and Z. Lian, *ICMC-90 High Temperature Superconductors,* Garmisch-Partenkirchen, FRG, May1990.
16. S. Jin, T. H. Tiefel, R. C. Sherwood, M. E. Davis, R. B. van Dover, G. M. Kammlott, R. A. Fastnacht and H. D. Keith, *Appl. Phys. Lett.* **53**(1988)2074.
17. M. Murakami, M. Morta, K. Doi and K. Miyamoto, *Jpn. J. Appl. Phys.* **28**(1989)1189.

18. K. Salama, V. Selvamanickam, L. Gao and K. Sun, Appl. Phys. Lett. **54**(1989)2352.
19. P. J. McGinn, M. A. Block and A. Valenzuela, *Physics C* 156(1988)57.
20. M. N. Rahman, L. C. DeJonge and M. Y. Chu, *J. Am. Ceram. Soc.* **71**(1988)C-237.
21. R. L. Meng, C. Kinalidis, Y. Y. Sun, L. Gao, Y. K. Tao, P. H. Hor and W. Chu, *Nature* **345**(1990)326.
22. P. A. Morris, B. G. Bagley, J. M. Tarascon, L. H. Green and G. H. Hull, *J. Am. Ceram. Soc.* **71**(1988)334.
23. J. P. Zheng, H. S. Kim, Q. Y. Ying, R. Barone, P. Bush, D. T. Shaw and H. S. Kwok, *Appl. Phys. Lett.* **55**(1989)10.
24. N. McN. Alford, J. D. Birchall, W. J. Clegg and K. Kendall, *J. Appl. Phys.* **65**(1989)2856.
25. N. Ozkan, B. A. Glowacki, E. A. Robinson and P. A. Freeman, *J. Mater. Res.* **6**(1991) No.9(in press).
26. B. A. Glowacki, N. Ozkan and P. A. Freeman, *in High Temperature Superconductors*, ed. P. Vincenzini (Elsevier Science Publishers B. V.,1991)p.663.
27. N. Ozkan and B. A. Glowacki (unpublished data).
28. P. A. Freeman and B. A. Glowacki, *in High Temperature Superconductors*, ed. P. Vincenzini (Elsevier Science Publishers B. V.,1991)p.915.
29. N. Sadakata, M. Sugimoto, O. Kohno and K. Tachik, *IEEE Trans. Mag.* **25**(1989)2180.
30. B. A. Glowacki, S. C. Gill, T. W. Clyne, J. E. Evetts and P. Chandler, *in Thermal Spray Technology-New Ideas and Processes*, ed. D. L. Houck (ASM International, Ohio, USA,1989)p.191.
31. J. Kase, K. Togano, H. Kumakura, D. R. Dietderich, N. Irisawa, T. Morimoto and H. Maeda, *Jpn. J. Appl. Phys.* **29**(1990)L1096.
32. K. Togano, H. Kumakura, H. Maeda, E. Yanagisawa, N. Irisawa, J. Shimoyama and T. Morimoto, *Jpn. J. Appl. Phys.* **28**(1989)195.
33. S. Jin, J. E. Graebner, T. H. Tiefel, R. B. vanDover, A. E. White and G. W. Kammlott, *Physica C* **177**(1991)189.
34. Wai Lo and B. A. Glowacki, *in High-Temperature Superconductors, Materials Aspects*, ed. H. C. Freyhardt, R. Flukiger and M. Peuckert (DMG Informationsgesellschaft mbH, Verlag, FRG,1991)p.1005.
35. A. P. Baker, N. Ozkan and B. A. Glowacki, *ICMAS-91* Paris, France, October 1991.
36. A. Oota, A. Yata, K. Hayashi and K. Ohba, *Supercond. Sci. Technol.* **4**(1991)S244.

37. Y. Yamada, B. Obst and R. Flukiger, *Supercond. Sci. Technol.* **4**(1991)165.
38. Wai Lo and B. A. Glowacki, *Supercond. Sci. Technol.* **4**(1991)S61.
39. L. R. Motowidlo, E. Gregory, P. Haldar, J. A. Rice and R. D. Blaugher, *Appl. Phys. Lett.* **59**(1991)736.
40. D. W. Johnson Jr. and W. W. Rhodes, *J. Am. Ceram. Soc.* **72**(1989)2346.
41. Wai Lo and B. A. Glowacki, *CEC/ICMC-91*, Huntsville, Alabama, USA, June1991.
42. B. A. Glowacki and Wai Lo, *ICMAS-91* Paris, France, October 1991.
43. M. Sekine, K. Ogava, H. Maeda and K. Numata, *Jpn. J. Appl. Phys.* **28**(1989)28.
44. T. Asano, Y. Tanaka, M. Fukutomi, K. Jikihara, J. Machida and H. Maeda, *Jap. J. Appl. Phys.* **27**(1988)L1652.
45. K. Ohba, T. Horio, K. Iwasaki and A. Oota, *6th International Workshop on Critical Currents,* Cambridge, England, July 991.
46. J. Tenbrink, M. Wilhelm, K. Heine and H. Krauth, *IEEE Trans. Magn.* **27**(1991)1239.
47. T. Hikata, M. Ueyama, H. Mukai, N. Shibuta, T. Kato and K. Sato, *The 3rd International Syposium on Superconductivity*, Sendai, Miyagi Prefecture, Japan, November1990.
48. M. Ullrich, W. Schaper and H. C. FGreyhardt, *Supercond. Sci. Technol.* **3**(1990)602.
49. Y. L. Chen, W. Lo, Y. S. Zhen and R. Stevens, *J. Mater. Sci.: Mater. Electronics*,**1**(1990)185.
50. S. X. Dou, K. S. Song, H. K. Liu, C. C. Sorrell, M. H. Apperley, A. J. Gouch, N. Savvides and D. W. Hansley, *Physica C* **160**(1989)533.
51. B. A. Glowacki, Wai Lo and J. E. Evetts, *The 6th International Workshop on Critical Currents*, Cambridge, England, July1991.
52. Y. Iye, S. Nakamura, T. Tamegai, T. Terashima, K. Ymamoto and Y. Bando, *Physica C* **166**(1990)62.
53. J. R. Clem and M. W. Coffey, *Phys. Rev. Lett.* **B 42**(1990)6209.

CONTRIBUTED PAPERS

DEFECTS OF STRUCTURE AND CRITICAL PARAMETERS OF HTSC

VLADISLAV V. KIRSANOV
*Computer Simulation Laboratory,
Tver Polytechnical Institute,
Pervomayskaya nab., 22, Tver 170026, USSR*

and

NARIMAN N. MUSIN
*Computer Simulation Laboratory,
Tver Polytechnical Institute,
Pervomayskaya nab., 22, Tver 170026, USSR*

ABSTRACT

Structural defects created by irradiation which may act as pinning centers in Y-Ba-Cu-O are investigated by using molecular dynamics method. Threshold energies of O, Y, Cu and Ba displacements are estimated. Features of collision cascades are described.

1. Introduction

One of the most important parameters of HTSC is the critical current density. Its value is influenced by many factors among which are structural defects occupying not the minima[1-3].

Irradiation of HTSC by fast particles causes in this materials the creation of different structural defects which act as strong centers of pinning resulting, as a rule, in increasing the current density at low fluences[2,4-10]. Such experiments indicate that defects of atomic scale (of order of coherence length, 1÷3 nm) act, to all appearance, as centers of pinning in HTSC.

Therefore considerable interest[11-15] is now being paid to the study of defects being formed under irradiation that can act as centers of pinning. Defect clusters (displacement cascades, interstitial phases, stacking faults, dislocations, etc.) with the size of 1-5 nm can serve as effective centers of pinning in irradiated high-temperature superconductors. For the present their microstructure is not clear. Some apparent evidence for defect location in Cu-O planes seem to correlate with low displacement threshold energies of these ions. Computer simulation is a useful tool for clearing up the mechanism of forming collision cascades and other defect clusters. The results given below were obtained by using the molecular dynamics method and they show what mechanisms of radiation defect formation are realized under low energy irradiation and the corresponding types of defects produced in each case. It is found established that oxygen ions possess, on the average, the least threshold energy of displacement and therefore, can be easily removed from their positions. Estimations of threshold energies of O, Cu, Ba, Y ion displacement of are presented, these estimations permit determination the number of radiation defects under different irradiation conditions.

2. METHOD

The computational experiment was carried out on model crystallite having the shape of rectangular parallelepiped. Computations were based on the molecular dynamics method. The appropriable computation algorithm used here has been described in detail elsewhere[18]. Potentials of interionic interaction were borrowed[17].

3. RESULTS AND DISCUSSION

Spreading of Cu-O collision chains along the axis z in the plane (010) and along the axis y in the plane (001) has been studied. As has been shown[18,19] running of dynamic crowdions can take place in such chains if the primary knocked-on atom (PKA) Cu has sufficiently high energies (E_{pka} > 40eV). Again one can see the process of oxygen defect formation. Oxygen ions move away from the position on the chain axis to an oxygen vacancy O(5). This mechanism of injection of stable oxygen displacements requires the transfer of a rather less energy to PKA than it would be during the formation of stable displacements of Cu ions. Threshold energies of Ba and Y ion displacements in the direction of Ba-Y rows is estimated to be about 25 eV and 60 eV, respectvely.

Fig. 1-2 show angular dependencies of threshold energy of oxygen displacement E_d^O. Positive angle corresponds to the direction of PKA escape at which the vector of initial velocity is turned clockwise from the Cu-O chain axis. Angular dependence E_d^O for PKA - O(1) ion is given in fig. 1 (plane (010)). Minimum value E_d^O constitutes 9 eV at angles 10° and 45°. Computed angular dependence of threshold energy of oxygen displacement is in agreement with experimental data. Thus if we will consider that degradation of structure and electrical properties of HTSC begins at electron energy equal to 100 keV[20,21], the recalculation in terms of maximum energy of oxygen ion recoil gives its value of 14.7 eV. If 100 keV is supposed to be the mean value at all angles of escape and lattice sites, the averaging of threshold energy of displacement over all angles of escape of PKA, investigated by us,(fig.1) gives the value of $\langle E_d^0 \rangle = 14$ eV. Computation results of threshold energy on Cu-O rows in basic plane are given in fig. 2. Oxygen ion O(4) was taken as PKA. On the average, threshold energy is lower than for the chain in the direction of [001]. Averaging on investigated angles gave the value $\langle E_d^0 \rangle = 10$ eV. In this case minimum value is equal to 5 eV at the knock-on angle 45°.

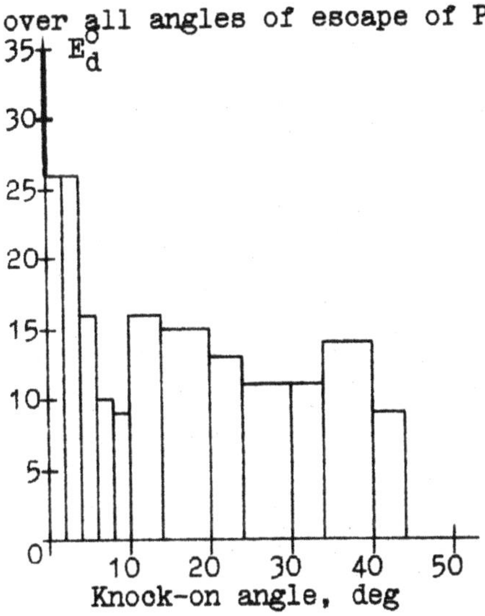

Fig.1 Angular dependence of E_d^0 if PKA - O(1) ion.

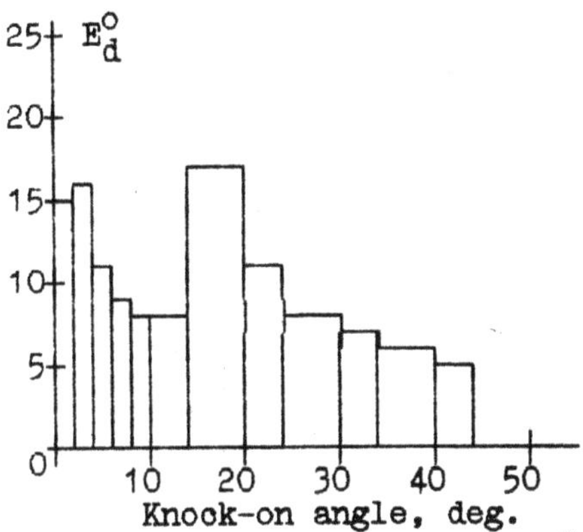

Fig.2 Angular dependence of E_d^0 if PKA - O(4) ion.

All the above points to the fact that the process of oxygen ion displacement is considered to be the one dominating on forming radiation defects. The latest works on electron - microscopy analysis of irradiated HTSC[11-13] provide support for the observation. Thus the formation of impurity phase Cu_2O[11,12], formation of defect strips on layers Cu(1)-O[14,15] prove that oxygen and copper sublattices are rather easily damaged than the rest. Besides, the disorder of the very layers Cu(1) - O rather than Cu(2) - O is also confirmed by molecular dynamics computations: oxygen ions on sites O(1) and O(4) are easily displaced than on sites O(2,3).

At present we are investigating more complicated cases of irradiation, namely, development of low-energy cascades of ion - ion collisions. The first results on studying collision cascades produced by 100 ÷ 300 eV PKA (Cu(2) ion) are given below. Initial direction of PKA motion was at an angle of 20° with an axis c. Fig. 3 - 5 show the cascade created by 200 eV PKA. Projection of model crystallite with 4 x 4 x 3 unit cells is presented in figs. 3 and 5. To perceive it better ions are not shown, only frames of unit cells are plotted. On the basis of analysis of this cascade geometry we can make the following conclusions:

1. Even though the initial motion of PKA really take place along the axis c cascade are noticed to dissociate to subcascades which are space separated by Y ion layers (fig.3). In fig. 3 one can see heavy ion displacements within one unit cell between Y planes, at the same time ion jumps over Y ion layers practically are not observed.

2. Furthermore, one can notice that the heaviest displacements take place near Cu(1)-O layers where the main displacements are concentrated. When passing the cascade damage of Cu(1)-O chains and oxygen redistribution between O(4) and O(5) positions is registered (fig.4)

3. Greater amount of PKA energy is dissipated on oxygen and copper sublattices that is easily seen in fig. 5 where only Ba and Y ion displacements are shown. Trajectories of these types are few in numbers (compare with fig. 3). Sublattice of light oxygen ions is a specific viscous medium in which the cascade development fades out. Therefore a well-known picture of branching-off

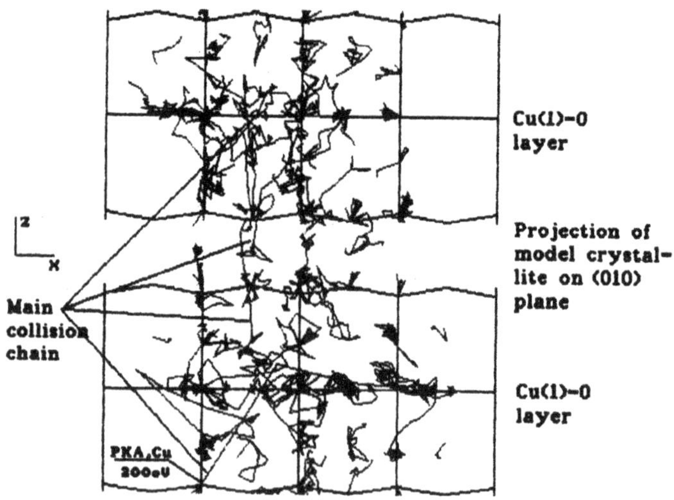

Fig.3 Projection of model crystallite on the (010) plane. PKA Cu(2) ion, E_{pka}=200 eV, angle 20° to the z axis.

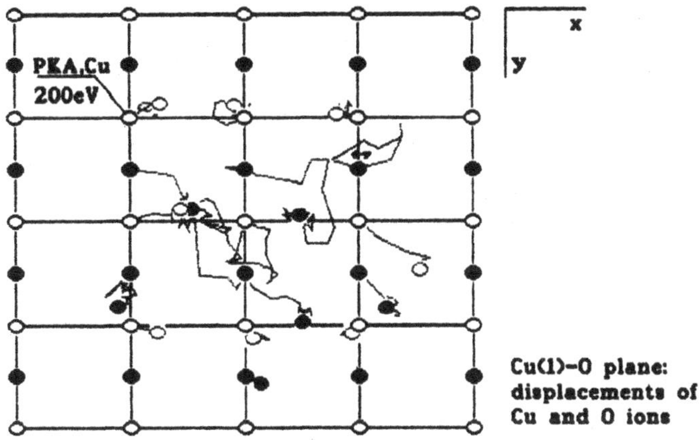

Fig. 4 Destruction of Cu(1)-O(4) chains in the Cu(1)-O(4) layer on developing 200 eV cascade.

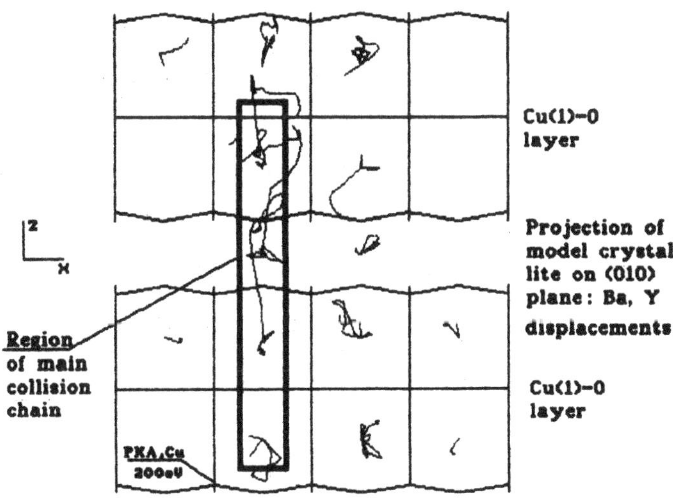

Fig.5 Projection of model crystallite on the (010) plane. Cascade as in Fig.3. Only displacements of Ba and Y ions are shown.

of main chain of cascade into the secondary branches is not observed. One can see only the primary branch of the cascade Cu(2)-Ba-Y-Ba-Ba, that is quickly dying down as in the case of earlier investigated chains of ion-ion collisions[18]. Near this branch "gas" of highly moving ions is formed which takes energy away from ions of the main collision chain.

The first obtained results reveal the formation of Cu(1)-O defect layers[14,15]. Due to this one can suggest that there is a tendency in the cascade development towards transfer of ion motion in the Cu(1)-O plane with the destruction of Cu(1)-O(4) chains which can lead to ortho-tetra transition.

4.REFERENCES

1. J.Piechota et al, *Phys. Scr.* **42** (1990) 109.
2. H.Kupfer et al, *Cryogenics* **29** (1989) S268.
3. D.Shi et al, *Supercond. Sci. Technol.* **3**(1990) 457.
4. A.Wisniewski et al, *Solid State Commun.* **65** (1988) 577.
5. F.Sauerzopf et al, *Phys. Rev.* **B43** (1991) 3091.
6. P.Hor et al, *Mod.Phys.Lett.* **B4** (1990) 703.
7. T.Kato et al, *Jpn. J. Appl. Phys.* **28** (1989) L766.
8. K.Shiraishi et al, *Jpn. J. Appl. Phys.* **28** (1989) L807.
9. S.Kohiki et al, *Appl. Phys.* **A50** (1990) 509.
10. R.van Dover et al, *Appl. Phys. Lett.* **56** (1990) 2681.
11. F.LeGoues at al, *Phil. Mag.* **A60** (1989) 525.
12. M.Ruault et al, *Phil. Mag.* **B60** (1989) 801.
13. M.Kirk et al, *Phil. Mag. Lett.* **62** (1990) 41.
14. D.Bourgault et al, *Nucl.Instr.Meth.Phys.Res.* **B42** (1989) 61.
15. R.Janam et al, *Solid State Commun.* **77** (1991) 259.
16. V.V. Kirsanov and N.N.Musin *Computer code for calculation of dynamics of radiation damage and point defect properties in high temperature superconductors* (Tver Polytechn. Inst. Report, Tver, USSR, 1990) (in Russian).
17. S.Chaplot, *Phys.Rev.* **B37** (1988) 7435.
18. V.V.Kirsanov, N.N.Musin, *Phys.Lett.* **A143** (1990) 420.
19. V.V.Kirsanov, N.N.Musin, *Phys.Lett.* **A153** (1991) 493.
20. F.Clinard, Jr. et al, *J.Appl.Phys.* **65** (1989) 5057.
21. M.Kirk et al, *Mat.Res.Soc.Symp.Proc.* **99** (1988) 209.

INFLUENCE OF NEUTRON IRRADIATION ON THE DISTRIBUTION OF ACTIVATION ENERGIES IN A $Bi_2Sr_2CaCu_2O_x$ SINGLE CRYSTAL

M. Reissner, W. Steiner
Institut für Angewandte und Technische Physik, T U Wien, Austria

W. Kritscha, F.M. Sauerzopf, H.W. Weber
Atominstitut der Österrreichischen Universitäten, Wien, Austria

ABSTRACT

We report on measurements of the time dependent magnetization of a BiSCCO-2212 single crystal irradiated to fast neutron fluences of 0.5 and $2.5\cdot10^{21}m^{-2}$. The results are analysed within a model of thermally activated flux creep with a distribution of activation energies, leading to a highly anisotropic relaxation time τ, which is found to be independent of neutron fluence. After irradiation the distribution of activation energies shifts to higher values. The results are related to a qualitative model of flux pinning by different defect structures.

1. Introduction

It has been shown that the analysis of measurements of the time dependence of the magnetization in terms of the model of Hagen and Griessen[1], which leads to a distribution of activation energies, allows some conclusions on the pinning behaviour in high T_c superconductors. From measurements on polycrystalline materials we have concluded that typical distributions exist for the different classes of superconductors, which emphasize the significance of strong intrinsic pinning[2]. Measurements on neutron irradiated single crystals offer the possibility to investigate the influence of changes in the defect structure and of the anisotropy on the distribution function. It is the aim of this work to determine the distribution of activation energies in a neutron irradiated single crystal of BiSCCO-2212 for both field directions and to compare it to results deduced from measurements of the field and temperature dependence of J_c and of the irreversibility line.

2. Experimental

The BiSCCO-2212 single crystal with dimensions of $2.5x1.45x0.4mm^3$ was irradiated by fast neutrons (E > 0.1MeV, sample temperature about 60°C) in two steps to total fluences of $5\cdot10^{20}m^{-2}$ and $2.5\cdot10^{21}m^{-2}$ in the TRIGA reactor in Vienna. Details of the sample preparation are given in Ref. 3. The transition temperature (83K) was not significantly influenced by the irradiation process.

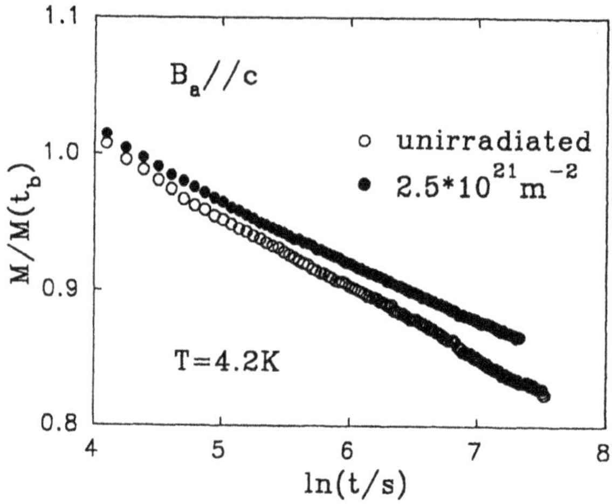

Fig. 1: Time dependence of the magnetization normalized to $M(t_b)$ for the unirradiated (O) and fully irradiated sample (●) measured at 1T.

Measurements of the relaxation of the magnetization were performed in a vibrating sample magnetometer in fields up to 6.7T and for temperatures $1.8K \leq T \leq T_c$. After zero field cooling to the measuring temperature, hysteresis cycles were recorded with a field sweep of 280mT/min. At certain fields, which were high enough to ensure full flux penetration into the sample, the field sweep was interrupted and the decay of the magnetization monitored for times up to 30min. The measurements were done for both field directions, $B_a \| c$ and $B_a \| (a,b)$.

3. Results

For both directions a broadening of the hysteresis loops is observed after irradiation. Although for low fields the width of the loops is smaller for $B_a \| (a,b)$ than for $B_a \| c$ the hysteretic behaviour extends to higher fields for $B_a \| (a,b)$ which indicates that the irreversibility lines, defined by the disappearance of the hysteresis, are different for the two directions. Whereas after the first irradiation step a shift of the irreversibility line to higher temperatures was measured only at low fields, a much more pronounced shift appears after the second irradiation step and is observable for all field values up to 8T[4].

Within the chosen time interval and the accuracy of the measurements the recorded decay of the magnetization is logarithmic irrespective of the irradiation of the crystal (Fig.1). The creep rate $dM/d\ln t$ versus temperature exhibits a maximum (T_{max}) which is shifted to lower values for higher external fields. Compared with the results

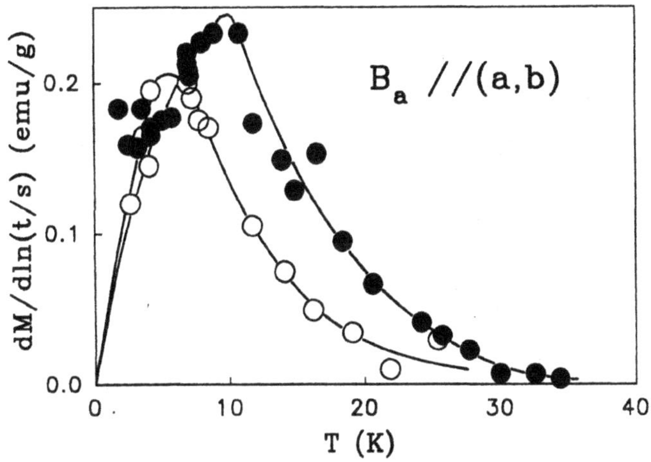

Fig. 2: Temperature dependence of the creep rate at 1T for fluences of $5*10^{20}m^{-2}$ (o) and $2.5*10^{21}m^{-2}$ (●).

obtained after the first irradiation step, T_{max} shifts to higher temperatures after the second irradiation step (Fig.2,3). Whereas at comparable temperatures the relaxation rate *increases* with increasing dose, the normalized relaxation rate *decreases* as one would expect for a material, into which additional defects have been introduced (Fig.1). After the second irradiation step a change in the relaxation behaviour is observed at low temperatures. Contrary to the results obtained after the first irradiation step the relaxation rate remains still high at the lowest available temperature (1.8K), which might point to nonvanishing flux creep at very low temperatures. However, it is quite unclear how defects which are introduced by irradiation could result in such a behaviour and measurements on other crystals are necessary to confirm this result. For the following analyses this contribution was therefore neglected and a relaxation behaviour according to thermally activated flux creep assumed.

The simplest form of a flux creep model is the single barrier height model, which leads to

$$M(t,T) = M(t=0,T)\cdot[1 - \frac{kT}{\langle E(T)\rangle} \ln(1 + \frac{t}{\tau})], \qquad (1)$$

where <E(T)> represents a mean effective activation energy. Using this model no influence of the irradiation is detectable (insets in Figs.4,5). The values of <E(T)> are slightly larger for B∥c, but the influence of the irradiation is smeared out in both directions. It should be noted, that the increase of <E> with temperature represents an "unphysical" result.

In a more refined model a distribution of pinning energies is included[1]. According to this model the time dependent magnetization is given by

$$M(t,T) = Mo\, f(T) \int_{E_o(t,T)}^{\infty} m(E)\, [1 - \frac{kT}{Eb(T)}\, ln(1 + \frac{t}{\tau})]\, dE \qquad (2)$$

with $Eo(t,T)=kT\,ln(1+t/\tau)/b(T)$ and $Mo=M(0,0)$.

The distribution of activation energies m(E), normalized by $\int_0^{\infty} m(E)dE=1$, can be calculated from

$$m(Eo) = \frac{1}{Mok(4/b(T)-3)}[\frac{d}{dT}(\frac{dM}{d\,lnt}) - \frac{1}{T}(\frac{dM}{d\,lnt})]. \qquad (3)$$

The temperature factor $f(T)$ was approximated by $1/b(T)$ with $b(T)=1-(T/Tc)^4$ as found in Ref. 1. For the analyses it was assumed that the relaxation time τ is independent of temperature and can thus be estimated from the measured data by

$$ln(\frac{t}{\tau}) = T\cdot\frac{dM}{dT}(\frac{dM}{d\,lnt})^{-1} \quad \text{in the limit } T \to 0. \qquad (4)$$

Since $M(t=0,T)$ cannot be determined directly in the experiment, $M(t_b,T)$ must be introduced, t_b being the time at which the analysis is started. For experimental reasons 60s was chosen for t_b. Mo can be obtained from the extrapolation of $M(t_b,T)$ to T=0 since the creep rate vanishes at T=0.

Because of the large change of $M(t_b,T)$ at low temperatures and the temperature limit of 1.8K in the present experiments the estimated value of τ is uncertain. However, it can be checked by the condition

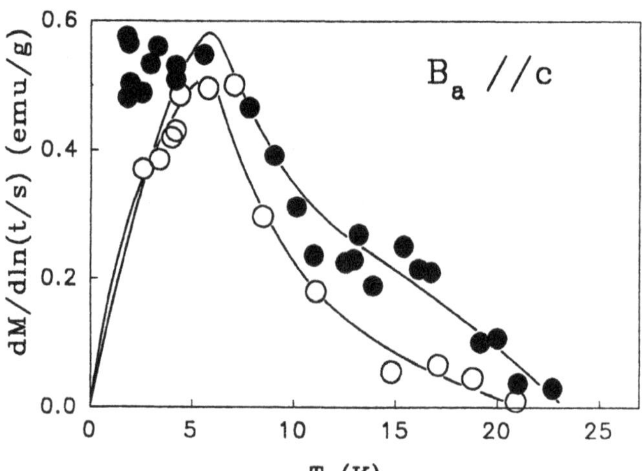

Fig. 3: Temperature dependence of the creep rate at 1T for fluences of $5*10^{20}m^{-2}$ (O) and $2.5*10^{21}m^{-2}$ (●).

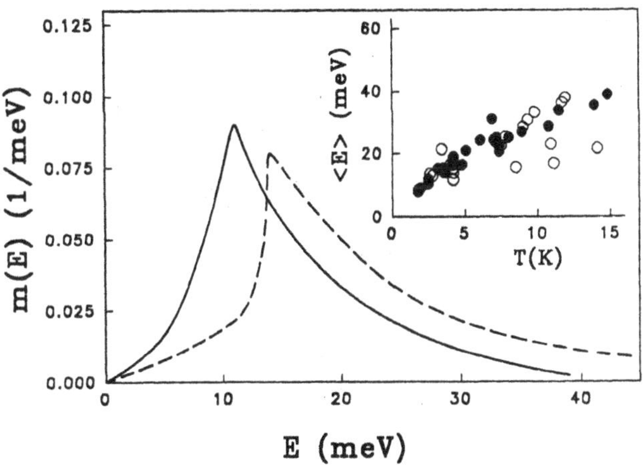

Fig. 4: Distribution of activation energies and temperature dependence of the mean effective activation energy for Ba‖(a,b) at Ba=1T for fluences of $5*10^{20}m^{-2}$ (———, O) and $2.5*10^{21}m^{-2}$ (– – – –, ●).

that m(E) is normalized to 1. For the measurements after the first irradiation step an estimation of the area under the distribution function was possible and leads to $\tau \sim 10^{-8}$s for Ba‖c and to $\tau \sim 10^{-6}$s for Ba‖(a,b)[5]. The same values were used for the analyses after the second irradiation step since no changes with increasing dose were detectable.

The resulting distribution of activation energies exhibits a smaller half width for Ba‖c (Fig.4,5). The influence of the irradiation in this direction is weak compared to the one for Ba‖(a,b) where a shift of the maximum and an increase of the high energy tail of the distribution appear.

4. Discussion

The observed anisotropic shift of the activation energy distribution as well as the difference in the influence due to irradiation for both directions are in good agreement with the results obtained for the irreversibility lines[3]. The absolute values of the energies are uncertain because of the strong simplification which enters into the analysis by the assumption of a temperature independent relaxation time τ, which is given by $\tau=2\mu_0 kT/\omega_0 VB_a^2$, where ω_0 is the hopping frequency and V the flux line bundle volume, which depends on the physical properties of the flux line lattice. Furthermore the model takes only the mean distance between the pinning barriers into account. Therefore misleading results can occur, if largely different defect structures exist in one sample. However, the distribution function is rather insensitive to changes of these

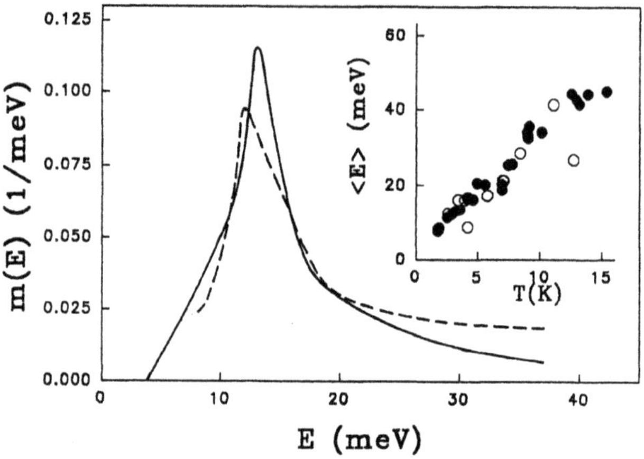

Fig. 5: Distribution of activation energies and temperature dependence of the mean effective activation energy for B∥c at Ba=1T for fluences of $5*10^{20}m^{-2}$ (———, ○) and $2.5*10^{21}m^{-2}$ (– – – –, ●).

factors because they enter only logarithmically via τ into the analysis. A spread in τ values over six decades leads only to uncertainties in E of 10 to 15%. In the special case of the investigated single crystal of BiSCCO-2212 the temperature dependence of these factors should be of no importance, because the measuring field is far below B_{c2} and the highest measuring temperatures are ~$0.25T_c$, i.e. the coherence length can be taken as nearly constant. From investigations on neutron irradiated YBCO single crystals (e.g. Ref. 6) two types of pinning centers, in addition to the anisotropic intrinsic pinning, are proposed to exist: firstly weak pinning centers, which are characteristic of the pre-irradiation defect structure (e.g. oxygen deficiencies) and which might be reduced by irradiation, and secondly large defect cascades, introduced by the irradiation with neutrons, which should act as strong pinning centers. The same behaviour is expected to appear in BiSCCO single crystals. This is in qualitative agreement with the distribution of activation energies obtained from the present analysis. The influence of intrinsic pinning is clearly seen by the observed anisotropy. Whereas the low energy part of the distribution, which corresponds to the weak pinning sites, is reduced (Fig.4), the number of pinning centers with high energies increases after irradiation. Electron microscopy investigations on YBCO show that the strong pinning centers are relatively rare (defect density $10^{22}m^{-3}$ for a neutron fluence of $2\cdot10^{21}m^{-2}$ in YBCO[6]). Therefore, no second peak but only an overall increase of the probability on the high energy side of the distribution function appear.

To sum up, measurements of the time dependence of the magnetization show clear evidence for enhanced flux pinning in a BiSCCO-2212 single crystal due to neutron irradiation. The observed anisotropies and shifts of the distribution of activation energies are consistent with a model of weak pre-irradiation defects and irradiation-induced strong pinning centers as well as strong intrinsic pinning.

5. Acknowledgement

Work supported in part by the Austrian Science Foundation (projects P8174 and P7970).

6. References

1. C.W. Hagen and R. Griessen, *Studies of High Temperature Superconductors,* Vol.3, ed. A.V. Narlikar (Nova Science Publ., New York, 1989) p.159.
2. M. Reissner and W. Steiner, *Supercond. Sci. and Technol.* in press.
3. W. Kritscha, F.M. Sauerzopf, H.W. Weber, G.W. Crabtree, Y.C. Chang and P.Z. Jiang, *Physica C* **179** (1991) 59.
4. W. Kritscha, F.M. Sauerzopf, H.W. Weber, G.W. Crabtree, Y.C. Chang and P.Z. Jiang, *Supercond. Sci. and Technol.*, in press.
5. M. Reissner, W. Steiner, W. Kritscha, F.M. Sauerzopf, H.W. Weber and G.W. Crabtree, *Physica C,* in press.
6. H.W. Weber, *this volume.*

NEUTRON IRRADIATION EFFECT ON MAGNETIC FLUX DISTRIBUTION IN Bi-Sr-Ca-Cu-O CRYSTALS

R. Szymczak, M. Baran, K. Piotrowski and H. Szymczak

Institute of Physics, Polish Academy of Sciences
Al. Lotników 32, PL-02-668 Warszawa, Poland

Abstract

Influence of fast neutron irradiation on the first critical field and the critical current as well as on the magnetic flux distribution in the mixed state of $Bi_2Sr_2CaCu_2O_x$ superconducting single crystals was studied by the magnetooptical method. In this method, the bismuth-doped iron garnet film was used as a magnetic flux sensor. The investigations of magnetic flux distribution in the mixed state, for zero field cooling and field cooling, were carried out in external magnetic fields parallel to the c-axis, and at temperatures from 4.2 K to 40 K.

In addition, the hysteresis curves were measured on the same samples and the obtained results were compared with those derived from the magnetooptical method.

1. Introduction

High temperature superconductors (HTSC) are characterized by high critical temperature T_c and simultaneously by rather low pinning forces and consequently - low value of the critical current density, J_c. One of the simplest and most effective methods to create additional pinning centers over the whole volume of the sample and to improve this way J_c is the neutron irradiation (for details and references see Wiśniewski et al., in this volume). From the experimental point of view, the easiest way to study the effect of neutron irradiation on HTSC are the magnetic and transport measurements. No one has examined this effect using recently developed[1,2,3] very effective magnetooptical method. In the version of this method used in our work some magnetic, uniaxial-garnet films were applied as magnetic flux sensors. The domain structure (usually stripe-like domains magnetized perpendicularly to the surface of the film) in a low-coercitivity garnet film strongly depends on the magnetic field acting on it. Therefore, the domain width in garnet film, placed directly on the superconducting sample, allows to estimate a local magnetic flux density in an investigated superconductor. Moreover, from the data provided by the magnetooptical method, it is possible to get the information on such fundamental parameters as the critical current J_c and the lower critical field H_{c1}.

In this paper magnetooptical method is used to determine the effect of fast neutron irradiation on the H_{c1} and J_c as well as on the magnetic flux distribution in the mixed state of $Bi_2Sr_2CaCu_2O_x$ superconducting single crystals. In addition, the d.c. magnetization measurements were performed on the same samples to compare the results obtained from both methods.

2. Experimental details.

The examined $Bi_2Sr_2CaCu_2O_x$ single crystals were obtained by float zone melting method[4]. The samples had the form of thin platelets of dimensions $1\times3\times0.15$ mm^3 (sample 1) and $2.6\times5.4\times0.9$ mm^3 (sample 2). The c-axis was perpendicular in both cases to the sample surface.

The visual observations of magnetic flux profile for zero field cooled (ZFC) samples and for samples cooled in magnetic field (FC) as well as d.c. magnetization measurements have been performed in magnetic fields parallel to the c-axis, at temperature range from 4.2 to 40 K, before and after fast neutron irradiation. As magnetic flux sensors, in the magnetooptical method, uniaxial bismuth-iron-garnet films, of low coercitivity, with measuring ranges up to 450 G were used. The experimental set-up and measurement technique were based on the magnetooptical system used previously to study the magnetic domain structure near phase transitions (see Ref.5 for details and references). The d.c. magnetization measurements were done by means of vibrating sample magnetometer (PAR 155).

The samples were irradiated by fast neutrons (of 1 MeV average energy) with fluence of $1.5\cdot10^{17}$ n/cm^2.

3. Results and discussion

Some series of experiments were made for both investigated samples. The samples were examined for some temperatures and for various values of magnetic field both in zero field and field cooling regimes.

First of all, the examined sample was cooled to chosen temperature in zero magnetic field. Next, increasing magnetic field it was possible to observe changes of the magnetic domain width beyond the area of the sample, whereas the magnetic domain structure in the area of the sample was unchanged. It is caused by the shielding effect of superconducting sample. From the domain structure observation, in the vicinity of superconducting sample, one can determine the real value of the magnetic flux, which is strongly concentrated near the sample edge. The value of magnetic flux near the sample edge, corresponding to first observable changes in the magnetic domain structure along the sample edge, was taken as the value of the lower critical field H_{c1}. It seems, that magnetooptical method is more correct method for determination of H_{c1}, than the evaluation of the beginning of nonlinearity in M(H) from the d.c. magnetization measurement. After exceeding the H_{c1} the magnetic flux starts to penetrate into the sample (see Fig.1). It should

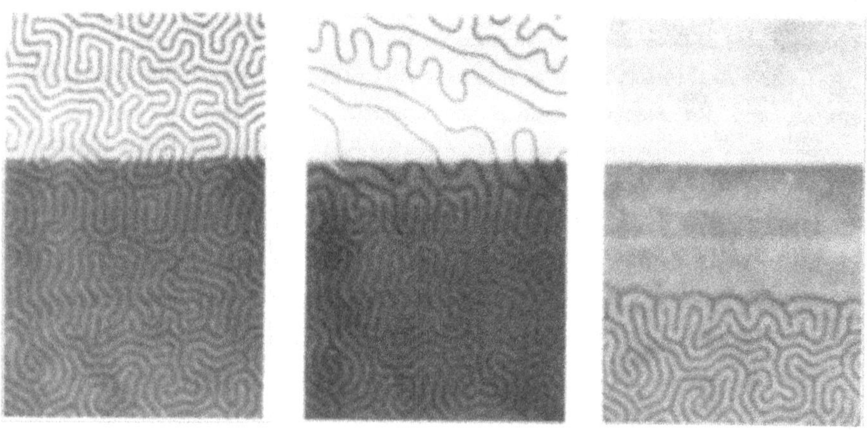

a b c

Figure 1.
Visualization of magnetic flux penetration into sample 1 cooled to 5.7 K in zero magnetic field. External magnetic field: a - 0 Oe, b - 185 Oe, c - 430 Oe.

be mentioned, that the front of the penetration is always parallel to the external shape of the sample. It reflects the homogeneous distribution of the pinning centers in the investigated samples. Analysis of the magnetic flux penetration patterns, similar to that given in Fig.1, gives possibility to estimate the flux profile in the superconducting sample. Examples of the flux profile determined for two values of external magnetic field (195 Oe and 300 Oe) after zero field cooling to 30 K are given in Figs.2a-2b. For comparison, on the same figures the flux profiles are presented for the neutron irradiated samples.

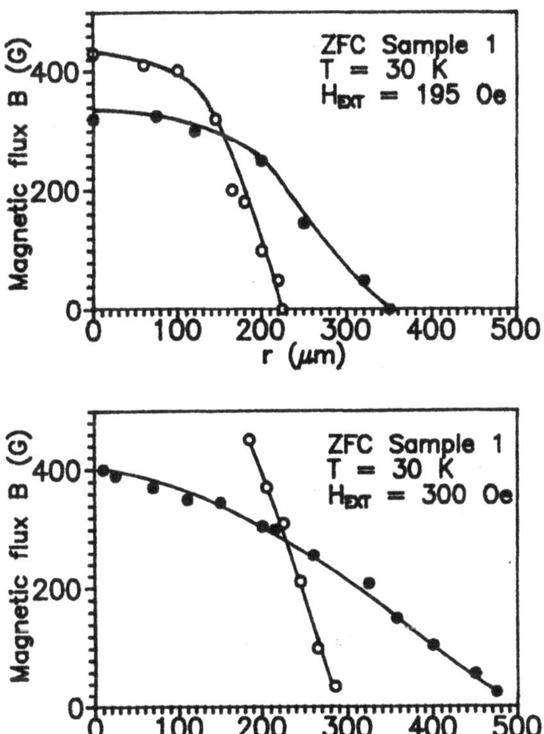

Figure 2.
Magnetic flux profile from the sample edge $r = 0$ to the sample center $r = 500\ \mu m$ (● - before and ○ - after irradiation).

a - $H_{EXT} = 195$ Oe
b - $H_{EXT} = 300$ Oe

It should be stressed that in both cases the flux distribution was determined in the same part of the sample. The similar experimental data are presented in another way in Figs.3a-3b, where the penetration distances versus applied magnetic field are given for fixed value of measured local field before and after neutron irradiation.

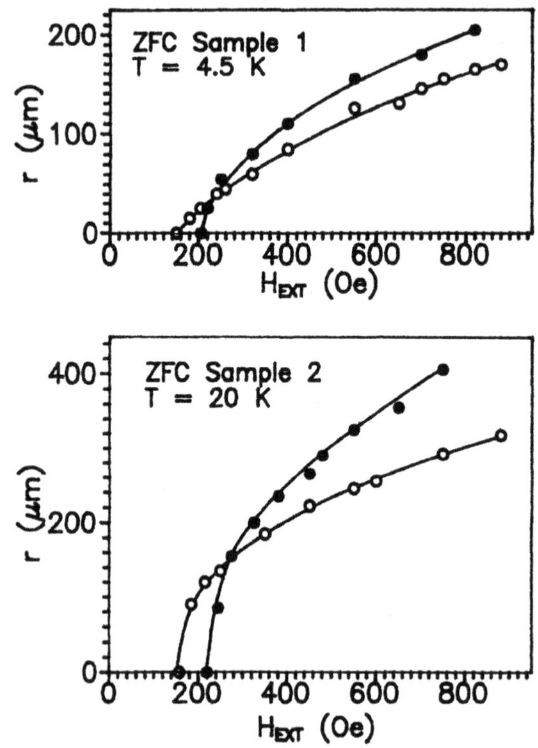

Figure 3.
Distance of penetration r by the magnetic flux B = 450 G versus external magnetic field (● - before and ○ - after neutron irradiation).

a - ZFC sample 1, T = 4.5 K
b - ZFC sample 2, T = 20 K

Experimental data presented in Figs.2-3 indicate that:
- neutron irradiation is responsible for a decrease (of about 20%) of the lower critical field H_{c1}.
- the slope of the flux profiles considerably increases after neutron irradiation.

The observed flux profiles allow to calculate the magnetic flux gradient and consequently - the critical current of each region of the sample. To determine the critical current from the flux gradient the Maxwell equation (in cgs units)

$$\left(\frac{4\pi}{c}\right) \boldsymbol{J} = \nabla \times \boldsymbol{H} \qquad (1)$$

is applied.

Assuming that cylindrical symmetry is appropriate to describe the geometry of the experiment and that the current flows in the plane of the sample ($J_r=0$ and $J_z=0$) one can calculate

$$J=J_\varphi=\frac{c}{4\pi}\left(\frac{\partial H_r}{\partial z}-\frac{\partial H_z}{\partial r}\right) \tag{2}$$

In this equation, the first term on the right, which can be referred to curvature term, for thick samples is usually neglected (see Ref.6). The most important, in our case, is the second term directly related to the experimentally observed magnetic flux profiles. It means, that the irradiation induced enhancement of the critical current can be described by the parameter p related to the flux gradient $\partial H_z/\partial r$:

$$p=\frac{J_c^{irr}}{J_c^{un}}=\frac{\left(\frac{\partial H_z}{\partial r}\right)^{irr}}{\left(\frac{\partial H_z}{\partial r}\right)^{un}} \tag{3}$$

The enhancement factor p has been also determined from d.c. magnetization measurements. If the sample is fully penetrated by the magnetic flux, the width of the M-H hysteresis loop is related to the critical current by the relation

$$J_c\sim|M_+-M_-| \tag{4}$$

where M_+ and M_- correspond to the magnetization measured in increasing and decreasing magnetic field, respectively. The example of the effect of irradiation on M-H hysteresis loop is presented in Fig.4. One can see, that magnetic irreversibility $\Delta M = |M_+ - M_-|$ is significantly influenced by the irradiation.

The values of the enhancement factor p, determined using the magnetooptical method as well as d.c. magnetization measurements, are given in Tab.I. The presented results indicate that both methods give similar results. One can see that the enhancement factor is rather low at 4.2 K and becomes significant at higher temperatures.

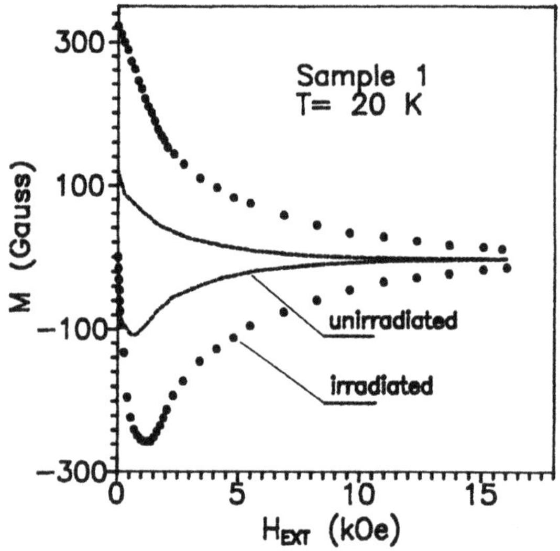

Figure 4.
The effect of the neutron irradiation on magnetic hysteresis loop.

Table I. The critical current enhancement p.

T [K]	D.C. magnetization		Magnetooptical
	H_{ext}=10kOe	H_{ext}=1kOe	H_{ext} = 200-900 Oe
4.2	1.3	1.1	1.2 - 1.4
10	3	1.7	1.6 - 1.8
20	10	3.2	2.2 - 2.4
30	-	-	4.5 - 6
40	-	7	-

The effect of neutron irradiation has been also observed in the field cooling regime. The flux profiles obtained for superconducting sample cooled down to 20 K in external field of about 800 Oe are presented in Fig.5. These profiles measured for different values of bias field indicate that a large amount of flux is trapped inside the sample, and that this effect becomes stronger after neutron irradiation. An interesting effect has been observed in remanent and near-remanent states. In these cases, because of the flux trapped in the center of the sample, the nonuniform magnetic state consisting of two Shubnikov phases (with opposite vortex polarity) and one Meissner phase is created. Similar effect was observed previously[7] in unirradiated samples, but in the case of irradiated samples the effect is significantly stronger.

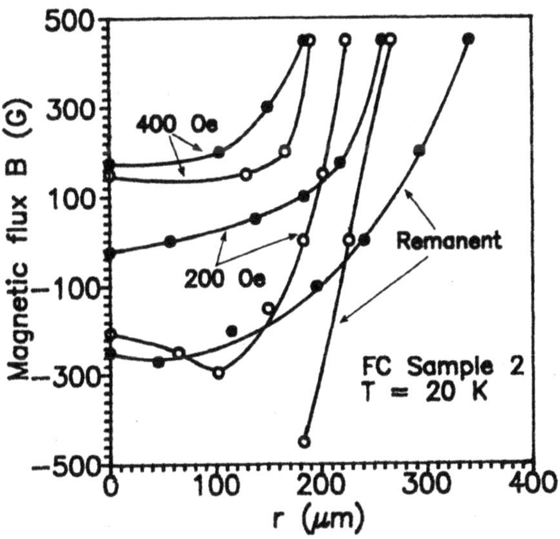

Figure 5.
Magnetic flux profiles B(r) for sample 2 cooled down to 20 K in external magnetic field 800 Oe (● - before and ○ - after irradiation).

The effect of neutron irradiation is not limited only to creating additional pinning centers. Another interesting effect, reported previously for ceramic samples[8], and observed also in the present work, is connected with a decrease of H_{c1} due to neutron irradiation. This effect is particularly surprising, since no considerable decreasing of T_c is observed in the investigated samples after irradiation. It is suggested to ascribe this effect to the irradiation suppressing of Bean-Livingston barrier[9] existing on the surface of investigated samples. Similar effect, but due to electron irradiation, was observed recently by Konczykowski[10].

4. Conclusions

Visual observations of the magnetic flux profiles have been performed for $Bi_2Sr_2CaCu_2O_x$ superconducting single crystals before and after fast neutron irradiation. It has been shown that neutron irradiation creates additional pinning centers and partly destroys the Bean-Livingston surface barrier. Additional d.c. magnetization measurements confirm the results of magnetooptical observations.

5. References

1. M.V.Indenbom, N.N.Kolesnikov, M.P.Kulakov, I.G.Naumenko, V.I.Nikitenko, A.A.Polyanskii, N.F.Vershinin and V.K.Vlasko-Vlasov, *Physica C* **166** (1990) 486.
2. A.Forkl, H.U.Habermejer, B.Leibold, T.Dragon and H.Kronmüller, *Physica C* **180** (1991) 155.
3. S.Gotoh and N.Koshizuka, *Physica C* **176** (1991) 300.
4. A.M.Balbashov, S.G.Karabashev and A.S.Nygmatulin, in *Proc. European Conf. on High-T_c Thin Films and Single Crystals*, eds. W.Gorzkowski et al. (World Scientific, Singapore, 1990) p.10.
5. R.Szymczak, *J.Magn.& Magn.Mat.* **35** (1983) 243.
6. L.T.Baczewski, K.Piotrowski, R.Szymczak, H.Szymczak and A.P. Malozemoff, *Physica C* **175** (1991) 363.
7. R. Szymczak, K. Piotrowski, H. Szymczak and A.M. Balbashov, *J.Magn.& Magn.Mat.* **92** (1990) L19.
8. A. Wiśniewski, M. Baran, P. Przysłupski, H. Szymczak, A. Pajączkowska, L. Dobrzyński, J. Waliszewski, M. Biernacka, K. Pytel, B. Pytel, in *High T_c Superconductors*, ed. H.W. Weber (Plenum, New York, 1988) p. 279.
9. C.P.Bean and J.D.Livingston, *Phys. Rev. Lett.* **12** (1964) 14.
10. M.Konczykowski, L.Burlachkov, Y.Yershurun and F.Holtzberg, *Phys. Rev.* **B43** (1991) 13707.

THERMOELECTRIC STUDY OF NEUTRON IRRADIATED $YBa_2Cu_3O_y$ SUPERCONDUCTORS

M. Pękała, A. Polaczek,
Department of Chemistry, University of Warsaw
Al. Zwirki i Wigury 101, 02-089 Warsaw, Poland

A. Pajączkowska,
Institute of Physics, Polish Academy of Sciences, Warsaw

B. Pytel and K. Pytel
Institute of Atomic Energy, Swierk, Poland

ABSTRACT

Thermoelectric studies of neutron irradiated high temperature superconductors $YBa_2Cu_3O_y$ reveal a gradual reduction of transition temperature and an increase of thermoelectric power, which are caused by changes in the defect structure during irradiation.

Introduction

Neutron irradiation, which redistributes the primary structure and introduces the new defects, is known to affect many properties of the high temperature superconductors (HTSC) such as the critical currents, the transition temperature, T_c, the electrical resistivity and other transport parameters[1]. It is of importance to note, that neutron irradiation leads to a uniform increase of defects in the sample. The present paper was aimed at a study of variations of thermoelectric power (TEP) and thermal conductivity under irradiation.

Experimental

The polycrystalline samples studied were prepared by standard methods. They were irradiated to fluences of 0.18×10^{18}, 0.61×10^{18}, 2.7×10^{18} and $3.3 \times 10^{18} cm^{-2}$ ($E > 1$ MeV). The influence of thermal neutrons was eliminated by means of a cadmium screen. The fluences were determined to an accuracy of 10 %. The thermoelectric power was measured by means of the differential method between 70 and 300 K. A temperature gradient of 0.5 to 3 K was generated by small heaters (2 to 40 mW) attached to the sample. The temperature was monitored by copper - constantan thermocouples. Copper wires were used to measure the voltage along the sample. The thermoelectric power of these copper leads was subtracted. The thermal conductivity was calculated from the power supplied to the sample and its dimensions (approximately 5 mm length and 3 mm diameter).

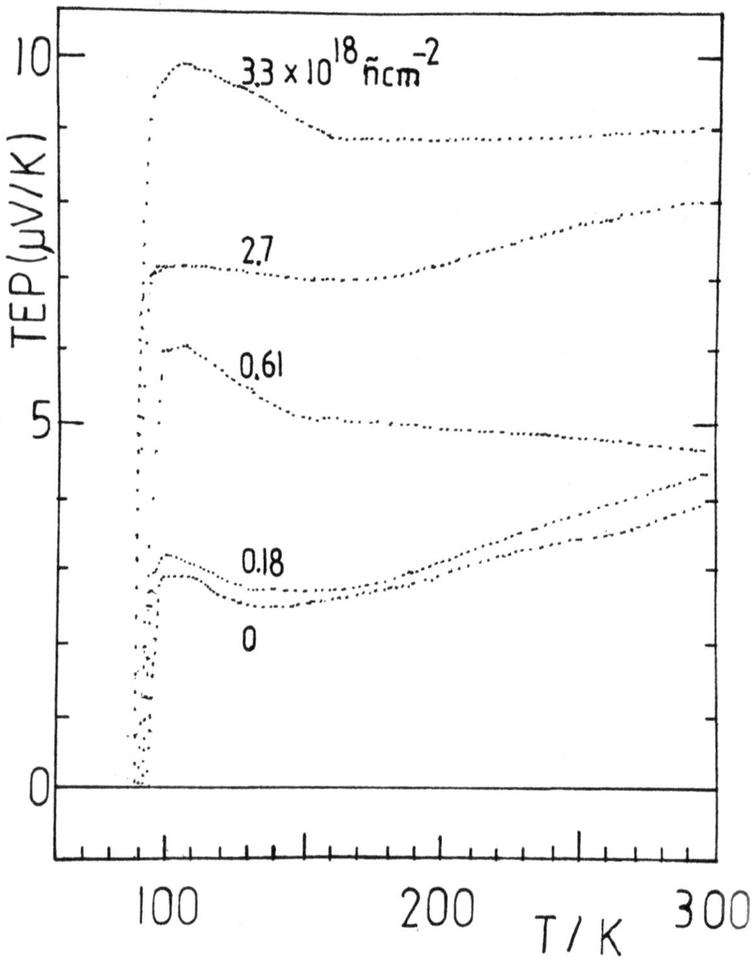

Fig. 1. Thermoelectric power of polycrystalline $YBa_2Cu_3O_y$ superconductors exposed to various neutron fluences.

Results

The measured thermoelectric power is plotted in Fig. 1 for the as prepared and the neutron irradiated samples. The temperature variation of thermoelectric power for both classes of samples is characteristic for high temperature superconductors[2]. The transition between the normal and the superconducting states is revealed by a disappearance of TEP in the latter one. In the normal state a broad maximum observed just above the transition temperature is followed by a weak temperature dependence of TEP at higher temperatures. For the unirradiated samples the thermoelectric power at the maximum just above T_c is equal to about 3 μV/K and increases to approximately 4 μV/K at room temperature. The thermoelectric power versus temperature curves of the neutron irradiated samples shift gradually towards higher values, but do not exceed 10 μV/K for the highest fluences (Fig. 1). This is accompanied by a shift in the transition temperature from 92 K for unirradiated to 87 K for samples subjected to the highest fluence. The width of the transition interval changes from 4 to 7 K. The thermal conductivity of the samples varies in the normal state and increases monotonically with temperature up to approximately 5 W/Km at room temperature. For the highest fluence the magnitude of thermal conductivity is reduced by irradiation by roughly 25 % as compared with the unirradiated samples.

Discussion

A comparison of the small and positive values of TEP of the unirradiated samples with published data on the relation between TEP and the oxygen content[3-6] allows us to conclude that the oxygen content in the samples studied is approximately 6.88 to 6.90. The TEP maximum appearing just above T_c is broad and its shape and magnitude are not strongly affected by the irradiation. There is a long lasting debate whether its origin is phonon drag[2,7] or fluctuation[8,9]. The transition to the superconducting state occuring for the unirradiated samples in a narrow temperature interval at 92 K confirms that samples are single phase and homogeneous. The influence of neutron irradiation on the transition temperature is moderate as may be seen from the rate of superconductivity degradation (1.5 K / 10^{18} cm^{-2}). This value is even smaller than that of 2.1 K / 10^{18} cm^{-2} reported by Uher and Huang[10] for the $YBa_2Cu_3O_y$ system with y very close to 7, which exhibits remarkably negative TEP values in the normal state. It is generally accepted that a suppression of the transition temperature should be ascribed to oxygen displacements and/or even some oxygen removal[1].

At the present state, where a consistent theory of thermoelectric effects in HTSC is still lacking, it is a difficult task to correlate precisely the shifts both in the magnitude of TEP and in T_c with microscopic phenomena. There are numerous factors which should be taken into account when analysing the influence of neutron irradiation onto HTSC materials. They include type, quality and preirradiation microstructure of HTSC on one hand and the conditions of the irradiation, such as the temperature of HTSC during irradiation and the energy spectrum of the neutron beam, on the other hand.

In polycrystalline HTSC the microstructure of weak links influences both the transport and the critical magnetic currents. Therefore, a modification of the weak links involves changes in both of these currents. For the present samples a considerable increase in critical currents was observed by Wiśniewski et al.[11] in agreement with other groups, who report irradiation induced increases of intragrain critical currents[12]. On the other hand, the transport properties are controlled predominantly by the intergrain phenomena. Thus, a weakening of the weak links has to lead to an enhancement of the electrical resistivity and the thermoelectric power in the normal state, as is seen in the system studied and in neutron irradiated BiPbSrCaCuO[13]. In order to evaluate the extent, to which the atomic structure is varied by neutron irradiation, one should also consider changes at the transition to the superconducting state. Since the transition temperature shifts under irradiation only by a few degrees and the width of the transition rises only slightly, one may conclude that the atomic structure is modified only in a very limited volume fraction of the samples. This conclusion agrees with the estimated number of displacements per atom[1], which is of the order of 10^{-3} for fluences of 10^{18} cm^{-2}.

The observed reduction in thermal conductivity is also consistent with the picture of the defects produced by neutron irradiation. However, it is impossible at present to confirm the presence of specific types of these defects, such as large cascades and some point defects[1,14] and their spatial and size distribution. However, taking into account that the heat transport in HTSC is dominated by phonons[15], one may infer that the suppression of the phonon mean free path is due to additional phonon scattering by the defects produced by neutron irradiation.

Acknowledgement

Work supported by G-MEN-183/90.

References

1. H.W. Weber and G.W. Crabtree, in *Studies of High Temperature Superconductors,* ed. A.V. Narlikar (Nova Science Publishers, New York, 1991), vol. 9, in press
2. A.B. Kaiser and C. Uher, in *Studies of High Temperature Superconductors,* ed. A.V. Narlikar (Nova Science Publishers, New York, 1990), vol. 7
3. P.J. Ouseph and M. Ray O'Bryan, *Phys. Rev.* **B41** (1990) 4123
4. H. Ishi et al., Physica **148B** (1987) 419
5. V.F. Kraidenov et al., *Fizika Nizkih Temperatur* (USSR) **16** (1990) 1076
6. Yu.M. Baikov et al., *Superconductivity: Physics, Chemistry, Technique* (USSR) **3** (1990) 254
7. J.L. Cohn et al., *Phys. Rev. Lett. 66* (1991) 1098
8. M.A. Howson et al., *Phys. Rev.* **B41** (1990) 300
9. A.J. Lowe, S. Regan and M.A. Howson, *Phys. Rev. Lett.* (preprint)
10. C. Uher and W.-N. Huang, *Phys. Rev.* **B40** (1989) 2694
11. A. Wiśniewski et al., *Solid State Commun.* **65** (1988) 577
12. F.M. Sauerzopf et al., *Phys. Rev.* **B43** (1991) 3091
13. A. Wiśniewski et al., *Physica* **C170** (1990) 333
14. M.A. Kirk et al., *Phil. Mag. Lett.* **62** (1990) 41
15. C. Uher, *J. Of Superconductivity* **3** (1990) 337

INCREASE OF PINNING IN Pb-Sr-(Y,Ca)-Cu-O SINGLE CRYSTAL BY NEUTRON IRRADIATION

S. Koleśnik, T. Skośkiewicz, J. Igalson,
Institute of Physics, Polish Academy of Sciences,
Al. Lotników 32/46, 02-668 Warszawa, Poland

K. Pytel, and B. Pytel
Institute of Atomic Energy, Świerk, Poland

ABSTRACT

We report on influence of neutron irradiation on the magnetic properties of Pb-Sr-(Y,Ca)-Cu-O single crystal. The results of the magnetization measurements show no significant change of the critical temperature T_c and the lower critical field (H_{c1}) in irradiated sample. Critical current density (J_c) and relaxation rate ($dM/d\ln t$) of the remanent magnetization of the crystal increases after irradiation. Our observations reflect the presence of stronger pinning in irradiated crystal.

Presence of magnetic flux pinning in high–T_c superconductors (HTSC) especially manifests itself in the irreversibility of magnetic properties. Magnetic irreversibility is observed in ceramic HTSC samples as well as in high–quality single crystals.[1–3] A short coherence length suggests possibility of an existence of an intrinsic pinning.[4] One of the useful methods to study the effects connected with the pinning is an introduction of new defects into the sample by fast–neutron irradiation.[5]

In this paper we report on the influence of neutron irradiation on the magnetic properties of Pb-Sr-(Y,Ca)-Cu-O single crystal. Preparation and characterization of this compound have been described elsewhere.[6] Transition temperature T_c of Pb-Sr-(Y,Ca)-Cu-O single crystal is 78 K.

Magnetization measurements were performed using superconducting quantum interference device (SQUID) magnetometer. Pb-Sr-(Y,Ca)-Cu-O single crystal was irradiated by the fast neutrons to a fluence of $1.1 \cdot 10^{17}$ n/cm^2. We observed no change of T_c after irradiation. The magnetization curves were measured in magnetic field parallel to a-b planes at some chosen temperatures before and after irradiation. The magnetization curves at $T = 60$ K are presented in Fig. 1. We observe weaker deviation of the magnetization from a linearity in $H > H_{c1}$ after irradiation. This reflects more difficult penetration of the magnetic flux into the crystal and, hence, stronger pinning after irradiation. While increasing magnetic field, magnetization

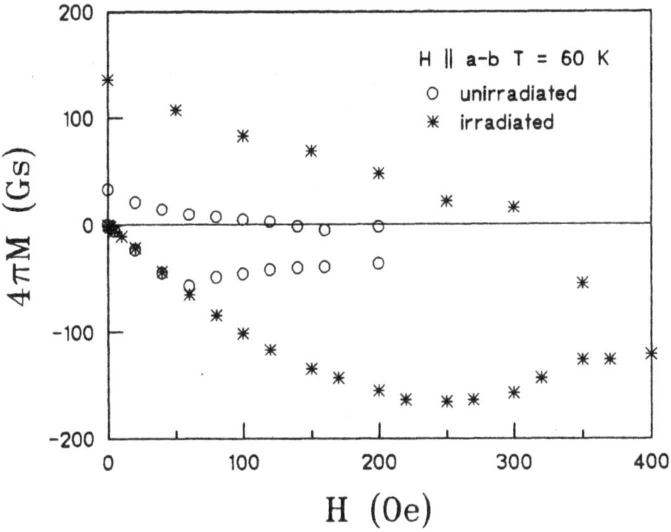

Fig. 1. Magnetization curves for Pb-Sr-(Y,Ca)-Cu-O single crystal measured at $T = 60$ K for **H** $\|$ a-b planes before and after fast–neutron irradiation.

passes through a shallow minimum and becomes almost independent on the magnetic field. This high–field value of magnetization is considered to be proportional to a critical current density J_c. According to the Bean model[7]

$$J_c = 20 \cdot M/d \tag{1}$$

where d is the thickness of the slab. Using this relation we calculate J_c for **H** $\|$ a-b. The results are given in following Table:

TABLE I. Critical current density of Pb-Sr-(Y,Ca)-Cu-O single crystal for **H** $\|$ a-b planes before and after irradiation.

Before irradiation		After irradiation	
T (K)	J_c (10^3 A/cm^2)	T (K)	J_c (10^3 A/cm^2)
60	3.4	60	10.6
67	2.4	64	8.4
73	1.6	73	1.1

We observe an enhancement of the critical current density at temperature range 60-70 K, while at $T = 73$ K J_c decreases after irradiation. (note, that critical temperature did not change during irradiation process). So far the reason for decrease of

J_c at higher temperatures is not clear. Time dependence of the remanent magnetization was studied before and after irradiation for magnetic field $\mathbf{H} \parallel$ a-b. The crystal was cooled down below T_c in certain magnetic field (200 Oe) to desired temperature. The field was then switched off and the remanent magnetization was measured as a function of time. We observe logarithmic decay of the magnetization according the formula

$$M = M_0 - S \ln t \tag{2}$$

The fitted parameters M_0 and S are given in Table II.

TABLE II. The initial value M_0 and the relaxation rate S of the remanent magnetization for Pb-Sr-(Y,Ca)-Cu-O single crystal for $\mathbf{H} \parallel$ a-b planes before and after irradiation.

	Before irradiation		After irradiation	
T (K)	M_0 (emu/cm^3)	S (emu/cm^3)	M_0 (emu/cm^3)	S (emu/cm^3)
67	2.51	0.121	7.11	0.174
73	3.1	0.065	3.59	0.108

It can be seen, that the relaxation rate S after irradiation increases both at 67 K and 73 K.

We also investigated an influence of the neutron irradiation on the lower critical field H_{c1} of Pb-Sr-(Y,Ca)-Cu-O single crystal. We used a modified method of determination of H_{c1} described in Ref. 6. In that method we measured a magnetization of the sample several times in given magnetic field H_M (say 10 Oe). Between the measurements the crystal was exposed to an increasing magnetic field $H_E > H_M$. For $H_E > H_{c1}$ magnetic flux penetrates the sample, reducing an absolute value of the magnetization in H_M. The results of these measurements at two temperatures 40 K and 65 K for two field orientations $\mathbf{H} \parallel$ a-b and $\mathbf{H} \perp$ a-b are presented in Fig. 2. The deviation of the magnetization from the initial value M_I is proportional to the square of H_E as can be seen from Fig. 2. The results show, that the introduction of new defects into the sample increases pinning in this material. On the other hand, the irradiation does not change significantly the lower critical field.

In our experiment, the neutron irradiation does not change critical temperature and lower critical field. An increase of the relaxation rate, critical current density and weaker penetration of the magnetic flux reflect an enhancement of the pinning force in the irradiated crystal.

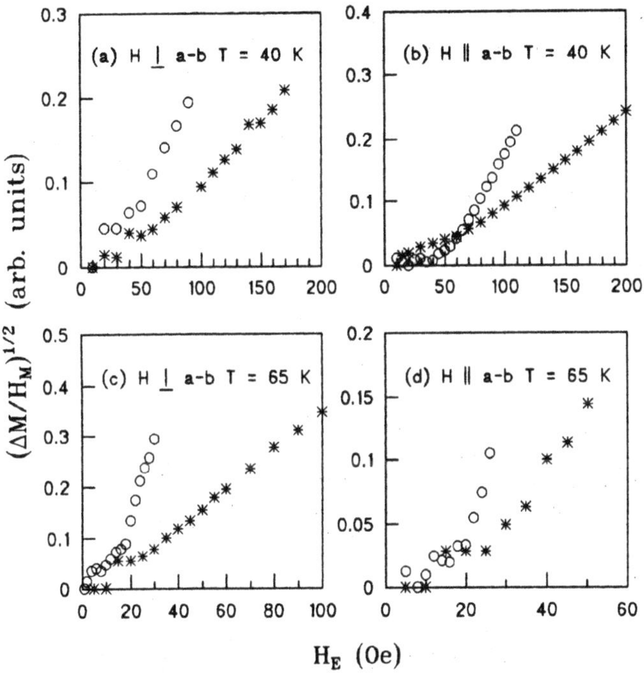

Fig. 2. Determination of the lower critical field of unirradiated (○) and irradiated (*) Pb-Sr-(Y,Ca)-Cu-O single crystal at $T = 40$ K and $T = 65$ K for two field orientations (see text). ΔM denotes the difference between the actual magnetization M and the initial magnetization M_I, both measured in H_M.

References

1. A.Wiśniewski, M. Baran, P. Przysłupski, H. Szymczak, A. Pajączkowska, B. Pytel, and K. Pytel, *Solid State Commun.* **65** (1988) 577.
2. T.R. Dinger, T.K. Worthington, W.J. Gallagher, and R.L. Sandstrom, *Phys. Rev. Lett.* **58** (1987) 2687.
3. J.P. Rice, D.M. Ginsberg, M.W. Rabin, K.G. Vandervoort, G.W. Crabtree, and H. Claus, *Phys. Rev.* **B41** (1990) 6532.
4. M. Tachiki and S. Takahashi, *Solid State Commun.* **70** (1989) 291.
5. H.W. Weber and G.W. Crabtree, to be published in *Studies of High Temperature Superconductors Vol. 9*, ed. A.V. Narlikar, (Nova Science Publishers, New York).
6. Z. Korczak, W. Korczak, S. Koleśnik, T. Skośkiewicz, and J. Igalson, *Supercond. Sci. Technol.* **3** (1990) 370.
7. C.P. Bean, *Phys. Rev. Lett.* **8** (1962) 250; *Rev. Mod. Phys.* **36** (1964) 31.

THERMOELECTRIC POWER AND THERMAL CONDUCTIVITY OF SUPERCONDUCTING SINGLE CRYSTALS $Bi_2Sr_2Ca_1Cu_2O_y$

M. Pękała, A. Polaczek,
Department of Chemistry, University of Warsaw
Al. Zwirki i Wigury 101, 02-089 Warsaw, Poland

A. Pajączkowska,
Institute of Physics, Polish Academy of Sciences, Warsaw

ABSTRACT

Studies of $Bi_2Sr_2Ca_1Cu_2O_y$ single crystals show that the layered structure leads to a moderate thermoelectric anisotropy and to a large anisotropy of the thermal conductivity parallel and perpendicular to the ab planes.

Introduction

The thermoelectric and thermal conductivity properties of bismuth based high temperature superconductors (HTSC) are less carefully investigated than the Y-Ba-Cu-O system. Moreover, the published data on thermoelectric power (TEP) scatter with regards to the sign, magnitude and temperature dependence[1]. The layered structure of the bismuth based HTSC leads to an anisotropy of all properties, including the transport parameters. These properties are also sensitive to the oxygen configuration. In view of these circumstances every opportunity to obtain new information on single crystals of this system should be explored.

Experimental

The investigated single crystals of the $Bi_2Sr_2Ca_1Cu_2O_y$ system had the approximate dimensions 1 x 3 x 4 mm. The thermoelectric power and the thermal conductivity were simultaneously measured using the experimental setup described elsewhere[2]. The absolute accuracy of the thermal conductivity was about 20 to 30 % due to the irregular shape of the samples.

Results and discussion

The thermoelectric results obtained in the present study coincide with the data typical for HTSC single crystals[1,3]. When passing the transition temperature T_c = 88 K, the TEP rises abruptly from zero in the superconducting state to a maximum just above T_c (Fig. 1). The transition, which occurs in a relatively narrow temperature interval of 3 to 7 K, proves that the samples contain only the Bi2212 phase.

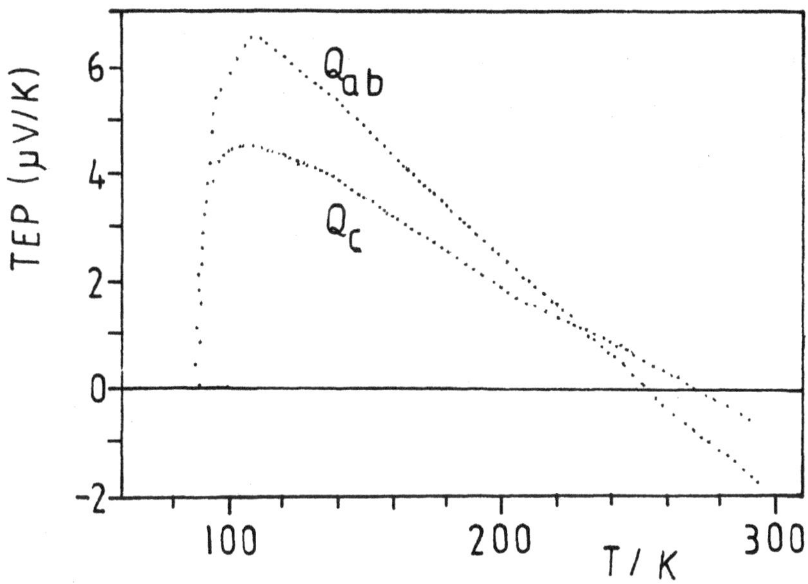

Fig. 1. Temperature dependence of thermoelectric power of $Bi_2Sr_2Ca_1Cu_2O_y$ single crystals for the (ab) and (c) orientations.

The anisotropy of TEP manifests itself by different magnitudes and temperature dependences of the TEP components, Q_{ab} and Q_c, in the ab plane and the c direction, respectively. Both the TEP components exhibit maxima at 110 - 115 K and have values of 6.5 and 4.5 μV/K for the ab and c orientations, respectively. At higher temperature they decrease proportional to temperature with different slopes of 0.045 and 0.025 $\mu V/K^2$, respectively. The behavior of Q_c observed here and for other Bi2212 single crystals[3-5] is in contrast to the roughly temperature independent or even positive slope of Q_c reported by Leon and Escudero[6] and by Crommie et al[7]. A comparison of the present thermoelectric and electrical resistivity data with those of Crommie et al.[7] for various oxygen concentrations suggests, that the Q_c behavior reflects the relatively high oxygen content of the crystals studied, which may be close to y = 8. The high oxygen content is additionally confirmed by the observed metallic like temperature dependence of the electrical resistivity. Unfortunately, no simple model of TEP in HTSC is capable to explain the anisotropic TEP in the crystals studied. Considering the proportionality betwen TEP and temperature, the relatively small values and the change in TEP sign below room temperature, one may presume, that a partially compensational mechanism with different scattering processes prevails.

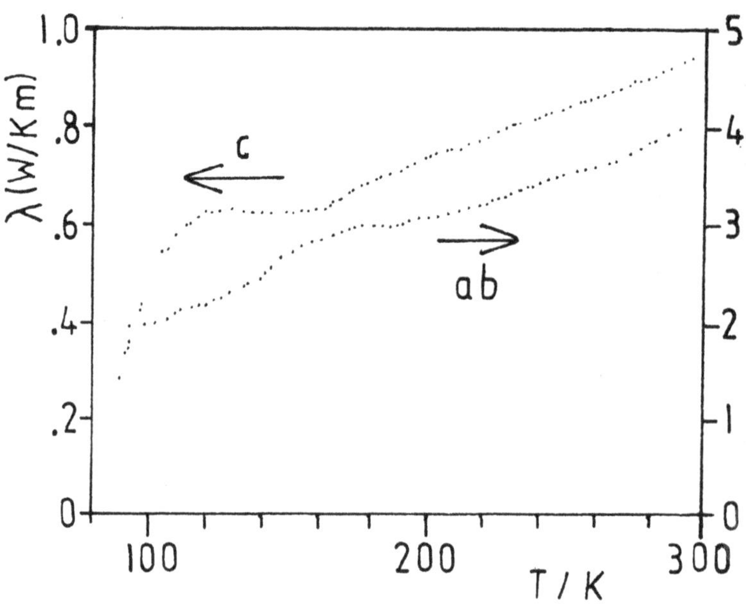

Fig. 2. Temperature dependence of thermal conductivity of $Bi_2Sr_2Ca_1Cu_2O_y$ single crystals for the (ab) and (c) orientations.

The anisotropy of the Bi based HTSC is much more pronounced in the temperature dependence of the thermal conductivity components, λ_{ab} and λ_c. As shown in Fig. 2, the ratio λ_{ab}/λ_c is about 5 between 100 and 300 K. Since the thermal conductivity consists of the phonon, λ^{ph}, and the electron, λ^e, contributions, one may estimate the electronic one by means of the Wiedemann - Franz relation

$$\lambda^e = L\ T\ /\ \rho \qquad (1)$$

where the Lorentz number is given by $L = 2.45 \times 10^{-8}\ V^2K^{-2}$. Such an estimation shows that the electronic thermal conductivity in the ab plane contributes about 6 to 10 % of the measured value which, is comparable with typical observations on HTSC[8]. The situation along the c direction is in a marked contrast to this result. Because of large anisotropy in the electrical resistivity ($\rho_c/\rho_{ab} \cong 1000$) the electronic thermal conductivity, λ^e_c, is negligibly small. Therefore, the low values of λ_c reveal strong scattering of phonons in the c direction. The similar observations of thermal conductivity anisotropies varying between 6 and 8 were reported by Crommie and Zettl[9].

References

1. A.B. Kaiser and C. Uher, in *Studies of High Temperature Superconductors*, ed. A.V. Narlikar (Nova Science Publishers, New York, 1990), vol. 7
2. M. Pękała et al., *this volume*
3. M. Pękała et al., *Solid State Commun.* **76** (1990) 419
4. I.M. Tsidilkovski et al., *Superconductivity: Physics, Chemistry, Technique* (USSR) **2** (1989) 45
5. L. Forro et al., *Solid State Commun.* **73** (1990) 501
6. L.M. Leon and R. Escudero, *Physica* **B165 & 166** (1990) 1211
7. M.F. Crommie et al., *Phys. Rev.* **B41** (1990) 2526
8. C. Uher, *J. Of Superconductivity* **3** (1990) 337
9. M.F. Crommie and A. Zettl, *Phys. Rev.* **B43** (1991) 408

HYSTERETIC TIME RELAXATION BEHAVIOR OF THE CRITICAL INTERGRAIN TRANSPORT CURRENT IN GRANULAR $YBa_2Cu_3O_7$

K. KWASNITZA and CH. WIDMER
Paul Scherrer Institute, CH-5232 Villigen, Switzerland

ABSTRACT

In sintered polycrystalline high T_c superconductors the intergrain transport I_c shows complex hysteretic time relaxation behavior. Depending on how the magnetic field was cycled before, the I_c-relaxation rate dj_c/dt can be positive or negative, can have different magnitudes, and, in certain cases, can be made to vanish. These experimental results can be explained at least qualitatively with the time relaxation behavior of the intragrain screening currents and the superposition of the field due to the grain magnetization and the applied magnetic field at the grain interfaces. These results could also be of importance for superconducting electronics applications.

1. Introduction

Bulk sintered high-T_c superconductors are usually of granular nature with large intra-grain- and small intergrain critical current densities j_c. At the grain interfaces the transport current flows through weak links and probably also through strong links of small cross section[1]. It is now well documented that in granular high-T_c superconductors the critical transport current is hysteretic in an applied magnetic field B_a [2-7]. This hysteresis can be explained by the vectorial superposition at the grain interface of applied magnetic field B_a and a hysteretic field component B_m created by the intragrain persistent currents[3,4]. I_c is however a monotonously decreasing function with the <u>true field</u> in the grain interfaces as is normal for superconductors. At 4.2 K where our measurements were done, high-T_c superconductors are of the hard type-II. The local distribution of the screening currents can then be described by the Critical State Model[8]. These screening currents are however not absolutely persistent, but decay slowly with a logarithmic time dependence[9]. This flux-creep is known to be a very pronounced effect in high-T_c superconductors compared to technical low-T_c superconductors but, apart from the size of the effect, the situation found in high-T_c superconductors at 4.2 K is analogous to that found in NbTi- and Nb_3Sn-wires[13]. In order to explain the time relaxation behavior of the intergrain transport I_c we have performed detailed flux-creep measurements in sintered $YBa_2Cu_3O_7$ samples as a function of prior magnetic field cycling.

Because it is difficult to directly measure transport-I_c continuosly in time we have measured the flux-flow voltage instead. We have published earlier[14], that there is a direct correspondence between the transport critical current and the flux-flow voltage which is given by $U = \frac{l \cdot \rho_f}{F} (I - I_c^*)$, where F is the sample cross section and l the length over which the voltage is taken. The flux-flow resistance ρ_f is the derivative $\partial E/\partial j$ with $E = U/l$ being the electrical field strength and j the transport current density. We have found experimentally[14], that ρ_f is practically constant with applied magnetic field in high-T_c superconductors at 4.2 K.

2. Experimental

Intragrain magnetization, flux-flow voltage and critical transport current were measured on several bulk granular $YBa_2Cu_3O_7$ samples as a function of applied magnetic field cycling. All experiments were done at 4.2 K. The material was fabricated in a three-stage sintering process[11]. The final sample density was 5.5 g/cm^3. The average grain diameter determined by SEM was about 18 µm. T_c was 92 K with a ΔT_c of about 2 K. The magnetization relaxation was measured on a cylindrical stack of seven pills with a diameter of 15.5 mm and a total height of 20.4 mm. The applied magnetic field was parallel to the cylinder axis. The magnetization was detected by the usual pick-up coil method. Flux-flow voltage and critical transport current was measured on platelet samples cut from a pill with a diamond saw. The length was 15 mm, the width was 2 mm and the cross section was 1.7 mm^2. The U and I_c measurements were done by a four point silver-contact method[8] over a length of 5 mm, with B_a being always perpendicular to the wide side of the sample. The criterion for j_c was a voltage drop of 100 nV/cm.

3. Results and discussion

3.1 Relaxation of intragrain magnetization

Fig. 1 shows the measured hysteretic intragrain magnetization M with virgin curve, upper and lower envelopes of the hysteresis area and transition curves - between the envelopes - which are measured after field sweep reversal. The arrows indicate the direction of cycling. At certain B_a-values the magnetic field increase or decrease was stopped and the subsequent change of M with time was recorded. The spikes on the curves indicate the change of M during a time interval Δt=60 sec after a stop of field sweep. On the upper and lower envelope curves we find a relatively large relaxation. On the transition curves after field sweep reversal and a small ΔB_a change, relaxation is strongly reduced. As ΔB_a becomes larger, one gets increasingly larger relaxation rates until the envelope is again reached.

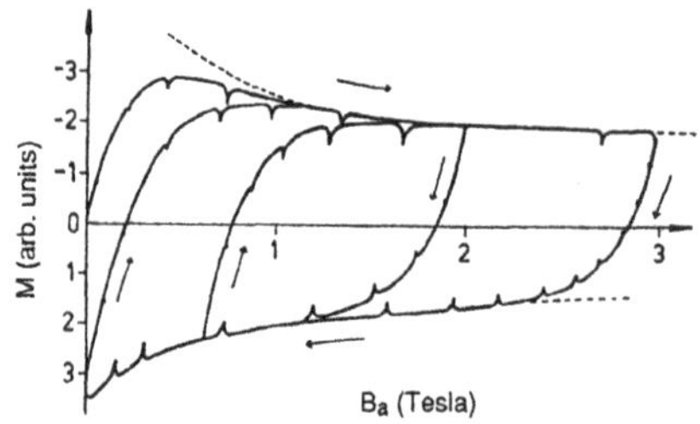

Fig. 1: Hysteretic intragrain magnetization curve. Intermediate field sweep stops of 60 s duration show relaxation of M due to flux creep.

3.2. Transport-I_c relaxation: Overview

Fig. 2 shows a schematical overview of the measured relaxation behavior of transport-I_c. On the lower envelope (for increasing B_a) the two components in the grain interface have the same orientation and are additive, so if B_m relaxes in time the total field becomes smaller in time and transport-I_c becomes bigger. On the upper envelope the opposite is true, the two field components are now subtracting. For the higher B_a-fields, B_a is bigger than B_m and therefore the B_m-relaxation leads now to a total field that is increasing in time. In point E, relaxation is reduced as the two field components cancel each other, and for even smaller fields the relaxation rate increases again but in the opposite direction as now $B_m > B_a$. On the transition curves I_c-relaxation reflects the behavior of the intragrain magnetization relaxation. Just after field reversal relaxation is nearly zero (points A-D in fig. 2) and then starts to increase in the opposite direction.

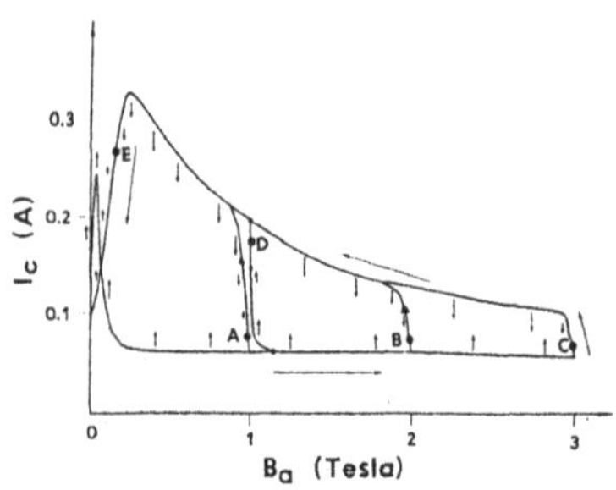

Fig. 2: Hysteretic intergrain I_c as function of field cycling. The vertical arrows show the tendency of I_c time relaxation. A-E are points where the time relaxation changes sign.

3.3. Increasing B_a-field

Figs. 3a and b show the flux-flow voltage-relaxation for B_a increasing from zero (sample in the remanent state). The relaxation in the remanent state itself is shown in fig. 3a for comparison (dashed curve). When the field is now increased again, the relaxation is immediately strongly reduced (arrow). Then, with further increasing B_a , the relaxation rate $|dU/dt|$ increases again. When the field is increased even further (fig. 3b), relaxation rate $|dU/_{dt}|$ increases again. When the field is increased even further (fig. 3b), $|dU/_{dt}|$ decreases again and approaches an approximately constant, small value. This behavior is exactly what one expects from the superposition model. In the remanent state only the strongly relaxing B_m-component is present, so also U is relaxing strongly. For the small B_a-increases in fig. 3a , the magnetization is on a transition curve, therefore B_m-relaxation is first strongly reduced but then starts to increase again. In the high field range in fig. 3b B_m relaxation would be again strong, but now the non-relaxing B_a-component is dominant, so the U-relaxation is small.

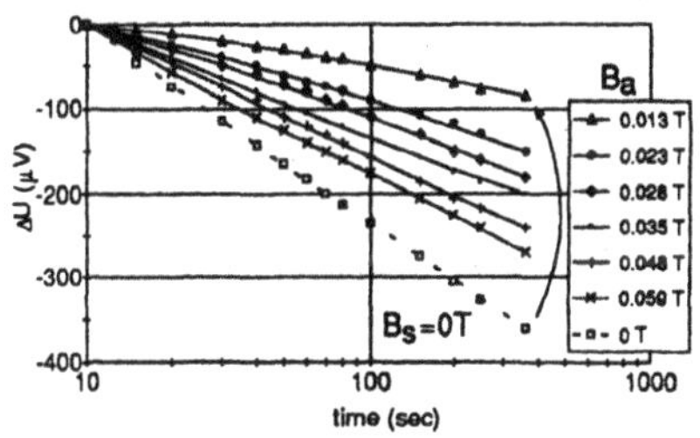

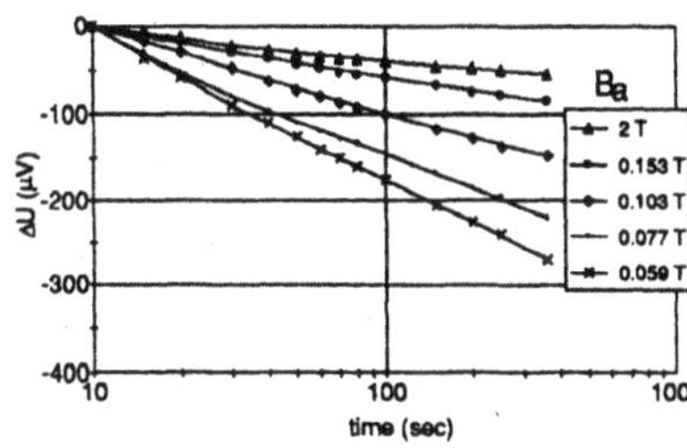

Fig. 3a Fig. 3b

Fig. 3: Flux-flow voltage relaxation for increasing B_a-field. Fig. 3a for the low field range and fig. 3b for the higher fields.

3.4. Decreasing B_a-field

Figs. 4a-c show the flux-flow voltage relaxation when the applied field is cycled from a maximum field B_s=3 T back to zero. On the I_c-curve in fig. 2 one is now on the upper envelope. Here the two components B_a and B_m have opposite sign. In fig. 4a, B_a is larger than B_m and as B_m relaxes the total field becomes larger in time. This results in a voltage increase. In fig. 4b B_m becomes comparable in size to B_a . This is the region near the peak in fig. 2. Here the effects of superposition can be seen most clearly. At 0.5 T, B_a is still bigger than B_m for all times. At 0.19 T, B_m is first bigger than B_a, then around t=60secs they become equal and as B_m relaxes further it is again becoming smaller than B_a . At B_a=0.11 T, B_m is then bigger than B_a for the whole time interval shown.

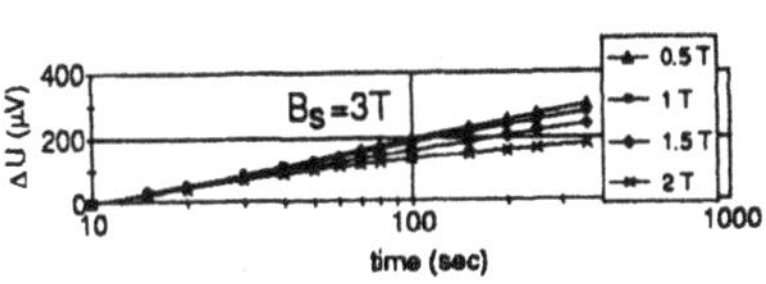

Fig. 4a

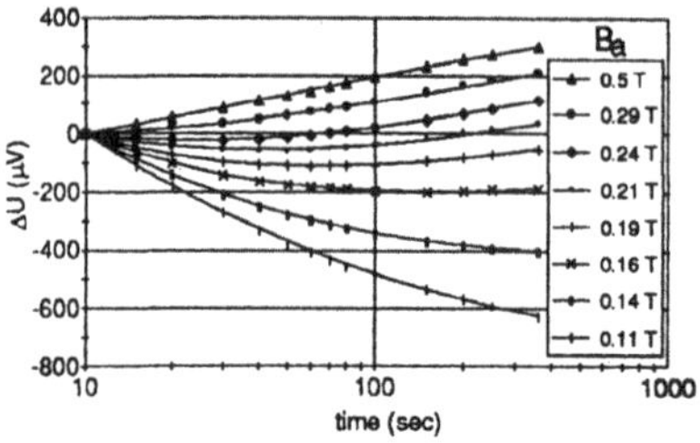

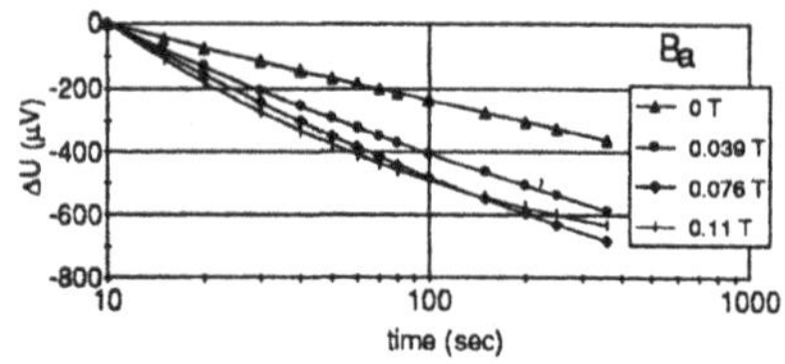

Fig. 4b

Fig. 4c

Fig. 4: Flux-flow voltage relaxation for decreasing B_a-field. Fig. 4a for large field ($B_a > B_m$), Fig. 4b for intermediate fields ($B_a \approx B_m$) and fig. 4c for small fields ($B_a < B_m$)

3.5. Transition curves

Fig. 5a shows the case when B_a is first cycled up to 1 T and then reversed (Curve with point A in fig. 2). The relaxation on the lower envelope at 1 T is shown dashed for comparison. One can see that the relaxation rate is almost zero after just very small field changes. For bigger reversal fields the flux-flow voltage starts again to relax, until the envelope is again reached.

Fig. 5b shows the opposite case when B_a is first cycled down to 1 T and then reversed (curve with point D in fig. 2). The measured relaxation is analogous to fig. 5a but with opposite signs. On the transition curves, the I_c-relaxation is completely determined by the relaxation behavior of B_m.

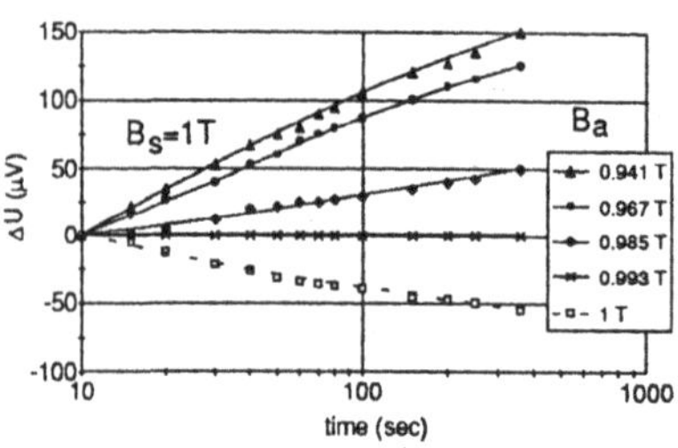

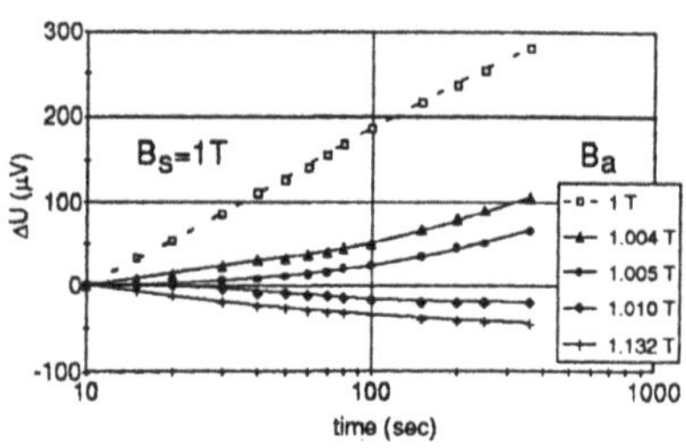

Fig. 5a

Fig. 5b

Fig 5: Flux-flow voltage relaxation on transition curves. Fig. 5a for cycling up to 1 T and then decreasing field. Fig. 5b for cycling down to 1 T and then increasing field.

Acknowledgements

The authors wish to thank R. Häfeli and F. Meier for their valuable help in sample preparation. The financial support of the Swiss Government (Schweizerischer Nationalfonds) is gratefully acknowledged.

References

1. J.W. Ekin, T.M. Larson, A.M. Hermann, Z.Z. Sheng, K. Tugano and H. Kumura, *Physica C* **160** (1989) 489
2. K. Kwasnitza, B. Jakob and G. Vécsey, *Proc. European Workshop on High T_c Superconductors and Potential Applications, Genova, Italy* (1987) 389
3. J.E. Evetts and B.A. Glowacki, *Cryogenics* **28** (1988) 641
4. K. Kwasnitza and Ch. Widmer, *Cryogenics* **29** (1989) 1035
5. K. Watanabe, K. Noto, H. Morita, H. Fujimori, K. Mizuno, T. Aomine, B. Ni, T. Matsushita, K. Yamafuji and Y. Muto, *Cryogenics* **29** (1989) 263
6. M.E. Mc Henry, M.P. Maley and J.O. Willis, *Phys Rev. B* **40** (1989) 2666
7. Y. Yang, C. Beduz and S.P. Ashworth, *Cryogenics* **30** (1990) 618
8. J.W. Ekin, T.M. Larson, N.F. Bergren, A.J. Nelson, A.B. Swartslander, L.L. Kazmerski, A.J. Panson and B.A. Blankenship, *Appl. Phys. Lett* **52** (1988) 1819
9. Y. Yeshurun and A.P. Malozemoff, *Phys. Rev. Lett* **60** (1988) 2202
10. C.W. Hagen and R. Griessen, *Phys. Rev. Lett.* **62** (1989) 2857
11. K. Kwasnitza and Ch. Widmer, *IEEE Trans. Magn.* **27** (1991) 2515
12. K. Kwasnitza and Ch. Widmer, *to be published in Physica C*
13. K. Kwasnitza and Ch. Widmer, *IEEE Trans. Magn.* **27** (1991) 1202
14. K. Kwasnitza and Ch. Widmer, *Physica C* **171** (1990) 211

FLUX PINNING ANISOTROPY IN $YBa_2Cu_3O_7$ FILMS

V.G. PROKHOROV, G.G. KAMINSKY, M.A. KUZNETSOV,
V.I.MATSUJ, Y.A. PETRUKHNO

Institute of Metal Physics, Vernadsky str. 36, Kiev, Ukraine, USSR

ABSTRACT

Experimental results of critical current density versus magnetic field at different orientations are given. From data obtained was shown that anisotropy of critical current density is due to microstructure features rather then the sample geometry.

1. Introduction

It is well known that twin domain or grain boundaries in films crossing a-b planes induce appreciable reduction of the critical current density in the absence of magnetic field [1] . Moreover, higher the misorientation angle of the adjacent grains or domains, the lowers $J_c(0)$. On the other hand the plane defects can serve as strong pinning centers and result in significant critical current density increase at high magnetic fields [2,3] . This contradiction could be reasonably explained taking into account that the energy of flux line creation at a plane defect provide more easy penetration of the vortices into the film (self current vortices are meant) determining a critical current density value in the absence of external field [4,5], and on the other hand the plane defect stimulates an interaction between flux line and defect [6].

The crystal structure anisotropy of the epitaxial quasi-single crystal or highly textured polycrystalline films results in anisotropy of a number of physical parameters (measured for the parallel and perpendicular to the film surface (FS) directions).This effect mainly is due to the anisotropy of the electron mass, coherence length, magnetic field penetration depth, etc. along and across a-b planes. Nevertheless, an appreciable difference between volume pinning force values for different orientations of the applied field relatively FS could have alternative nature.

2. Experimental results

All measurements were made for $YBa_2Cu_3O_7$ prepared by laser ablation onto $SrTiO_3$ substrates [7]. The grain size was ⟨D⟩ = 50 - 100 nm along c axis and about 500 nm in the basal plane. The misorientation (of c axes) of adjacent grains ranged from 1 - 2° up to 30° and was controlled by the film deposition regime. The grain boundaries in the specimens under investigation consisted mainly of misfit dislocations (dislocations walls) and were approximately 4 - 5 nm width [8]. The superconducting transition temperature was T_c = 89 - 93 K; ΔT_c = 1.0 - 1.5 K; the residual resistivity was ρ_o = 100 - 200 $\mu\Omega$cm; the resistivity ratio was $\gamma = \rho_{300}/\rho_o = 3$.

The geometry required for measurements was made by photolitography, 200 - 300 nm thickness, 50 - 100 μm width, and 2 - 10 mm distance between potential probes. The silver contacts were prepared by vacuum vapor deposition. The current corresponding 1 μV/cm sample voltage was considered as critical. To avoid an influence of the thermal electric and contact effects the I - V curves were measured by dc as well as by pulse - lock - in device.

For pulse - lock - in measurements rectangular double polar pulses were passed through the sample, after then relaxation of transition processes were integrated and stored in computer memory. The difference signal from both devices was monitored with X - Y recorder, the pulse duration was about 36 μs, and the sequence frequency was approximately 870 Hz.

Fig. 1 shows that the behavior of the critical current density versus magnetic field is essentially different depending on the applied field orientation relative to the film surface. The maximum J_c value can be reached in the case of the so called "forceless" measurement configurations, B || FS, B || I, that is usual for non-ideal type II superconductors. However, the discrepancy observed for J_c(B) behavior at B ⊥ FS and B || FS (B ⊥ I in both cases) has to be discussed separately. As a rule the volume pinning force anisotropy in this case could be due to two circumstances. The first is a specific shape of defects that are able to be pins. The second is an influence of the free film surfaces that could be regarded as additional pins. In the present work we shall not discuss a possibility of the intrinsic pinning which is due to the peculiarities of HTSC film crystal structure, because there are no assured experimental data confirming its real existence.

The insert in Fig.1 shows that the dependence of the volume pinning force on magnetic field obeys the law $F_p \propto b^{1/2}$ in the low field region ($b = B/B_{c2}$). As it has been noted in [2,3] such F_p(b) behavior could be due to flux line pinning at the grain boundaries because of superconducting parameters anisotropy, while the interaction of flux lines with dislocations being able to result in similar effect was not excepted [9]. Moreover, it should be noted that the F_p(b) behavior is the same at the alternative film orientation relatively applied magnetic field. It could be an argument for that the flux line pinning mechanism does not change when the geometry changes, and the anisotropic F_p(b) behavior cannot be explained taking into account only surface pinning. The same conclusion can be made analyzing the data shown in Fig.2. Because the film sizes in this case are approximately equal (width and thickness), the surface contribution to the net volume pinning force has to be approximately equal for different samples too.Nevertheless,

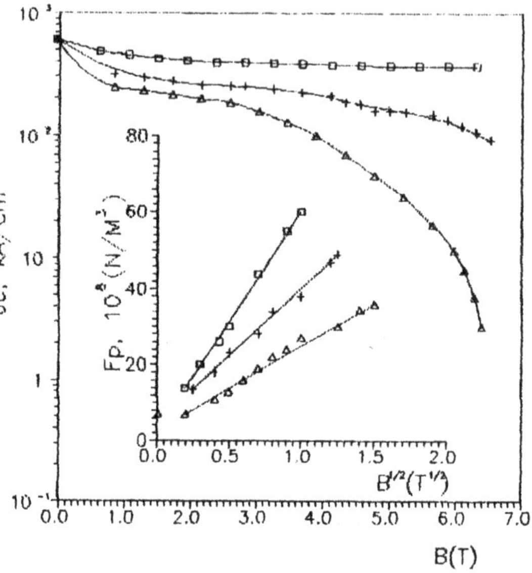

Fig. 1: Critical current density vs applied magnetic field at 77 K. □ - forceless configuration; + - B || FS; Δ - B ⊥ FS.

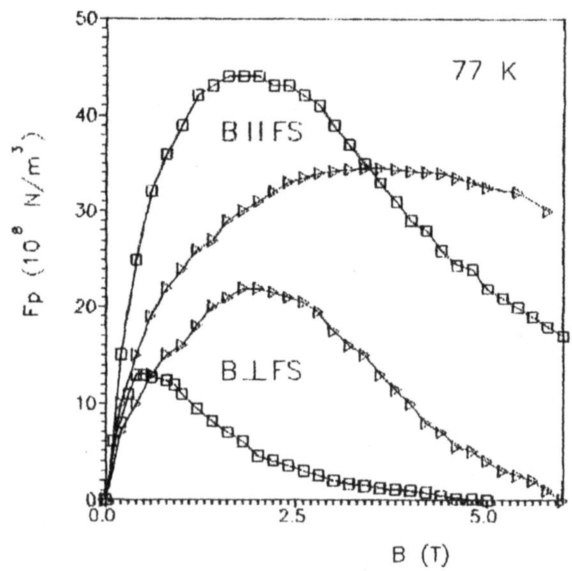

Fig.2: Volume pinning force vs applied magnetic field at 77 K. □ - sample 1; Δ - 2

the contrary effect is observed experimentally. The maximum pinning force value F_p^{max} at B || FS proved to be higher for the sample with the smaller F_p^{max}(B ⊥ FS).

The electron microscopy has shown that grains (or domains) in the investigated samples were tablet - like, which thickness were 5 - 10 times smaller than diameter. Therefore a number of plane defects in a unit volume, at which vortices can be pinned, several times higher at B || FS than at B ⊥ FS. Using the expression for the volume pinning force due to the vortex interaction with grain boundaries [9] the expected F_p (b) discrepancy for different orientations can be estimated.

$$F_p \approx N^{1/2} f_p / a_o^2, \qquad (1)$$

where N is a number of planes (grain or domains boundaries), f_p is the elementary pinning force, a_o is the flux line lattice period. Choosing a constant a_o value (e.g., $B = B_m$ when $F_p = F_p^{max}$ at B ⊥ FS) and supposing the f_p value to be almost constant for different B orientations, one can obtain F_p(B || FS)/F_p^{max}(B ⊥ FS) = 2 - 3. The obtained results well agree with the data for the films with the c misorientation angle is the smallest, 1 - 2°(sample 1 in Fig.2). An increase of the c axis misorientation up to 10° for adjacent grains results in its spheroidization and a decrease of the volume pinning force anisotropy (sample 2 in Fig.2).

3. Conclusion

In summary it can be concluded that the F_p(b) anisotropy observed in $YBa_2Cu_3O_7$ is determined by microstructure features rather than by the sample geometry.

4. Acknowledgements

This work was supported by Scientific Council on the problem of HTSP and carried out within frameworks of project 90229 of State Program of High Tc superconductivity.

5. References

1. P. Chaudhari, J. Mannhart, D. Dimos et al., *Phys. Rev. Lett.* 60 (1988) 1653.
2. V.M. Pan, S.V. Gaponov, G.G. Kaminsky et al., *Cryogenics* 29 (1989) 392
3. S.V. Gaponov, G.G. Kaminsky, E.B. Kluenkov et al., *JETP (Sov.)* 95 (1989) 392.
4. A.I. Larkin, Yu.N. Ovchinnikov, *JETP (Sov.)* 61 (1971) 1221.
5. L.G. Aslamazov, S.V. Lempitsky, *JETP (Sov.)* 84 (1983) 2216.
6. V.G. Prokhorov, C.G. Tretiatchenko, G.G. Kaminsky, *Fiz. Nizk. Temp.(Sov.)* 10 (1984) 878.
7. S.V. Gaponov, F.V. Garin, V.N. Golubev et al., *JETP (Sov.)* 95 (1989) 1086.
8. N.A. Kiselev, A.L. Vasiliev, O.V. Uvarov et al., *Proc. of EUREM 88* York, England, Inst. Phys. Conf. Ser. 92 (1988) 223.
9. E.H. Brandt, *Phys. Rev. Lett.* 57 (1986) 1347.

FLUX DYNAMICS AND PINNING IN HIGH - Tc FILMS

V.G. Prokhorov, G.G. Kaminsky, A.L. Kasatkin,
M.A. Kuznetsov, Y.A. Petrukhno

Institute of Metal Physics, Vernadsky str. 36, Kiev, Ukraine, USSR

ABSTRACT

The results of experimental investigation of vortex dynamics in mixed state of $YBa_2Cu_3O_7$ films are presented. A phenomenological theory describing the influence of planar defects on electrodynamics of superconductors with small coherence length, taking into account a local suppression of order parameter near planar defects due to the proximity effect, has been developed as well. Analyses of the creep and viscous flux flow regimes using current - voltage curves for the films under investigation leads to the conclusion that the one - dimensional "easy slip channels" along which the vortices preferably move under the Lorentz force, play an important role in the formation and properties of the resistive state (at least at high enough temperatures and rather weak fields).

1. Introduction

A magnetic field in a type II superconductor exists in the form of lattice of quantized flux vortices, each containing one flux quantum ϕ_o. The motion of these vortices under a variety of driving forces produces an electric field which is proportional to their velocity. This movement is opposed by viscous forces and by interactions with defects and inhomogeneities in the superconductor [1]. Therefore the shape of current - voltage characteristic can reflect some correlation with the microstructure peculiarities of studied superconductors [2]. The investigation of vortex motion in dynamic mixed state could give a lot of information about pinning processes and thermally assisted flux creep (TAFC) in high temperature superconductors (HTSC). On the other hand, most of the theoretical models do not take into account an influence of the spatial inhomogeneity of the superconducting order parameter, which we feel , is essential. In the present work we used specially developed pulse technique for the investigation of dissipation processes which arise because of flux creep and viscous flux flow in HTSC films. We interpret the results taking into account the specimen inhomogeneity.

2. SAMPLE PREPARATION.

The $YBa_2Cu_3O_7$ films prepared by laser ablation were used for measurements. The cross - beams scheme of deposition was realized with using two YAG (1068 nm) lasers. The film thickness was $d \propto 200 \div 300$ nm. The superconducting transition temperature was $T_c = 89 \div 93K$, the transition width was $\Delta T_c = 1.0 \div 1.5K$, the residual resistivity was $\rho_o = 100 \div 200\ \mu\Omega cm$, the resistivity ratio was $\gamma = \rho_{300}/\rho_o \propto 2.6 \div 3.0$ $SrTiO_3$ and ZrO_2 single crystals were used as substrates. The samples were prepared by mask technique to achieve the following geometry: $0.3 \div 1$ mm film width, $2 \div 10$ mm distance between potential probes. The silver contacts were vacuum deposited. The films were quasi - single - crystalline. Most of them had low angle boundaries between regions of coherent scattering. The mean misorientation angle in a - b plane determined by X - ray studies was $4 \div 5^\circ$ and one in c axis was $< 0.5^\circ$. A typical grain (regions of coherent scattering) size was $\langle D\rangle = 0.5 \div 4.0\ \mu m$ in basal plane and could be regulated by the temperature of substrate during deposition. The critical current densities of the studied films in the absence of magnetic field were $J_c(77K) \propto 1 \div 3\ 10^6 A/cm^2$.

3. EXPERIMENTAL RESULTS

The current - voltage measurements of $YBa_2Cu_3O_7$ films were performed by pulse technique to avoid the sample heating. Fig. 1 shows the first derivatives of I - V curves for film at low temperature. It is seen from the figure that the experimental dV/dI curves could be divided into three specific current ranges with different V(I) dependencies. The first is connected with the initial part of non - linear I - V behavior at $I \propto I_c$ and well described with known empirical expression

$$J(E) = J_c + J_1 \ln(E / E_o) \qquad (1)$$

where E is electric field, $J_1 \propto 0.01\ J_c$, and $E_o = 10^{-6}$ V/cm. To explain such I - V curve behavior the thermally assisted flux creep model [3] is usually used. At this approach $J_c = kT/a_o\phi_o d$, where a_o is the flux line lattice period, E_o is the electric field value at which all vortex leave pins, and d is the flux line length (the film thickness in the present case). The J_1 estimation in the frameworks of this model is in a good agreement with experimental results.

In the second current range a power dependence $V(I) \propto (I-I_c)^{n(H,T)}$ is the case. Similar dependencies were often observed earlier for non - ideal type II superconductors [2]. For our samples studied at different temperatures n was about 2 ($2.2 \div 2.4$ within the whole field range).

At last the steep growth of incremental resistivity at the end of non - linear part of I - V curve is due to the sample overheating and transition to the normal state.

At the high temperature ($T \geq 0.9T_c$), sometimes flat parts of dV/dI (I) curves were observed (see Fig. 2). We suppose they are due to the viscous flux flow (VFF).

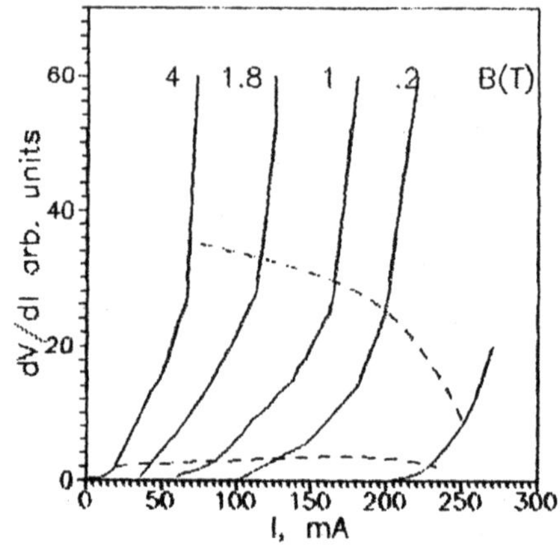

Fig. 1: First derivatives of I - V curves at different values of the magnetic field. T = 56 K, B is normal to film surface.

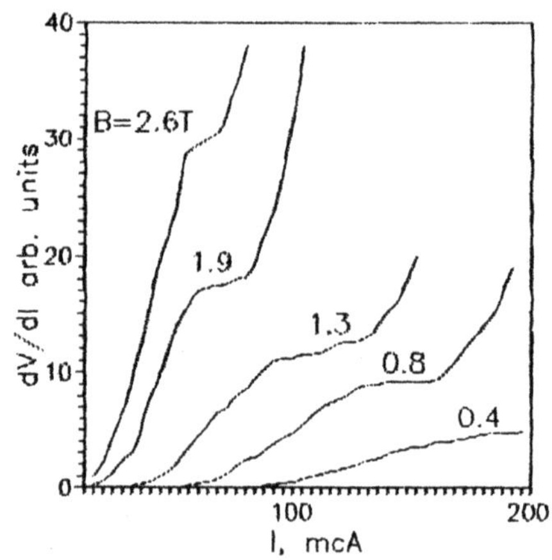

Fig. 2: First derivatives of I - V curves at different values of the magnetic field. T = 83 K, B is normal to film surface.

The field dependence of the VFF resistance is shown in Fig. 3. It is seen that this dependence could be divided into two parts with different $R_f(B)$ behavior, where R_f - resistance of VFF.

The initial part is well described by the $R_f \propto B^{1/2}$ law (see the insert in Fig. 3). At higher fields the VFF $R_f \propto B$ being usual for conventional type II superconductors. It should be also noted that the linear dependence does not start from zero R_f.

Fig. 4 shows dependencies of volume pinning force vs applied magnetic field at different temperatures. It is seen on the insert that the temperature dependence of of elementary pinning force $f_p(t)$, where $t = T/T_c$, is not monotonous. At $t \leq 0.9$ $f_p \propto (1 - t)$ and at $t \geq 0.9$ $f_p \propto (1 - t)^{5/2}$. The values of elementary pinning forces obtained from the $F_p(B)$ slopes are $f_p \propto \phi_o dF_p(B)/dB$, supposing that the vortices are in so - called insulator regime, when the interaction between vortices is negligibly weak.

4. Discussion

As we have shown earlier [4] $R_f(B)$ dependence cannot be described by Bardeen - Stephen model, because in the low field range $R_f(B) \propto B^{1/2}$. Such field dependence can be explained assuming that vortices move along one - dimensional easy slip channels rather then within the whole volume. These channels are formed along planar defects, near which the superconducting order parameter is locally suppressed, and the activation energy and pinning force are lower compared with the rest volume of the sample.

According to the results of structure studies such channels could be formed along the low angle grain boundaries. Under these conditions $R_f(B)$:

$$R_f(B) = \beta R_n n_\perp \xi^2 = \beta R_n \frac{\xi^2}{La_o} = \beta R_n \frac{(\phi_o B)^{1/2}}{LB_{c2}} \qquad (2)$$

where $\beta \propto 1$, $n_\perp = 1/La_o$ is the concentration of vortices in channels, L is the mean distance between channels (grain size).

Assuming these channels are not identical and there is some distribution of pinning forces (and critical currents) for channels, an increase of transport current induces into increase a number of effective channels. For the I - V curve the following can be written:

$$U = \int f(J_c) R_f(B)(J - J_c) dJ_c$$

$$\int f(J_c) dJ_c = 1 \qquad (3)$$

where $f(J_c)$ is the critical current distribution function of channels. Using the simplest rectangular stair - like distribution function $f(J_c) = 1/(J_{c2} - J_{c1})$, where J_{c2} and J_{c1} are maximum and minimum values of the

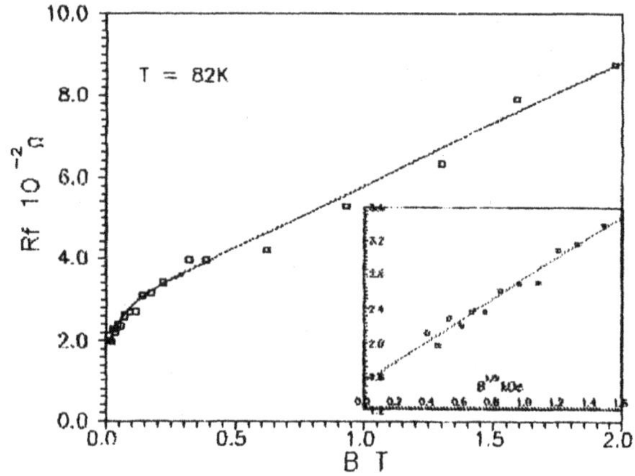

Fig. 3: The R_f (B) at flux flow regime. T = 82 K.

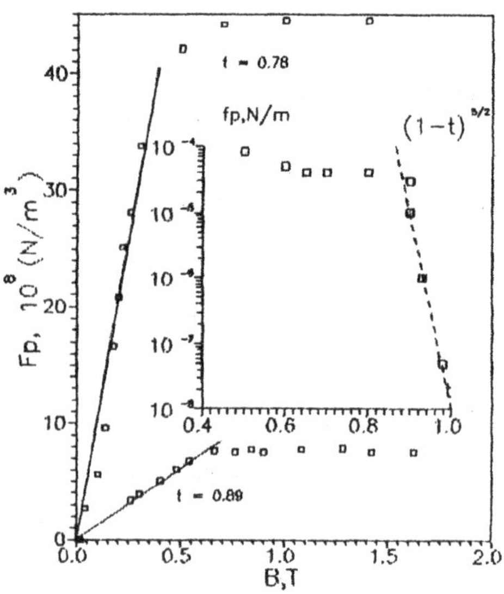

Fig. 4: Temperature dependences of volume F_p and elementary f_p pinning forces.

critical current in channel, it is easy to derive for $J_{c1} < J < J_{c2}$:

$$U = \frac{R_f(B)}{2(J_{c2}(B) - J_{c1}(B))}(J - J_{c1}(B))^2 \qquad (4)$$

that corresponds to experimental data with $J_{c1}(B) \equiv J_c$ and $n(H,T) = 2$. By the way, similar results were earlier obtained for conventional type II superconductors [1,2].

The relevant temperature dependence of the the elementary pinning force $f_p(t)$ at $t \to 1$ (Fig. 4) are also well agree with the concept of the vortex motion along easy slip channels. Really, for the elementary pinning force in the channel the following expression could be written:

$$f_p \propto U_o = \beta \frac{H_{c2}^2(t)}{8\pi}\left[\frac{\Delta_i}{\Delta_o}\right]^2 V_c \qquad (5)$$

where U_o is the vortex activation energy in channel, $V_c \propto \xi d a_o \propto (1 - t)^{-1/2}$ is the correlation volume of flux line lattice in a single thermal activation process, Δ_i/Δ_o is the ratio of the superconducting order parameters in channels and out of them, which obeys the $(\Delta_i/\Delta_o)^2 \propto (1 - t)$ law [5], and $H_c(t) \propto (1 - t^2)$ is the thermodynamic critical field. It is worth to notice that $H_c(t) \to (1 - t)$ at $t \to 1$.

Thus, in the framework of present concept $f_p \propto (1 - t)^{5/2}$ near T_c which in a good agreement with the experimental results.

5. Conclusion

The main peculiarity of the HTSC films in the mixed state is the existence of easy slip channels (ESC) along planar crystal defects, where flux lines could easily move. The experimental results are well described by ESC theory taking into account the spatial inhomogeneity of superconducting order parameter.

6. Acknowledgements

This work was supported by Scientific Council on the problem of HTSP and carried out within frameworks of project 90229 of State Program of High T_c superconductivity.

7. References

1. M.G. Blamire, J.E. Evetts, *Phil. Mag.* 51 (1985) 421
2. V.G. Prokhorov, G.G. Kaminsky, C.G. Tretiachenko, *Fiz. Nizk. Temp. (Sov.)* 12 (1986) 675
3. P.W. Anderson, *Phys. Rev. Lett.* 9 (1962) 309
4. V.G. Prokhorov, G.G. Kaminsky, A.L. Kasatkin et. al., *Fiz. Nizk. Temp. (Sov)* 17 (1991) 467
5. G. Deutscher, K.A. Muller, *Phys. Rev. Lett.* 59 (1987) 1745

THE INFLUENCE OF ADDITIONAL PINNING CENTERS ON THE CRITICAL CURRENT IN WEAKLY COUPLED PERCOLATIVE STRUCTURES

JANUSZ JACKIEWICZ and JAN LESZCZYNSKI
Departament of Materials Science
Institute of Principles of Electrical Power Engineering I-12
Technical University of Lodz
ul. B.Stefanowskiego 18/22, 90-924 Lodz, Poland

ABSTRACT

It is shown that the "internal" loop of magnetization hysteresis of weakly coupled high-Tc superconductors is due to the superposition of magnetic moments of superconducting ring structures with different critical shielding currents. The influence of the superconducting rings on the transport current Ic occurs only by agency of the dissipation fluxes of the shielding currents.

The effect of introduction of additional pinning centers on the global magnetic moment of the sample and on its critical current is investigated assuming that the transport current flows through a percolative network of weak junctions.

1. Experiment

The sample of ceramic superconductor of the type $Y_1Ba_2Cu_3O_{7-y}$ with the dimensions 3x3x14 mm^3 was placed in a system of two measurement coils, push-pull coupled.

In the figure 1a the magnetization M of the sample is shown, during magnetization by alternating external field H_e with the amplitude H_{em} = ±2.3 kA/m, in LN temperature, and in the figure 1b - the intergranular magnetization $M_I = -(M + \mu_o \cdot H_e)/(1 - \gamma)$, after taking into account the diamagnetism of the grains remaining in the Meissner state, with relative volume γ.

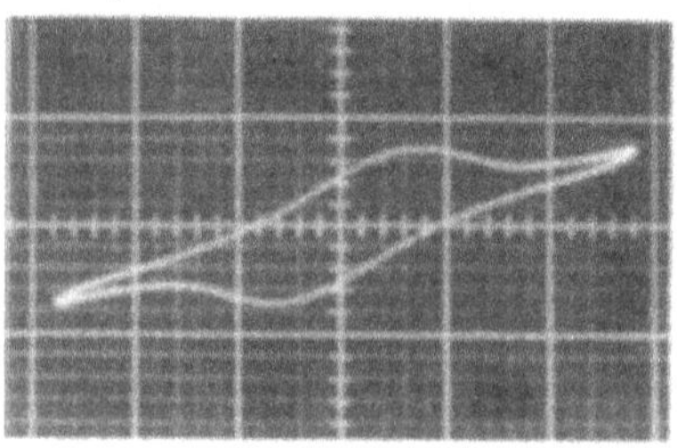

a.

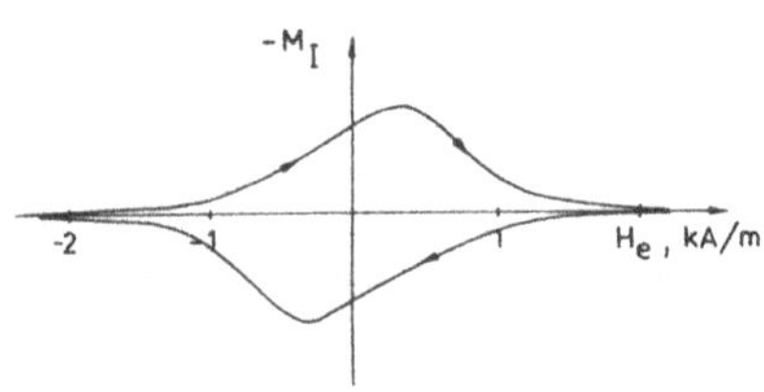

b.

Fig. 1. The magnetization of the sample of weakly coupled granulated superconductor in alternating field with amplitude $H_{em} < H^G_{c1}$, where H^G_{c1} - the first critical field of the grains (a), and the magnetization of the intergranular region (b).

In the effect of a measurement of the critical transport current carried out by the DC four probe method with the use of diffusion electrodes of Au wire, in the field with amplitude ±2.3 kA/m the hysteresis like dependence presented in the figure 2 was obtained.

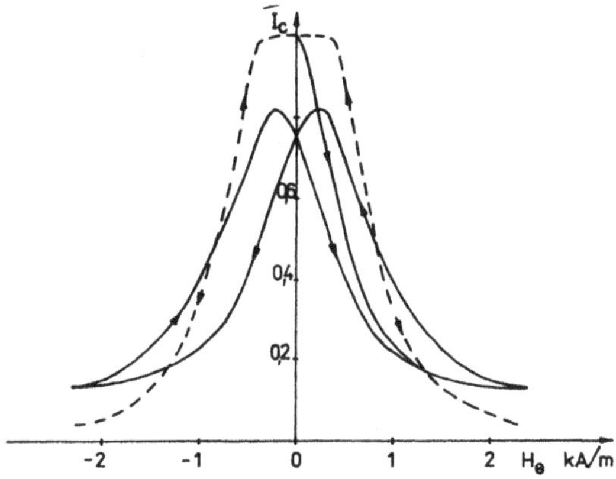

Fig. 2. The critical transport current of weakly coupled granulated superconductor in alternating magnetic field; the arrows show the direction of the external field changes (the dotted line shows the dependence obtained from the Bean's model).

According to the Bean's model [1], the density of the macroscopic critical current is proportional to the pining of the magnetic flux in superconducting material:

$$J_c = 3/4 \cdot (M^+ - M^-)/\mu_o R \qquad (1)$$

where: M^+ and M^- are the sample magnetizations for the same field value, during its increase and decrease correspondingly, R - the sample radius, μ_o - the magnetic permeability of vacuum.

The critical current dependences, presented in the figure 2, obtained by the transport current measurements and by the magnetization of the intergranular region are not in qualitative agreement, which points at a dissimilarity of the mechanisms limiting superconducting current in weakly coupled high-Tc superconductors and in homogeneous type II superconductors.

2. The properties of circulating screening current

The intergranular area magnetization, visible in the figure 1b, may be due to the superposition of the magnetic moments m^R, corresponding to the screening currents I_s^R of diverse structures of the superconducting rings.

In weak magnetic fields, in granulated superconductors the influence of the screening current flow becomes important in two types of the rings:

1 - those formed by superconducting grains coupled by weak junctions (figure 3);

2 - those formed by the grains themselves (figure 4), in this case the screening current may be the Meissner current causing the grains diamagnetism or the effect of Abricosov's vortexes superposition.

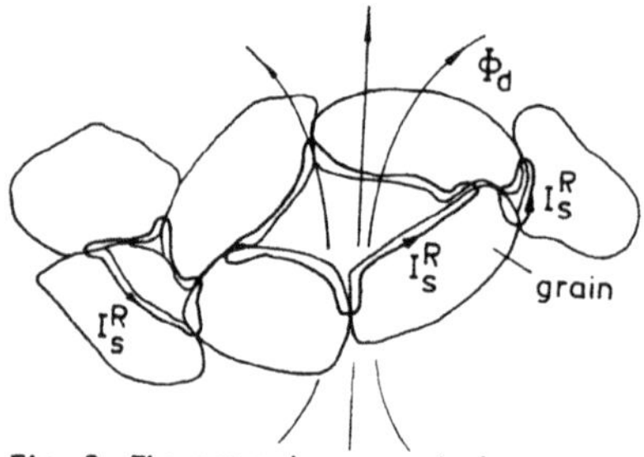

Fig. 3. The screening current in a ring formed by coupled superconducting grains.

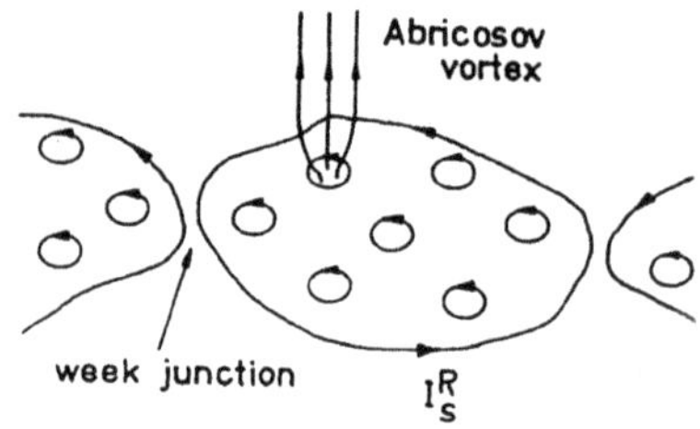

Fig. 4. The screening current in superconducting grains.

In stronger magnetic fields in high-Tc superconductors, the formation of screening currents in the superconducting rings formed at the locations of any defects of the crystal structure (microdislocations in thin monocrystalline films, twinnings etc.) is possible.

The superconducting ring may be characterized by critical fields H_{c1}^R and H_{c2}^R, the H_{c1}^R field corresponds to the beginning of the magnetic flux penetration through the walls (the weak junctions) of the ring, in the H_{c2}^R field the ring (the weakest junction in the ring) loses its superconducting properties. The screening current of the ring cannot exceed the critical value $I_c^R = \alpha \cdot H_{c1}^R$, where α - some proportionality coefficient.

The value of the screening current in the ring depends on the value of the external magnetic field H_e and on maximum value of the magnetizing field H_{em}. In the figure 5 the hysteresis of the screening current of a superconducting ring is presented, assuming that I_c^R = const.

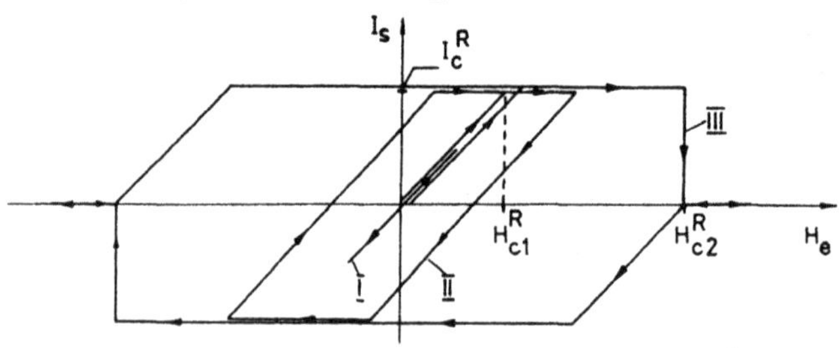

Fig. 5. The dependence of the screening current I_s^R of the superconducting ring on the external field for diverse values of the maximum magnetizing field Hem: Hem < H_{c1}^R (I), H_{c1}^R < Hem < H_{c2}^R (II), Hem > H_{c2}^R (III).

3. The influence of the screening current of the superconducting rings on the transport current in percolative structures

The screening current of the superconducting rings influences the percolative network of the transport current of the weakly coupled superconductor only by dissipation fluxes $\Phi_d = LI_s^R$, where L - inductivity of the ring, and the interaction depends stochastically on relative location of the weak junctions of the precolative network of the superconducting rings.

The magnetic field $\vec{H}$ in the area of every weak junction of the percolative network is the effect of superposition of the external field $\vec{H}_e$ and some number of dissipation fields $\vec{H}_d$:

$$\vec{H} = \vec{H}_e + \sum_i \vec{H}_d^{(i)}. \tag{2}$$

And so, the magnetization of the sample of a granulated superconductor (weakly coupled) corresponds closely to the total flux "frozen in" in the superconducting rings ($H \sim \sum^R I_s$), while the transport current depends on the properties of the weak junctions of the percolative network in local field H, and, in general, on the structure of the network.

The figure 6 presents the result of a computer simulation of dependence of the critical current of percolative network (51x19) of weak junctions on alternating magnetizing field with the amplitude $\pm H_{em}$, carried out accordingly to the proposed superconducting rings model (SRM) [2]. In the figure 7 there is a result of summing up the magnetic moments of the screening currents of superconducting rings, influencing, by means of the dissipation fluxes Φ_d, the weak junctions of the analysed percolative network.

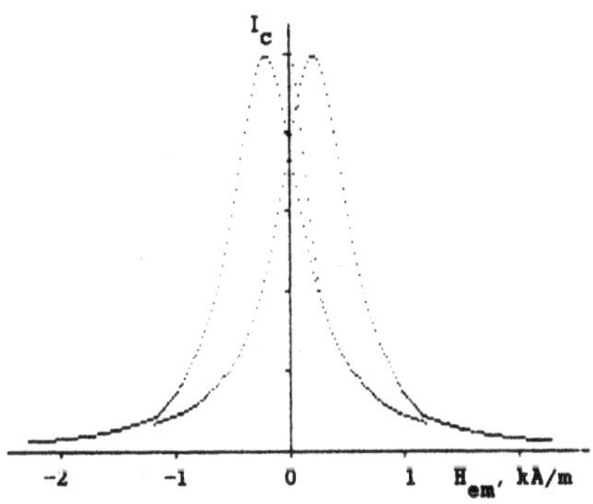

Fig. 6. The result of the RSM analysis of the dependence of the critical current of the percolative network of weak junctions on alternating magnetizing field.

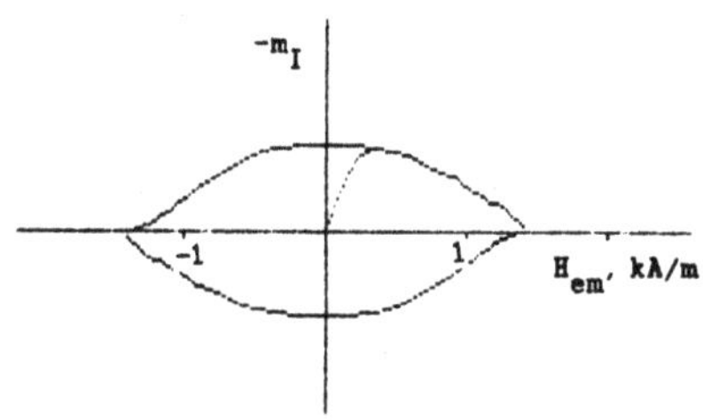

Fig. 7. The global magnetic moment of the screening currents of the rings, influencing the critical current presented in the figure 6.

The dependences of the critical current of the percolative network, accordingly to SRM [2], and of the sum magnetic moment, are in qualitative agreement with the experimental dependences of the transport current I_c (figure 2) and of the magnetization of the intergranular areas M_I (figure 1b) on the alternating magnetizing field H_e.

4. The influence of additional pinning centers on the critical current of the percolative network

The figures 8a and 8b present the influence of additional pinning centers (superconducting rings), formed at expense of the existing percolative network, on global magnetic moment, the figures 9a and 9b present the influence of these centers on the critical transport current of the analysed network.

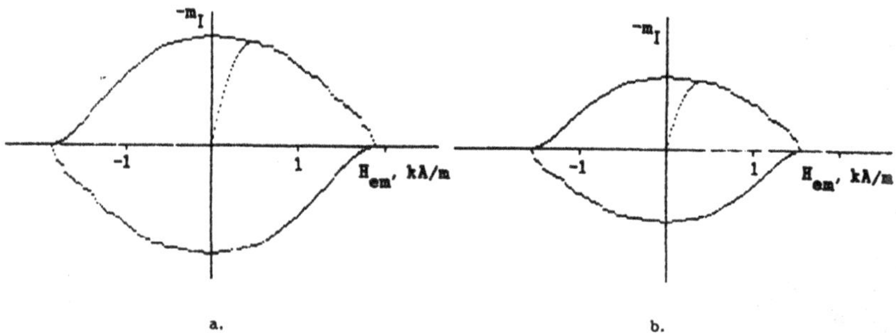

Fig. 8. The changes of the magnetic moment in the case of formation of additional pinning centers at the expense of the percolative network of the transport current.

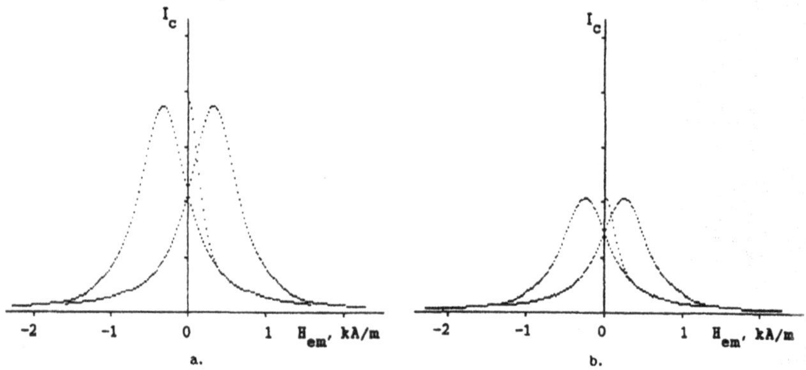

Fig. 9. The influence of the additional pinning centers, formed at the expense of the percolative network, on the critical current of the network.

In the case of the formation of additional pinning centers at the expense of the existing percolative network, initially there occurs an increase and then a decrease (because of the imposibility of the formation of circular intergranular junctions) of the sample magnetization. The pinning centers of this type always cause a decrease of a total critical current, both because of the influence of additional dissipation flux and because of a decrease of the number of the superconducting junctions in the percolative network.

In the case when additional pinning centers are located outside the percolative network of weak junctions (e.g. in the superconducting grains) they cause a decrease of the critical current (figures 11a and 11b), in spite of the monotonic increase of the structure magnetization (figures 10a and 10b). In this case the influence of additional pinning centers occurs by the dissipation fluxes of the screening currents.

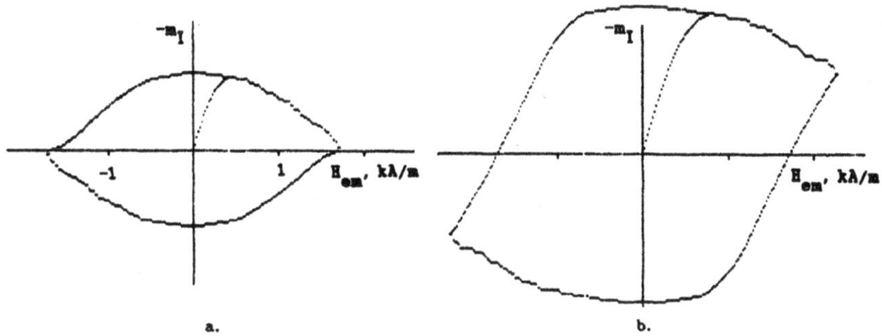

Fig. 10. The influence of the pinning centers, formed in superconducting grains, on the magnetization of the sample.

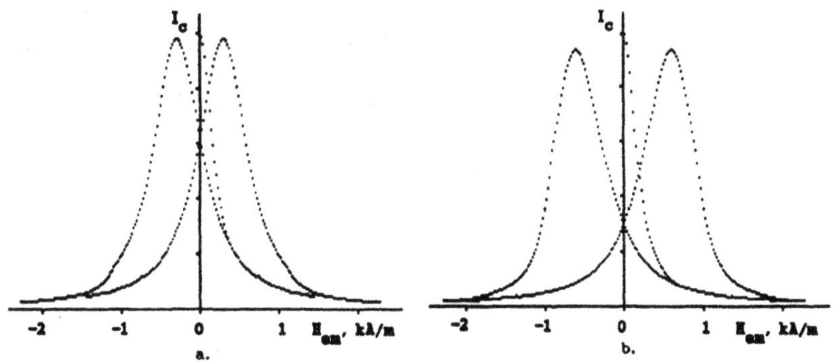

Fig. 11. The changes of the critical current of the percolative network as the effect of the additional pinning centers formed outside the network.

The increase of the critical transport current I_c in percolative structures may occur in effect of additional pinning center formation only in the following cases:

a) when these centers are localized in the weak junctions of the percolative network,

b) when the critical current I_c^J of these junctions is an effect of the magnetic flux creep, and not of the uncoupling of the superconducting carriers.

5. References

1. C.P.Bean, *Rev. Mod. Phys.* **36** (1964) 31-39.
2. J.Jackiewicz and J.Leszczynski, *Proc. 6th Int. Workshop on Critical Currents (Cambridge, 1991)* (to be published in *Supercond. Sci. Technol.*).

THE INDUCTANCE OF A GRANULATED SUPERCONDUCTOR OF $Y_1Ba_2Cu_3O_{7-y}$ TYPE

JAN LESZCZYNSKI and JANUSZ JACKIEWICZ

Departament of Materials Science
Institute of Principles of Electrical Power Engineering I-12
Technical University of Lodz
ul. B.Stefanowskiego 18/22, 90-924 Lodz, Poland

ABSTRACT

A method for measuring the equivalent magnetic permeability of high-Tc ceramic superconductors is presented. The method consists in measuring the changes of the inductance of a normal conductor placed along the axis of a superconducting cylinder occurring during the transition from the normal to the superconducting state. The results presented here concern superconductors of the type Y1Ba2Cu3O7-y in their powdered and solid forms with different contents of the superconducting phase. Thus obtained values of the equivalent magnetic permeability μrs were used to verify the Clem's model [1].

1. Introduction

The alternating current flow through a superconducting conductor induces the electric field with a component depending on the time variations of the magnetic field and a component which corresponds to the carriers inertia. The overall voltage corresponds to both these components:

$$u = u_{in} + u_k \tag{1}$$

The voltage u_{in} induced by the changing magnetic field may be expressed as a function of the rate of the current change:

$$u_{in} = L \frac{di}{dt} \tag{2}$$

The coefficient L appearing in the above formula is the geometric inductance of the conductor, depending on its dimensions, distribution of the current density j(x) and on the magnetic characteristics of the conductor and medium. The geometric inductance may be treated formally as the sum of the, so called, internal inductance L_{in} and external inductance L_{ex}.

$$L = L_{in} + L_{ex} \tag{3}$$

The external inductance corresponds to the magnetic field in the medium around the conductor and in accordance with the principle $\oint H\,dl = J$ (where J is the current in the conductor) it does not depend on the distribution of the current density j(x) in the transverse cross section of the conductor.

The inductivity L_{in} depends on the j(x) distribution. The value of this component changes with the j(x) distribution while shifting from the normal into the superconducting state.

The kinematic voltage u_k is the effect of the acceleration of the superconducting carriers by the electric field in a resistanceless medium (with no energy dissipation).

The kinetic inductivity is described by formula:

$$L_k = \frac{\mu_0 l}{S}\lambda^2 \qquad (4)$$

where: λ - the London's penetration depth,
S and l - transverse cross-section and the length of the conductor respectively

The total inductance is given by:

$$L_t = L + L_k \qquad (5)$$

And so the transition from the normal state into the superconducting state is accompanied by the disappearance of the resistance R, corresponding to dissipative movement of the carriers in the crystal lattice of the normal conductor, and the appearance of the kinematic inductance L_k. At the same time the L_{in} component of the geometrical inductance changes because of the change in the current density distribution. There is also some small component of the voltage on the resistance appearing as the effect of the movement of normal electrons in the induced electric field. It is possible to show that the value $L_k \sim (\lambda/D)^2$ is insignificant in comparison to L_{in}.

The measurement of L_{in} may be used to determine substitute parameters of the superconductor, especially to determine the substitute magnetic permeability μ_{rs}. The system producing magnetic field with cylindrical geometry is presented in figure 1. A thin wire made of normal conductor is placed along the axis of a superconducting cylinder.

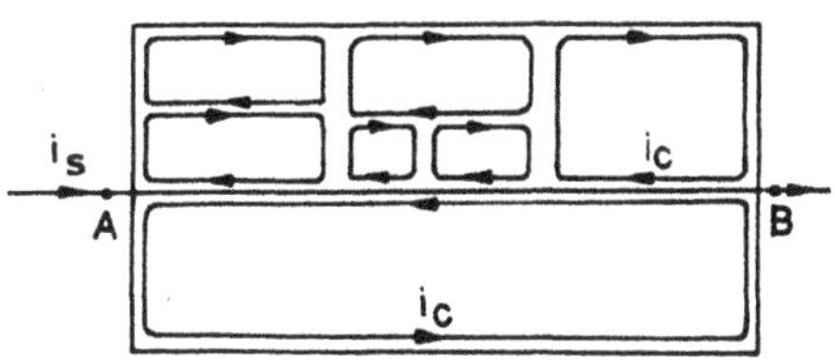

Fig. 1. A simplified schematic of a transformation towards the surface when a transport current is flowing through the central current filament inside of the superconducting cylinder.

In the normal state of the cylinder, the distribution of the magnetic field corresponding to the central conductor is a "natural" distribution of the type 1/x (x - a distance from the cylinder axis). In the superconducting state, because of shielding of the cylinder inside, there appears shielding current, corresponding to the "central" current transformation towards the surface.

The efficiency of the transformation, and at the same time also the depth of the zone occupied by the shielding current and the magnetic field, depend on the quality of the superconductor, i.e. on its shielding capability.

If one accepts that in the normal state of the cylinder $\mu_r = \mu_{rn}$ and that in the superconducting state $\mu_r = \mu_{rs}$, then, during the transition from the normal into the superconducting state, the voltage change Δu_{in}, induced between the A and B points of the Cu wire, will be:

$$\Delta u_{in} = \frac{\mu_o l}{2\pi} \ln\frac{D_{ex}}{D_{in}} \left(\mu_{rn} - \mu_{rs}\right) \frac{di}{dt} \tag{6}$$

where l, D_{ex}, D_{in} are respectively: length, external and internal diameters of the cylinder.

Accepting $\mu_{rn} = 1$ we get:

$$\mu_{rs} = 1 - \frac{2\pi}{\mu_o l}\cdot\Delta u_{in}\cdot\left(\frac{di}{dt}\cdot\ln\frac{D_{ex}}{D_{in}}\right)^{-1} \tag{7}$$

The quantity μ_{rs} is the equivalent relative magnetic permeability of the superconducting cylinder.

The degree of shielding of the magnetic field corresponding to the central current i_s may be determined also by finding the D_s - the diameter of the equivalent cylindrical current path which, for the current i_s, would create the same magnetic field in the medium as the current in the central conductor, being surrounded by superconducting cylinder. And so the D_s diameter is the diameter of the current path "seen" by the u_{in} voltage measurement.

From Eq.(7) it follows that:

$$\frac{D_s}{D_{ex}} = \frac{D_{in}}{D_{ex}}\cdot\exp\left(\left(1 - \mu_{rs}\right)\cdot\ln\frac{D_{ex}}{D_{in}}\right) \tag{8}$$

For ideal shielding, i.e. for $\mu_{rs} = 0$, we get:

$$D_s = D_{ex} \tag{9}$$

2. Experiment A - the overall voltage change

The ceramic material of the type $Y_1Ba_2Cu_3O_{7-y}$, being the starting substance for the production of all the samples used in subsequent investigations, was obtained in the form of superconducting powder with high content of the orthorhombic phase (sample P300) accordingly to the widely known technique [2]: precisely 1/2 mole of Y_2O_3, 2 moles of $BaCO_3$ and 3 moles of CuO were mixed; the obtained mixture was subjected three times to the following processes: pressing the tablets, sintering in pure oxygen atmosphere in the temperature 1230 K during 5 hours, slow cooling to the room temperature and grinding to powder with granule dimensions of 5 - 20 μm.

Using the superconducting powder (the material of the sample P300) and a special die a cylindrical samples with 4 mm diameter and 30 mm length were prepared with four connections made of Au wire with 0.4 mm diameters. They were placed symmetrically along the axis of the samples. The distance between the middle connections was 22 mm (the voltage electrodes) and between the extreme ones - 26 mm (the current electrodes).

Such a samples were sintered in the temperature 1230 K in O_2 atmosphere during 5 hours, and then they were slow cooled at a rate 2 K/min to the temperature 830 K (sample S830) and 300 K (sample S300).

The critical currents of the samples S830 and S300, measured by the four-probe method, were respectively 2.8 A/cm^2 and 64 A/cm^2 (T = 77 K) (the critical current of the P300 sample found by the same method was equal to zero).

The scheme of the system for measurement of the overall voltage is presented in the figure 2.

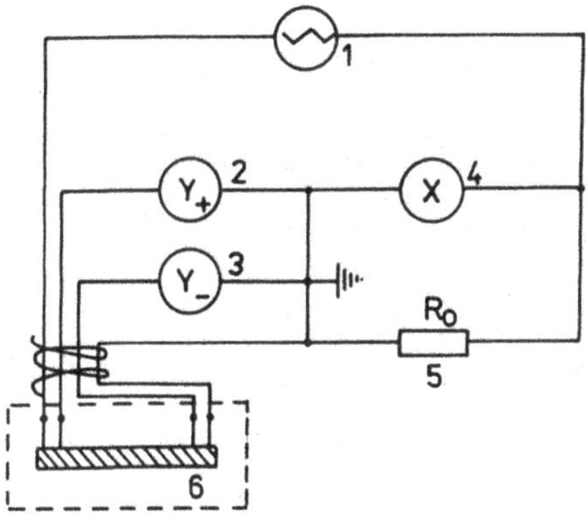

Fig. 2. The schematic of the system measuring the inductance of the fragment of the Y-Ba-Cu-O conductor; 1. - triangular signal generator (Imax = 50 mA, f = 300 kHz), 2. - positive input "Y" of the oscilloscope, 3. - negative input "Y" of the oscilloscope, 4. - the "X" input of the oscilloscope, 5. - measurement resistance (R_o = 42.2 Ω), 6. - the investigated sample.

A typical time dependence of the $u = u_R + u_{in}$ voltage between the voltage electrodes is presented in the figure 3, and the value of the u voltage as a function of the sample current - in the figure 4 (in all the cases triangular current signal was used).

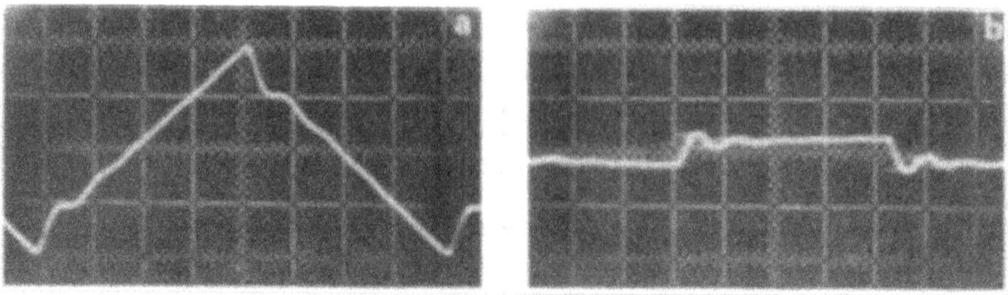

Fig. 3. Oscillogram of u(t) of the S300 sample; u axis - 1 cm = 1 mV; T = 300 K (a), T = 77 K (b).

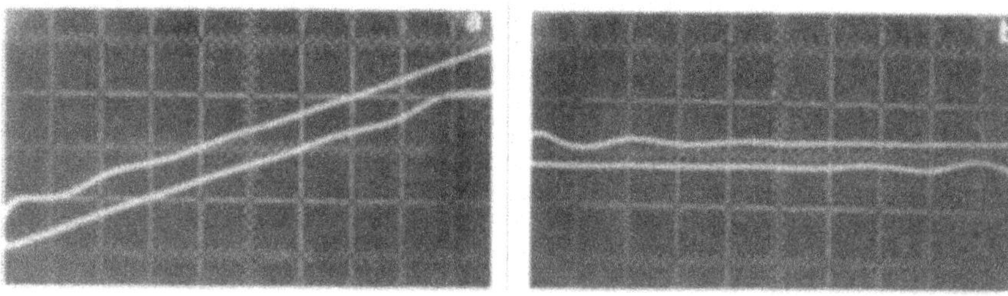

Fig. 4. Oscillogram of u(I) of the S300 sample; u axis - 1 cm = 1 mV; T = 300 K (a), T = 77 K (b).

3. Experiment B - the magnetic permeability measurements

For the measurements of the magnetic permeability μ_{rs} we used Y-Ba-Cu-O cylinders with D_{ex} = 10 mm, D_{in} = 0.5 mm and 20 mm long made of:

1. material from the P300 sample - superconducting powder, pressed under 10^7 Pa pressure, covered by ARALDIT epoxy resin;
2. ceramic sinter S830;
3. ceramic sinter S300.

The u_{in} measurements were carried out in the LN temperature using measurement system presented in the figure 2. The results of the measurements of the relative equivalent magnetic permeability μ_{rs} and the equivalent diameter D_s are gathered in the table 1.

Table 1.

The superconductor sample	I_c A/cm^2	μ_{rs}	$\frac{D_s}{D_{ex}}$
P300	0	0.645	0.198
S830	2.8	0.245	0.527
S300	64	0.037	0.911

The figure 5 presents the dependence of μ_{rs} on the intensity of constant external magnetic field parallel to the sample axis, determined on the basis of u_{in} measurement.

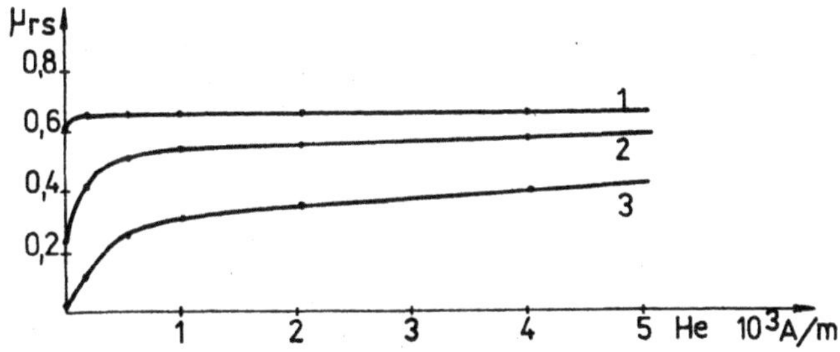

Fig. 5. The influence of the constant magnetic field on relative magnetic permeability μ_{rs} of the samples: P300 (1), S830 (2) and S300 (3); T = 77 K.

Relatively strong influence of the H_e field on the μ_{rs} value proves that the destructive interaction between the field and the intergranular junctions occurs already at low values of H_e ~ 100 A/m. This idea is confirmed by notable absence of the magnetizing field influence on μ_{rs} of the P300 sample with a structure without intergranular junctions.

In [1] a typical value of the active permeability of $Y_1Ba_2Cu_3O_{7-y}$ μ_{eff} = 0.3 was obtained. This value corresponds to the μ_{rs} permeability of the sample in the form of isolated superconducting granules (P300 sample) and the permeability μ_{rs} of the sample S830, where the granules are connected by the lattice of junctions with particularly low interaction energies and (probably) with somewhat exceeded percolation threshold.

The value μ_{rs} = 0.037 for the S300 sample with high content of inter-granular superconducting phase (maximum I_c current) proves that in the fields $H_e \to 0$ the granules together with the superconducting inter-granular regions considerably enlarge the region of internal shielding.

4. Conclusions

1. We describe a method for the measurement of the inductivity of cylindrical ceramic superconductor, which makes possible the determination of substitute magnetic permeability μ_{rs} in the case of cylindrical symmetry of the magnetic field.
2. The above method enables investigation of the magnetic properties in very weak specific fields so it is possible to reveal the presence of very weak intergranular junctions with interaction energies much lower than the mean energy in the structure of a massive ceramic superconductor .

5. References

1. J.R.Clem, *Physica C* **153-155** (1988) 50-55.
2. M.K.Wu et al. *Phys. Rev. Lett.* 9 (1987) 908-910.

THE CONTACT RESISTANCE IN THE YBaCuO/Ag SYSTEM

V.M. SVISTUNOV, A.I. D'YACHENKO, V.YU. TARENKOV
Physico-Technical Institute, Ukrainian SSR Academy of Science
340114 Donetsk, USSR

and

J. LESZCZYNSKI, J. JACKIEWICZ
Department of Materials Science
Institute of Principles of Electrical Power Engineering I-12
Technical University of Lodz
ul. B.Stefanowskiego 18/22, 90-924 Lodz, Poland

ABSTRACT

The characteristics of the YBaCuO/Ag contact have been investigated at high measurement current densities. It is found that high hydrostatic pressure does not change the transition SC ceramics-metal resistance. This fact points at the absence of dielectric transition layer in the Ag/ceramics contact region. The thesis was accepted that relatively high electric resistance of the contact silver-ceramics may be due to high difference of the characteristics of electron subsystems of SC ceramics and normal metal.

1. Introduction

There is considerable interest to understand the physical processes which define the transition resistance at the SC ceramics - normal metal boundary [1,2]. For practical applications, the choice of the contact metal is very important, because it determines the possibilities of its application in high current devices and in the production of weakly coupled weak-junction Josephson structures with high exploitation parameters.

There exist several opinions on the physics of the processes in the transition layer with relatively high resistivity depending on different contact preparation methods [1,2]. There is a question whether the resistance is connected with the appearance of the barrier as the effect of high chemical activity of the HTSC ceramics, or whether the high resistance is caused by considerable difference in the electron structure of the contact materials. The answer to this question may give advice for the search of optimum technological solution.

It seems that the problem may be solved by investigating the contact resistance reaction to high hydrostatic pressure. It is well known that in the case of tunneling between two metals, the tunneling current increases exponentially at small changes of the barrier thickness [3]. This property is the basis of the experiments on the electrical properties of the contact here described here.

2. The samples preparation and their characteristics

The superconducting ceramics plates with dimensions 1x0.7x10 mm^3, obtained by pressing the 1:2:3 phase powder at 40 kBar pressure, underwent rapid heating up to 1040 °C. The samples were kept in this temperature for 15 min, and then they were placed in a region of 980 °C and annealing for 6 hours with slow cooling afterwards.

The contacts for critical current measurements and the transition resistivity of the Ag/YBaCuO ceramics were made of pure (99.99%) silver pieces fused into the ceramics surface by narrow flame of an alcohol burner. The contact surface of the current and voltage electrodes was 1 ÷ 1.5 mm^2, the probes for contact properties measurements were $(50)^2$ ÷ $(100)^2$ μm^2 of surface. There were 2 ÷ 3 semicircular microcontacts with necessary minimum dimensions produced on every individual sample. The current and voltage electrodes and the investigated microcontacts displacement are presented in the insertion in the figure 1.

The samples with fused in contacts were again annealed in 920 °C during 0.5 hour and subsequently kept in pure O_2 atmosphere in 460 °C during 6 hours. The second annealing temperature was chosen to guarantee minimum silver diffusion into the ceramics.

The measurements leads were soldered to the silver contact droplets with indium; the indium - ceramics resistance was not taken into account because of its value, very high in comparison with the one of the silver - ceramics contact.

3. Experiment

The figure 1 presents the R(T) characteristics of the ceramics sample measured by the four-probe method and (by the contact silver droplet) by the three-probe method. Some widening of the transition curve, notable in the case of the three-probe method, is due to the increased measurement current density in the case of a contact with much lower surface (~$(50)^2$ μm^2).

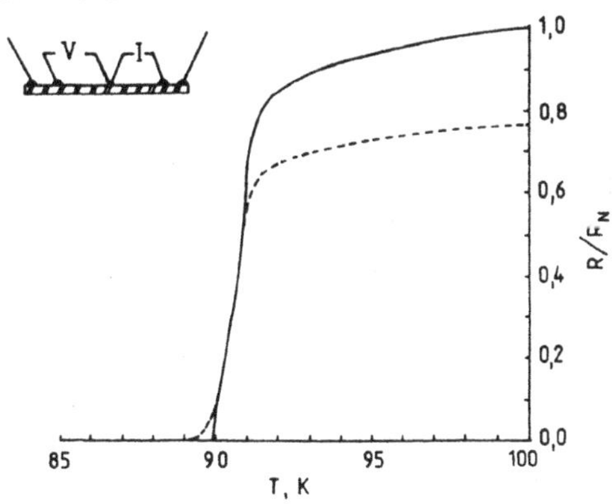

Fig.1. The resistance-temperature characteristics of the YBaCuO ceramic sample, obtained by four- (solid line) and three- (dotted line) probe method. The insert presents the displacement of the contact electrodes.

It is self evident, that the temperature dependence of the contact resistance reflects the changes of the contact specific resistivity and the resistivity of the adjacent ceramics layers, i.e. that the four-probe measurement method principle is violated. And so, correct measurement of the contact resistance may be conducted only in temperatures $T < T_c$.

In the figure 2 the I-U characteristics of the contact, obtained in LN temperature, are presented.

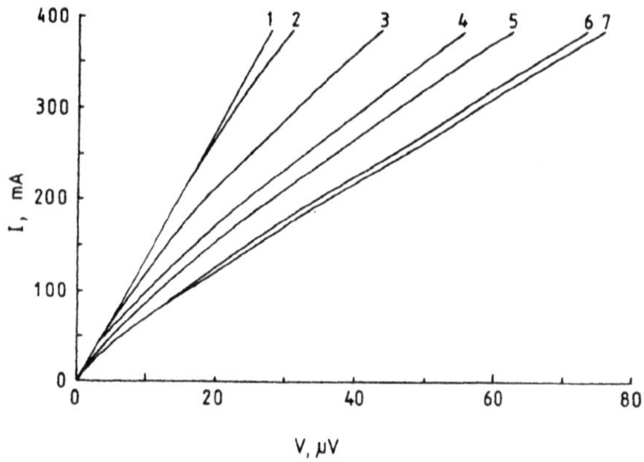

Fig.2. The current-voltage characteristics of the Ag/YBaCuO contact in various magnetic fields: He = 0 (2), He = 200 Oe (3), He = 300 Oe (4), He = 400 Oe (5), He = 1 kOe (6), He = 2 kOe (7); the tangent, obtained during the determination of the current using the I(V) dependence, deviates from the Ohm law (1).

There may be seen a straight portion in the curves, reflecting the real resistance of the transition layer. Starting from some current value, the linearity of the characteristics is violated, this violation is connected with the appearance of the resistance zones in the ceramic regions immediately adjacent to the silver contact.

The contact resistance value R_c calculated on the basis of the visually determined surface of the Ag/YBaCuO contact, equals to $R_c = 1.5 \cdot 10^{-9}\ \Omega cm^2$. This value was obtained at the measurement current density $j = 10^4\ A/cm^2$.

So high current density appears only in the immediate vicinity of the contact, its part inside the ceramics rapidly decreases due to the current spreading. The deviation from the linearity of the I-U characteristics of the contact is caused by the appearance of resistive areas in near-contact part of the ceramics, where the measurement current density exceeds the critical value of the investigated sample ($j_c = 5.7 \cdot 10^3\ A/cm^2$).

The magnetic field influence decreases the critical current of the ceramics $j_c(H_e) < j_c(0)$ and, correspondingly, leads to an increase of the sum resistance of the contact layer and the near-contact area at measurement currents values $j > j_c(H_e)$. At $j < j_c(H_e)$ the Ag/YBaCuO contact resistance does not depend on external magnetic field.

It follows that the Ag/YBaCuO transition resistance does not depend on the magnetic field $H_e \leq 2$ kOe.

In LN temperatures there is no visible proximity effect, because the $R_c = 1.5 \cdot 10^{-9}\ \Omega cm^2$ value is sufficiently high. The $R_c = 10^{-8} \div 10^{-5}\ \Omega m^2$ value, most often obtained during experiments [4 - 6], is sufficiently high as compared to the typical resistance values of metals and may be due to the formation of dielectric layer at the boundary ceramics - metal [1,2]. Assuming the existence of such a layer, the current flow should display tunnel characteristics. In our case, the lack of significantly thick layer with a semiconductor properties is proved by the observed weak temperature dependence of the contact resistance.

One of the methods for determining the transition layer character is an experiment investigating the contact resistance reaction to pressure. The measurements of tunnel transitions of the type Al - Al_2O_3 - metal at 10 kBar pressure shown a resistance decrease by 2.5 ÷ 3 orders of magnitude [3].

Assuming the compression permeabilities of the dielectric layers in the ceramics - metal contact and in Al_2O_3 are of the same order, than, in the case of a dielectric layer formation, there should be a strong pressure influence on the metal - ceramics contact observed.

High pressure, 10 ÷ 12 kBar, was produced in a beryllium bronze chamber. The pressure and temperature sensors were placed in the chamber in immediate vicinity of the sample.

In the figure 3 the dependence of the contact resistance on the magnetic field, for the hydrostatic pressures 1 and 10 kBar, is presented.

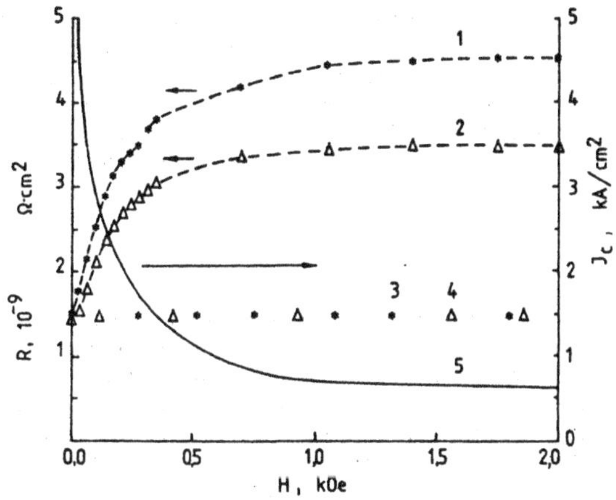

Fig.3. The $R_c(H_e)$ dependence of the contact for P = 1 kBar (*) and for P = 10 kBar (>). The curves 1 and 2 were obtained for the ceramics currents $j > j_c(H_e)$, the curves 3 and 4 for $j < j_c(H_e)$; the curve 5 presents the $j_c(H_e)$ dependence of the ceramics.

As may be seen, the pressure does not cause any changes of the contact resistance, if the measurement current density does not exceed the critical value $j_c(H_e)$ of the ceramics. If the measurement current density exceeds the critical value, the $R_c(H_e)$ curves corresponding to various pressures "diverge". It is a consequence of the critical current j_c increase under the pressure influence [4]. Hence, $R(H_e)$ has a constant value in those magnetic fields, for which j_c of the sample also remains constant (figure 3).

4. Results

The conducted experiments point at the absence of a tunnel barrier in the area of the contact SC ceramics - metal, i.e. the contact displays metallic character. The high value of the transition area resistance may correspond to substantial difference between the characteristics of the electron subsystems in ordinary metal and in a ceramics of the type YBaCuO. If the difference between the Fermi velocities of the conductors being in contact, v_{f1} and v_{f2}, is considerable, it causes high value of the coefficient of electron reflection on the contact boundary:

$$R = (v_{f1} - v_{f2})^2/(v_{f1} + v_{f2})^2 \cong 1 \tag{1}$$

and corresponding decrease of the coefficient D of the transmission through the contact $D = 1 - R \ll 1$. In effect, the two metal boundary assumes the contact resistance:

$$R_c = (\hbar/e^2)(2\pi/k_f)^2\langle\cos(\theta)D(\theta)/R(\theta)\rangle^{-1} \tag{2}$$

where: k_f is a Fermi impulse of the metal with lower Fermi velocity, θ - the angle between the velocity vector of an electron in this metal and the normal to the contact surface. The formula (2) for the resistance of the contact in normal state holds also for the contact of superconductor with metal in normal state [5]. The physical sense of the contact resistance, according to (2), is more clear if the formula is applied to the resistance "quantum" $R_Q = \hbar/e^2 \cong 4.1\ k\Omega$ and to the length of the de Broigle wave of the Fermi electron $\lambda_f = 2\pi/k_f$. Then, for $D \ll 1$, the contact resistance becomes:

$$R_c \cong R\cdot\lambda_f^2\langle D\cos(\theta)\rangle^{-1} \tag{3}$$

If the contact contains dielectric layer with thickness d and barrier height φ, the transmission coefficient is $D \sim \exp(-d(2m\varphi/\hbar^2)^{1/2})$. Usually $D \sim 10^{-3} \div 10^{-6}$. Following these assumptions and taking $k_f = 10^8\ cm^{-1}$, from (3) we get the characteristic resistance of the tunnel contacts $R_c \cong 10^{-8} \div 10^{-5}\ \Omega cm^2$. The value of $R_c \cong 10^{-8}\ \Omega cm^2$ is close to the values observed for Ag/YBaCuO contacts. However, in a tunnel contact, even at low changes of the thickness d of the dielectric layer, the transmission D increases exponentially causing a decrease of R_c, which was not observed in experiments.

Let's consider the possibility of explaining the high value of the contact resistance within the framework of the model taking into account only the high difference of the Fermi velocities in the SC ceramics and in normal metal. In silver $V_f^{Ag} \cong 1.4\cdot10^8$ cm/s, while in the YBaCuO ceramics, because of low carriers density, $V_f^{YBCO} \cong 3 \div 6\cdot10^7$ cm/s. Finally, using (2), we get a minimum Ag/YBaCuO contact resistance:

$$R_c^{min} \cong (R_Q/4)(2\pi/k_f)^2 V_f^{Ag}/V_f^{YBCO} \cong 10^{-10}\ \Omega cm^2 \tag{4}$$

(k_f was assumed to be $\sim 3\cdot10^7\ cm^{-1}$).

Comparison of the obtained R_c^{min} with the experimental values $R_c = 10^{-9} \div 10^{-10}$ Ωcm^2 shows that practically ideal silver - ceramics boundary is technologically achievable. At the same time the minimum contact resistance of the boundary is achieved, being determined by substantial difference between the electron subsystems of silver and copper oxide ceramics (especially in the ab plane, normal to the c axis).

In some studies extensive data may be found concerning the contact resistance in the range $10^{-11} \div 10^{-12}$ Ωcm^2 [6,7]. It should be noted that the technological conditions described in these works favored formation of widely branched contact surfaces, being effect of intensive silver diffusion in elevated annealing temperatures. In this case the effective contact surface could be much greater than the one observed visually. We are of the opinion that this is the cause of the difference between the R_c values obtained in [6,7] and in the presented R_c^{min} estimation. In our case ($R_c \sim 1.5 \cdot 10^{-9}$ Ωcm^2) the production conditions of the contact were defined in such a way that the diffusion coefficient cause minimum correction to the effective transition resistance calculation.

5. Conclusions

Based on the response of the silver - ceramics contact resistance to the influence of magnetic field and pressure it is found that in technologically achievable conditions it is possible to obtain contact junctions with metallic character with resistance $R_c \cong 10^{-9}$ Ωcm^2. This value corresponds to the estimation of the minimum resistance R_c^{min} of the contact of two metals with strongly differing Fermi velocities (of the copper oxide ceramics and silver). The lack of the contact resistance susceptibility to the influence of strong magnetic fields points at the technological possibility of producing resistive high - current composites basing on SC ceramics in metal matrix.

6. References

1. J.Talvacchio, *IEEE Trans. on Comp. Hybrids and Manufact. Technol.* 1 (1989) 21-31.
2. M.Yu.Kuprianov and K.K.Likhariev, *Usp. Fiz. Nauk* 5 (1990) 49-87 (rus).
3. V.M.Svistunov, M.A.Belogolovskij and O.I.Chern'yak, *Usp. Fiz. Nauk* 1 (1987) 31-66 (rus)
4. V.M.Svistunov, V.Yu.Tarenkov and A.I.D'yachenko, *Mod. Phys. Lett.* 10 (1990) 645-650
5. A.V.Zajcev, *Zur. Eksp. Teor. Fiz.* 5 (1984) 1742-1758
6. S.Jin, M.E.Davis, T.H.Tiefel et al., *Appl. Phys. Lett* 25 (1989) 2605-2607
7. Y.P.Liu, K.Warner, C.Chan et al., *J. Appl. Phys.* 11 (1989) 5514-5517

MAGNETIC IRREVERSIBILITY AND CRITICAL CURRENT IN YTTRIUM BASED HIGH TEMPERATURE SUPERCONDUCTING COMPOUNDS

J. Sosnowski, R. Clad [1/], R. Kuentzler [1/]

Electr. Inst., 04-703 Warszawa, Pozaryskiego 28, Poland

[1/] - IPCMS, Strasbourg University, Strasbourg, France

ABSTRACT

Magnetic irreversibility in high T_c yttrium based compounds prepared in air atmosphere has been investigated. The measurements of full and minor hysteresis loops of magnetization curves, flux trapping and flux creep effects have been performed in liquid helium temperature using sensitive digital magnetometer of the integrating type. From minor hysteresis loops measurements has been determined critical current density using developed numerical method. In the flux trapping experiments we have observed the anomalous behaviour of the remanent moment for the samples doped with iron impurity. Lower critical or first penetration field varying between 4 and 20 mT at liquid helium temperature has been estimated for the samples with different composition. The magnitude of pinning potential (of the range 13-18 meV) as well as activation energy for the flux creep effect was also determined.

1. Introduction

Great attention is devoted recently to the investigations of magnetic irreversible properties of the high temperature oxide superconductors. This is connected with specific features of the granular structure of these materials as well as increasing importance of magnetic methods in determining the superconducting critical current of these ceramical materials. Resistivity measurements of the critical current do not bring in the case of bulk samples satisfactory results, so inductive methods are developed.

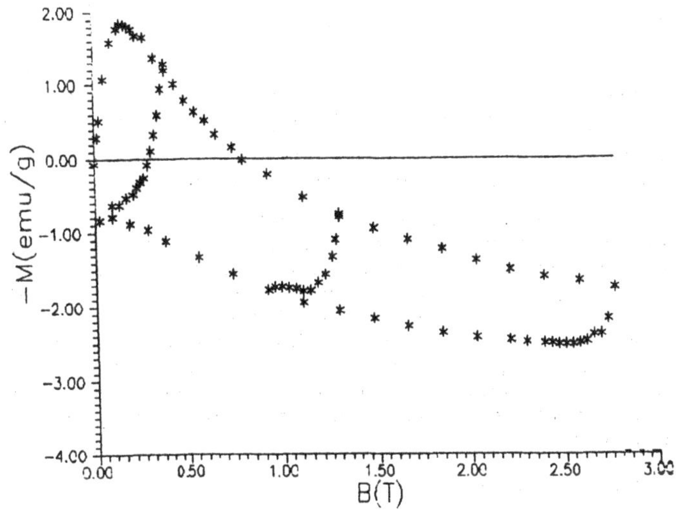

Fig. 1: The magnetization hysteresis measured at 4.2 K on $YBa_2Cu_{2.85}Ag_{0.15}O_{7-\delta}$ ceramics

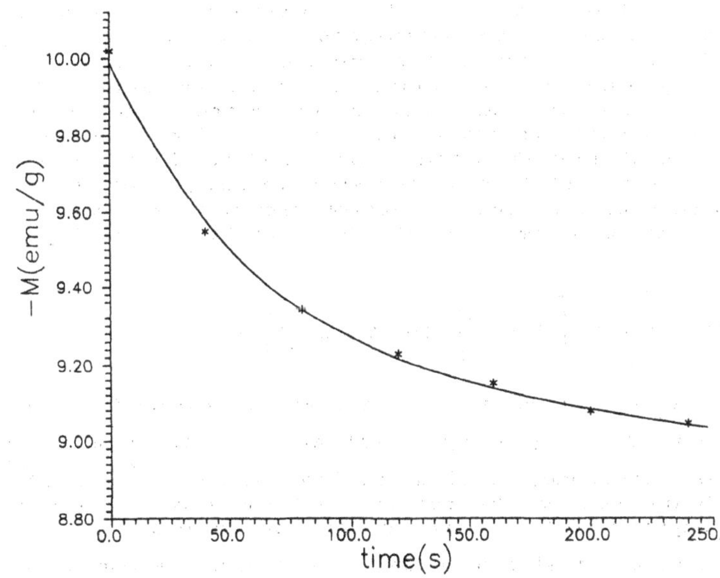

Fig. 2: Time dependence of magnetization decay for $YBa_2Cu_3O_{7-\delta}$ sample at B=1. T, T =4.2 K.

2. Sample preparation and experimental procedure

The measurements have been performed on the yttrium based samples of the type $YBa_2Cu_3O_{7-\delta}$, on Ag doped material and on samples doped with iron impurity prepared by firing the ceramics in air atmosphere. The ternary $YBa_2Cu_3O_{7-\delta}$ compound has been presesed into pellets, sintered in flowing oxygen at 1190K or under air at 1240K. On the above samples magnetization irreversibility has been measured at the temperature of 4.2 K using the sensitive digital magnetometer of the integrating type, connected with two oppositely wound pick up coils. The external magnetic field of the amplitude up to about 5 T has been generated using solenoidal superconducting magnet. In that equipment also flux creep experiments and flux trapping investigations have been performed.

3. Experimental results

The results of measurements full and small hysteresis loops of magnetization curve $YBa_2Cu_{2.85}Ag_{0.15}O_x$ ceramical sample at 4.2 K are shown in Fig. 1. As it follows from that Figure the sample indicates significant irreversibility, which denotes large critical current values. In Figure 1 is observed also transition of magnetic moment into paramagnetic side for large magnetic fields, which is probably connected with existence of silver impurity. The measured magnetic moment decreases with time, what is related to flux creep effect. An example of variation magnetic moment with time is shown in Fig. 2 for the case of Y-Ba-Cu-O sample annealed at air in 1240 K. From above experimental relations we have determined the value of pinning potential and activation energy investigated samples according to Griessen et al. formulas [1] of the type:

$$M(t) = M(0)\left[1 - \frac{kT}{U}\ln(1+\omega t_B)\right] \qquad (1)$$

where $E_a = U/kT$ denotes the activation energy for flux creep process, U - is pinning potential, k - Boltzmann's constant while ω frequency varying usually between 10^8 and 10^{12} s^{-1}, which depends on the geometry of the probe, efficiency of the hopping mechanism and others parameters. t_B - denotes the time of first measurement. In present research we have obtained the following values of these parameters at 4.2 K assuming $\omega = 10^9$ s^{-1}, t_B = 20 s: E_a = 40, U=14.5 meV at

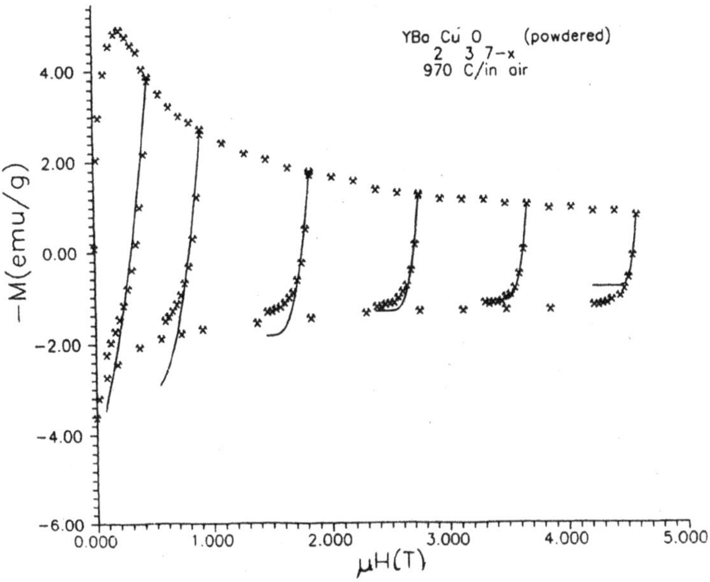

Fig. 3: Comparison of theoretical (solid line) and experimental shape of small hysteresis loops for YBaCuO powdered sample annealed at 1240K in air atmosphere, T=4.2K.

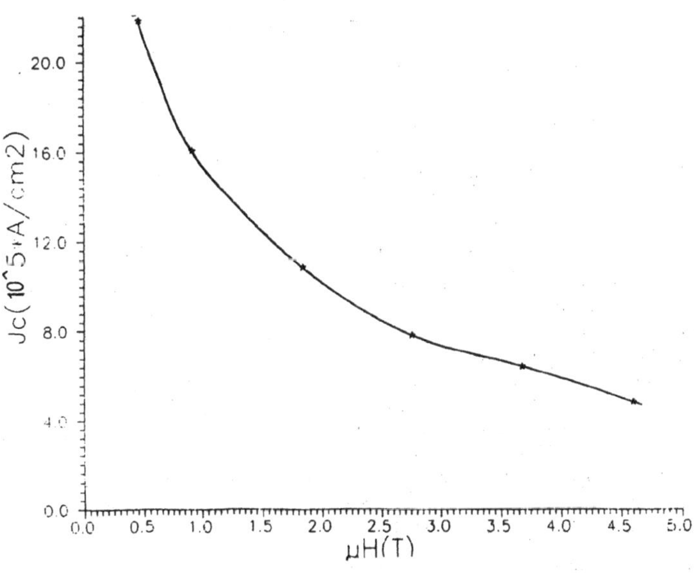

Fig. 4: Magnetic field dependence of the critical current density of $YBa_2Cu_3O_{7-\delta}$ annealed at 1240K in air powdered cermics obtained from small hysteresis loops investigations

B=3.7 T for sample annealed at 1190 K and E_a = 53, U = 18.2meV at 1.1 T for sample annealed at 1240 K.

Since flux creep modyfies the shape of magnetization curves so also influences small hysteresis loops and critical current values to which subject is devoted this paper. As it is well known the resistivity measurements do not bring frequently the satisfactory critical current values in bulk materials what is connected with granular structure of ceramics. From other side inductive methods of determining critical current from full magnetization curves are based usually on Bean's critical state model, which in fact neglects any details of magnetic induction distribution inside the superconductor assuming linear profiles. Therefore we have developed another method of determining critical current density and also pinning forces, which is based on small hysteresis loop analysis. An advantage of this method is that it allows to consider more precisely the real magnetic induction profile in the ceramical material treating the various phenomenological parameters as constant only in the thin region of magnetic field variation of the dimension the small hysteresis loop.

The comparison of the shape minor hysteresis loops obtained from theoretical developed model based on Yrie-Yamafuji approach with measured data is shown in Fig. 3. This fitting procedure has been performed using especially prepared numerical program, which assures least square deviation of theoretical model to experimental data. In that approach we do not need precise exactly the nature of this current, for instance its relation to Josephson's junctions array, but only approximate final magnetic field dependence with power-like relation. Basing on above fitting procedure various parameters describing pinning mechanism have been obtained for each minor hysteresis loop, which next allowed us to determine the critical current density of ceramic. The dependence of critical current density normalized to grain dimension on magnetic field at 4.2 K is shown in Fig. 4 for YBaCuO sample annealed at 1240 K. As it follows from Fig. 4 critical current density reaches the value of $1.5*10^6$ A/cm^2, which is very promising result from the point of view of applicational properties of these ceramical materials.

Important part of performed investigations is connected with flux trapping experiments. As we have shown in previous papers [2-3] these investigations can be useful for determining lower critical field or rather first penetration field. This value is determined from appearance of initial flux trapping when magnetic field decreases to zero, from maximal value B_m or as we have it done in present paper as the value of B_m for which flux trapping reaches 1% of maximal trapped flux. The results of flux trapping

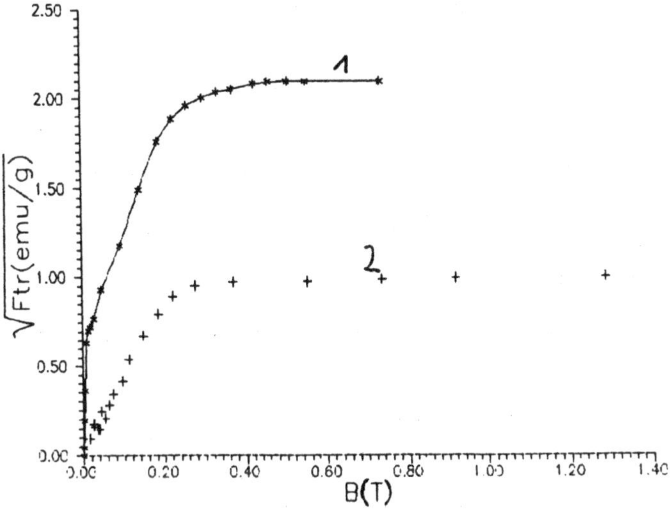

Fig. 5: The results of flux trapping experiments performed on bulk YBaCuO sample (2) and Fe doped sample (1) at 4.2 K

investigations shown in Fig. 5 in the root scale for yttrium and iron doped ceramical samples indicate however on anomalous behaviour of flux trapping in iron doped sample. Values of first penetration field are equal to about 4 mT for bulk iron doped samples, 7 mT for powdered while 20 mT for pure powdered or bulk yttrium samples, at 4.2 K.

Experimental part of that work has been fulfilled during author's stay at Strasbourg University, France.

4. References

1. R.P. Griessen, C.F.J. Flipse, C.W. Hagen, J. Lensink, B. Dam, G.M. Stollman, J. Less Common Metals 151 (1989) 39.
2. J. Sosnowski, Phys. Stat. Sol. (a) 60 (1980) K73.
3. J. Sosnowski, R. Clad, R. Kuentzler, Phys. Stat. Sol. (a) 87 (1985) 575.

NORMAL AND INVERSE PEAK EFFECT IN HIGH T_C OXIDE SUPERCONDUCTORS

SOSNOWSKI J.

Electrotechnical Institute, Warszawa, Poland

and

DATSKOV V.I.

Joint Institute for Nuclear Research, Dubna, USSR

ABSTRACT

In the paper have been presented the results of investigations dynamical current voltage characteristics of the high temperature oxide superconductors. The existence of peak effect has been well established. The investigations of peak effect we have performed in the function of current amplitude and velocity of magnetic field sweep rate at different temperatures. The inverse peak effect has been also discovered while its dependence on current amplitude and magnetic field sweep velocity is presented. Theoretical model describing existence of peak effect based on analysis of asymmetry of magnetic induction distribution caused by transport current flow has been proposed for various pinning forces mechanisms.

1. Introduction

The applications of high temperature oxide superconductors are presently strongly dependent on the efforts in the achievment of higher critical transport currents densities. Current densities measured resistively on presently prepared samples do not bring else frequently satisfactory results. For improving our knowledge on transport current related phenomena in these materials is devoted just this paper.

2. Experimental procedure

The investigations of dynamical current-voltage characteristics of high Tc oxide ceramics described in present paper have been performed on various yttrium based samples in the form of parallelepiped of the length varying between 10 and 14 mm and with thickness 1-4 mm. The voltage

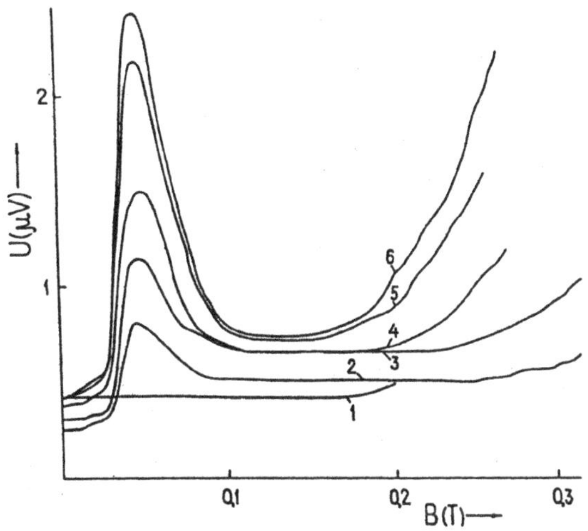

Fig. 1: Experimental dependence of peak voltage on Y-Ba-Cu-O ceramical sample on the flowing transport current amplitude at T =60.5 [K], $\dot{B}$ = 53.3 [G/s], (1) I=160, (2) 180, (3) 190, (4) 200, (5) 210, (6) 215 [mA]

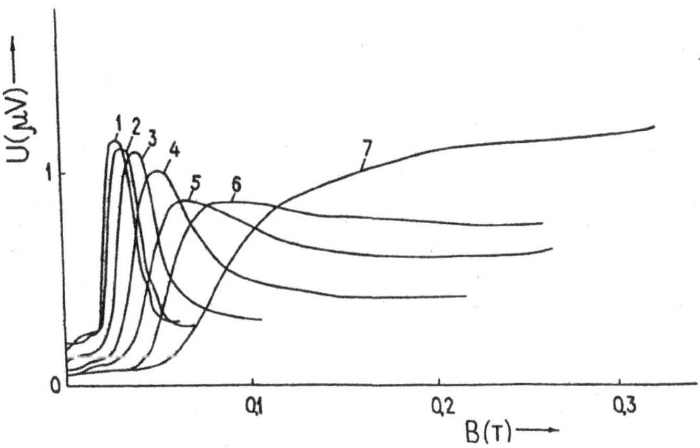

Fig. 2: Experimental dependence of peak voltage on Y-Ba-Cu-O ceramic on the velocity of magnetic field sweep rate for I=190 [mA] at T=60.5 [K]: (1) $\dot{B}$ = 10.67, (2) 16, (3) 32, (4) 64, (5) 106.7, (6) 160, (7) 320 [G/s]

was detected using four probe method, while the contacts were performed using silver paste. The magnetic field was generated with help of superconducting magnet of great homogeneity, which allowed to vary magnetic field in controlled way. The sistem of temperature control allowed to obtain and stabilize temperature in the region between liquid helium and nitrogen temperature.

3. Experimental results

The results of current-voltage characteristics measurements on Y-Ba-Cu-O ceramics are shown in Figs. 1-4. Fig. 1 presents the dependence on transport current amplitude induced voltage for temperature of measurements equal 60.5 K. While external magnetic field perpendicular to the surface of superconducting slab increased with rate 53.3 G/s then in some range of magnetic field amplitudes in superconducting slab arised rapidly the voltage, which amplitude next was decreasing. This voltage occured while transport current was flowing only. As it follows from Fig. 1 peak voltage is increasing function of current amplitude. Another unexpected result is seen in Fig. 2 which presents the dependence of induced voltage on the velocity of magnetic field sweep rate. The temperature of measurements was equal to 60.5 K. As it follows amplitude of peak effect and its shape is strongly influenced by magnetic field sweep rate. For rapid variation of magnetic field peak effect becomes more wide, while its amplitude slowly decreases and finally effect disappears. Also maximum of peak effect shifts then into the region of higher magnetic fields. Peak effect is also influenced by the temperature of measurements. For higher temperatures peak voltage increases [1,2]. When magnetic field is stopped voltage disappears what indicates on dynamical features of phenomenon. This effect therefore can be the base of high temperature superconducting device for detecting velocity of rotation superconducting machines. For proper work of such device it is however necessary to recognize the nature of discovered I-V anomaly, to which subject will be devoted next part of present paper. The results of measurements current-voltage characteristics at T= 61 K on other investigated YBaCuO sample are shown in Figs. 3-4. For that sample transport current has initiated origin of normal zone in the sample. For better leglibility the curves shown in Figures are shifted at the voltage axis. As it follows from Fig. 3 in that case we have observed different behaviour of peak voltage with current amplitude. Induced voltage has opposite direction now. This phenomenon can be therefore treated as inverse peak effect. The dependence of inverse peak effect on magnetic field sweep rate is shown in Fig. 4. This dependence is in fact very similar to the behaviour of

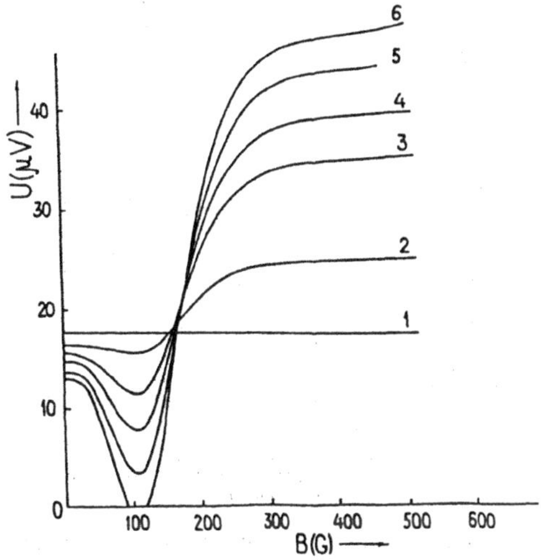

Fig. 3: Experimental dependence of peak voltage on other sample of Y-Ba-Cu-O ceramic on the flowing transport current amplitude at T= 61.5 [K], B=32 [G/s], (1) I=0, (2) 20, (3) 30, (4) 35, (5) 40, (6) 45 [mA]

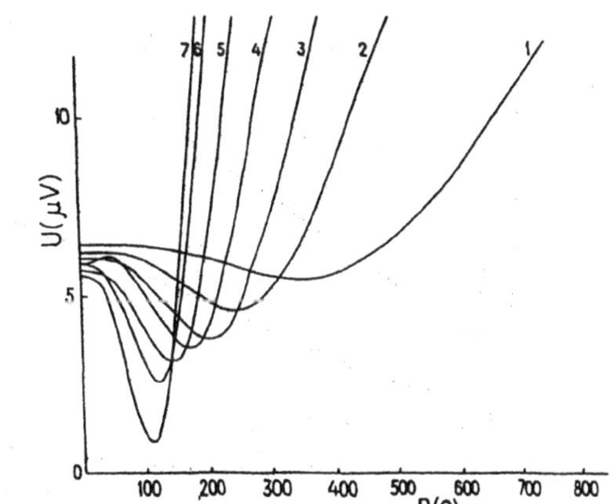

Fig. 4: Experimental dependence of peak voltage on Y-Ba-Cu-O ceramic on the velocity of magnetic field sweep rate for I=30 [mA] at T = 60.7 [K] for: (1) B = 320, (2) 160, (3) 106.7, (4) 80, (5) 53.3, (6) 40, (7) 32 [G/s])

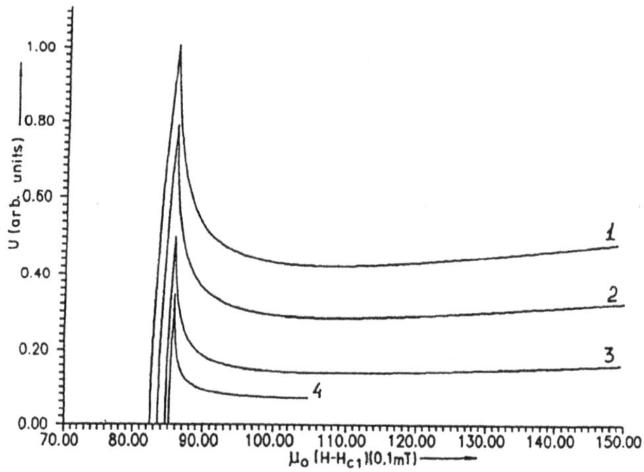

Fig. 5: Dependence of generated voltage in arbitrary units on transport current amplitude for pinning force mechanism described by Irie-Yamafuji (I-Y) type model for following values of material parameters: $\alpha = 0.6$ $[T^{1.7}/m]$, $\gamma = 0.7$, $x_m = 2*10^{-4}$ [m], $\dot{B} = 16*10^{-3}$ [T/s], (1) I = 0.3, (2) 0.2, (3) 0.1, (4) 0.05 [A].

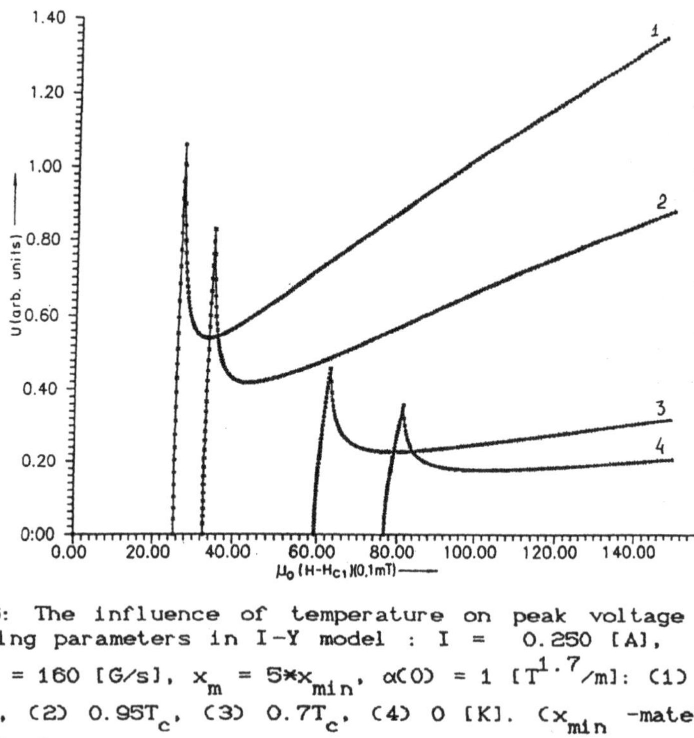

Fig. 6: The influence of temperature on peak voltage for following parameters in I-Y model : I = 0.250 [A], γ = 0.7, $\dot{B}$ = 160 [G/s], $x_m = 5*x_{min}$, $\alpha(0) = 1$ $[T^{1.7}/m]$: (1) T = $0.97T_c$, (2) $0.95T_c$, (3) $0.7T_c$, (4) 0 [K]. (x_{min} -material parameter).

normal peak effect but the polarization of effect is now opposite.

4. Theoretical model of the peak effect

Developed model is based on an analysis of asymmetrical penetration of magnetic induction at both sides of superconducting slab during transport current flow. For the case without transport current flux does not cross the middle of the sample, so measured then electric field disappears. If transport current flows magnetic flux penetrates into second side of the slab and electric field is generated in that case. It appears then also the region in which the increase of magnetic induction inside the sample is more slow, what brings just the peak-like behaviour of generated voltage. For pinning forces model of the power-like form $\partial B/\partial x=\alpha/B^{\gamma}$ generated at the depth x electric fields for nontotal E_n and total E_t penetration of the magnetic induction are:

$$E_n = \dot{B}\,[(B+\Delta B)^{1+\gamma}-\alpha(1+\gamma)x]^{1/(1+\gamma)} * (B+\Delta B)^{\gamma}/\alpha \tag{1}$$

$$E_t = \dot{B}\,[(B+\Delta B)^{\gamma}\,B(x) - [(B+\Delta B)^{\gamma} + (B-\Delta B)^{\gamma}] * B_{av}]/2$$

where it has been introduced notation:

$$B_{av} = \{[(B+\Delta B)^{1+\gamma} + (B-\Delta B)^{1+\gamma}]/2 - \alpha(1+\gamma)x_m\}^{1/(1+\gamma)} \tag{2}$$

Temperature dependence of electric field is connected then with variation of parameter $\alpha(T)$. x_m - is halfwidth of sample, B external magnetic induction, while ΔB the shift of magnetic induction on the surface connected with transport current flow. Fig. 5 presents in arbitrary units the dependence of peak voltage on transport current amplitude calculated according to Eqs. (1-2), while Fig. 6 on temperature. Similar behaviour has been observed in experiment.

References

1. I.V. Velichkov, V.I. Datskov, M.M. Marinova, Communic. JINR, Dubna, P17-89-704 (1989) 1-7.
2. J. Sosnowski, V.I. Datskov, Electr. Inst. Works. 165 (1991) 51-70.

CRITICAL FIELDS OF HIGH-TEMPERATURE SUPERCONDUCTORS $REBa_2Cu_3O_7$ WHERE RE = Y AND Dy

J.CHMIST, A.KOŁODZIEJCZYK, R.ZALECKI AND T.ŚCIĘŻOR
Department of Solid State Physics, Academy of Mining and Metallurgy, 30-059 Cracow, Poland

ABSTRACT

The lower and upper critical fields as a function of temperature of $DyBa_2Cu_3O_7$ and $YBa_2Cu_3O_7$ were measured. The former was elucidated from the a.c. susceptibility versus temperature and magnetic field measurements for the fine powdered specimens and it turned out to be an intrinsic feature of the materials which can be compared with other experimental and theoretical findings. The latter was determined from the standard magnetoresistance measurements for bulk specimens and the values of $H_{c2}(T)$ were always drastically too small in comparison to the other estimations and measurements. The reason is the so-called granular effect and the penetration of very small magnetic field to an interior of inter-grain nonsuperconducting regions. We observed the linear temperature dependence of $H_{c1}(T)$ close to the transition temperature which is predicted by theory.

1.Introduction

The important role played by the so-called granular effect in different physical characteristics of the high-temperature superconductors (HTS) has been emphasized almost since their discovery. In these materials the short coherence length and the presence of insulating inter-grain regions or the so-called weak links are reasons of the granular characteristics. The granular characteristics include rather larger resistivity in the normal state, low critical currents and critical fields which are affected very much by the granular structure of HTS. In order to compare experimental results with theory one has to extract from experiment intrinsic properties of HTS or one has to include the granular effect to theory [1].

A goal of the present paper is to measure the critical fields of HTS and to compare them with the existing theories in order to conclude on possible pairing mechanism. We have focused our attention on the effect of granularity. In this purpose a method of determination of the intrinsic lower critical field $H_{c1}(T)$ from the a.c. susceptibility measurements has been applied. Also we have measured the magnetoresistance in order to find the temperature dependance of the upper critical field $H_{c2}(T)$. At the end of paper some possible limitations of the critical fields and currents due to granular effect has been discussed.

2.Experimental

The specimens were prepared by the standard sintering procedure but a special case was undertaken to regrind the material a few times between subsequent sinterings in the agate ball-mill in order to reach as homogenized material as possible. Dy- and Y-specimens chosen to the present experiments were ones among the best quality specimens whenever we managed to prepare in our sintering technology. The transition temperatures measured resistively and by a.c. susceptibility are T_c=91.2 K and 91.0 K for $DyBa_2Cu_3O_7$ and T_c=90.25 K and 91.0 K for $YBa_2Cu_3O_7$ respectively. The width of transitions are very small $T=T_{90\%}-T_{10\%}\approx 0.7$ K.

The oxygen contents were estimated from our thermogravimetric calibration curve [2] as being x=6.95 ± 0.05 for the both Dy- and Y- specimens. The resistance and magnetoresistance were measured by the standard four-points a.c. method, at the frequency of 6.72 Hz, for the applied current of 2 to 3 mA rms, for rod-like shaped specimens with dimensions 2x2x15 mm^3, mainly within the range of liquid nitrogen temperature close to T_c in the helium cryostat with the Pt-100 thermometer, and with about 1% accuracy.

The in-phase a.c. susceptibility was measured by the standard mutual inductance bridge operating at the frequency of 237 Hz with sensitivity $4x10^{-6}cm^{-3}g^{-1}$ μV^{-1}. The magnetic field was applied by the conventional copper liquid-nitrogen cooled solenoid for the small fields up to 3.5 kOe (~45 Am^{-1}).

3.Results and their analysis

In order to determine the lower critical field as a function of temperature the a.c. susceptibility as a function of applied magnetic field and temperature was measured for the bulk as well as the powdered specimen of $DyBa_2Cu_3O_7$ .The results are shown in Fig.1.The method of determination of H_{c1} from a.c. susceptibility is as follows.The Meissner state diamagnetic magnetisation is linear function of applied magnetic field, M=-H, up to or very close to the lower critical field H_{c1} and then the a.c. susceptibility, $\chi'=\delta M/\delta H$, should be constant.So, the H_{c1} is defined as the maximal field for which the constant χ'-value does not exist further or equivalently as a maximal field for which the χ'-value is still constant.The constant value does exist for properly powdered specimen but not for the bulk specimen (c.f. the Fig.1).Also for crush powdered specimen the constant value of χ' does not correspond to the lower critical field yet. This is due to the granular effect. The plot of $H_{c1}(T)$ is shown in Fig.2.The values of H_{c1} are always about two times larger than the values obtained for bulk specimen or even single crystals by the standard magnetisation measurements. This is because the lack of granular effect for powdered specimen and then always H_{c1} is larger than for bulk specimen with granular effect in.The superconductors consist of superconducting grains separated by effectively nonsuperconducting or weakly superconducting intergrain barriers due to short coherence length. That is why even very small applied magnetic field flux penetrates easily to the bulk specimen through the intergrain boundaries and the field is much lower than the intrinsic inside-grain lower critical field H_{c1}. For powdered specimen the intergrain structure is destroyed and then it is possible to measure intrinsic H_{c1}-value.

In order to determine the upper critical field, the magnetoresistance-like measurements were carried out for the bulk $YBa_2Cu_3O_7$ specimen. We put our attention on the region very close to T_c because a primary purpose was to compare the experimental data of $H_{c2}(T)$ and $-(dH_{c2}/dT)_{Tc}$ with theory, for example with the Ginzburg-Landau-Abrikosov-Gorkov (GLAG) theory [3].

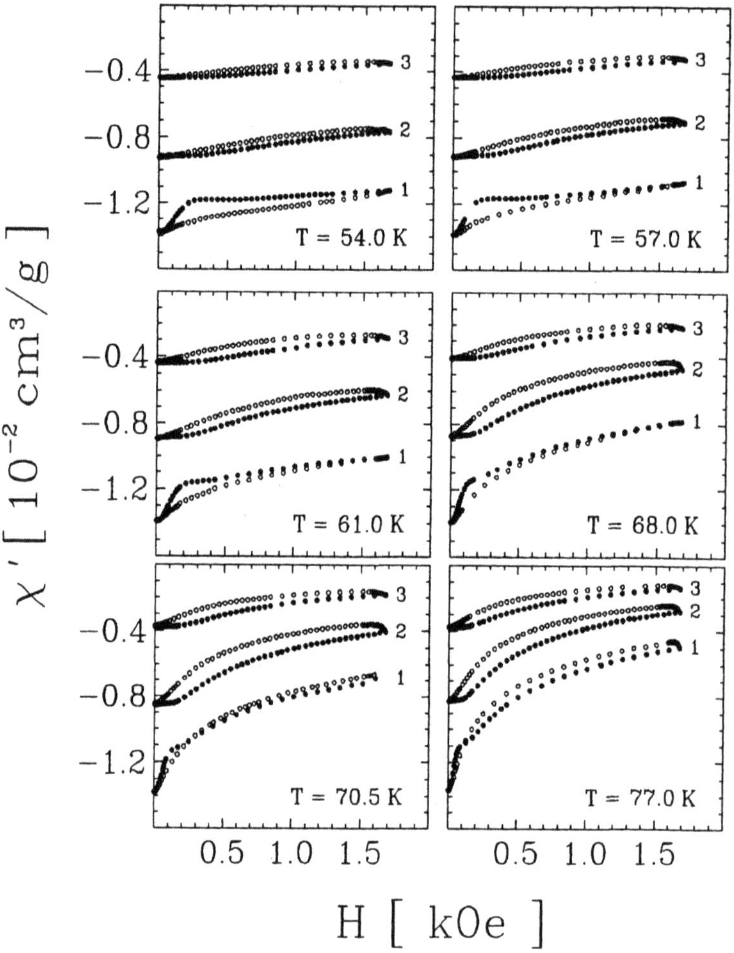

Fig.1. The field dependencies of a.c. susceptibility χ' of $DyBa_2Cu_3O_7$ for the bulk specimen -1, for the crush powdered specimen -2, for the fine powdered specimen -3 and with the increasing field (●) and the decreasing field (O).

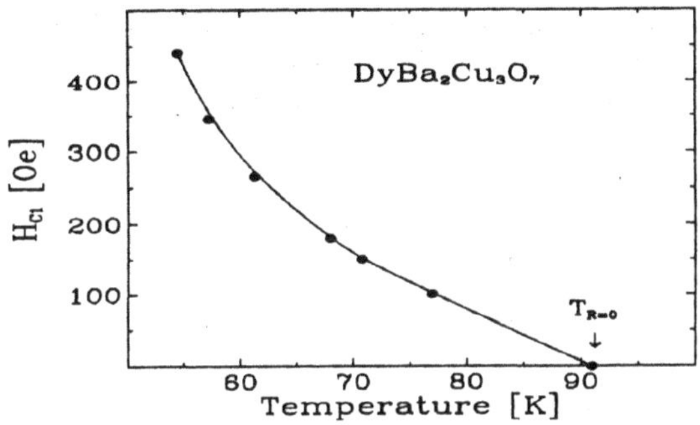

Fig.2. The temperature dependence of lower critical field H_{c1} of $DyBa_2Cu_3O_7$.

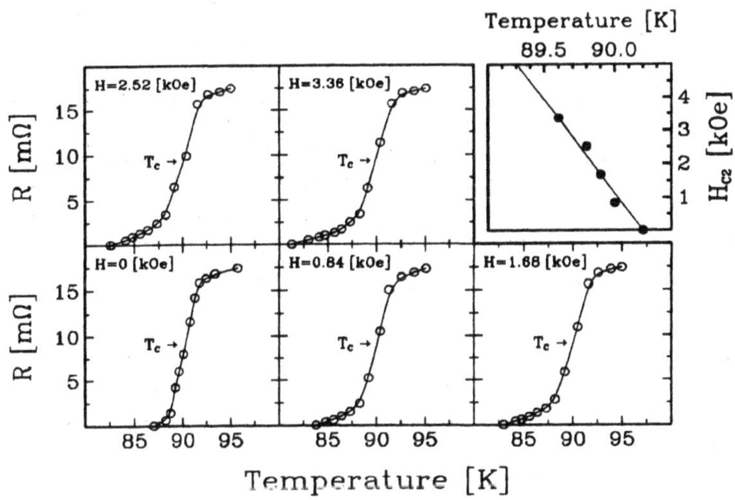

Fig.3. The temperature dependencies of resistance R at the given magnetic field H and the temperature dependence of upper critical field H_{c2}, (upper right corner), of $YBa_2Cu_3O_7$.

The Fig.3 shows the magnetoresistance results and the $H_{c2}(T)$ dependence but the resistance was measured as a function of temperature at the given magnetic field. According to the GLAG theory the $H_{c2}(T)$ dependence:

$$H_{c2}(T)=2^{1/2}\ \kappa\ H_c(T) \qquad (1)$$

as well as the $H_{c1}(T)$ dependence :

$$H_{c1}(T)=H_c(T)\ \ln\kappa\ (2^{1/2}\ \kappa)^{-1} \qquad (2)$$

are a linear function of temperature, because :

$$H_c(T)=4.23\ \gamma^{1/2}\ T_c\ (1\text{-}T/T_c)\ \text{in Oe}, \qquad (3)$$

where the specific heat coefficient γ is in (erg cm^{-3} K^{-2}) and in our case $\gamma=2710$ erg cm^{-3} K^{-2} [4]. This is really the case for our $H_{c1}(T)$ and $H_{c2}(T)$ dependencies (c.f. Fig.2 and Fig.3). One can fit these dependencies in order to calculate the Ginzburg-Landau parameter κ and later on a number of electronic superconducting and normal state quantities as well. But it is impossible to reconcile the κ-values extracted from (1) and (2) being respectively $\kappa=16.2$ and 91.5. We believe that the second value is right and intrinsic one but the first is affected again by the granular structure of the specimen, and the values of H_{c2} are much smaller than found for example in [5]. Sintered bulk sample consists of randomly oriented and loosely connected grains which in combination with very short intragrain superconducting coherence length forms weak superconducting and nonsuperconducting links among the grains. Thus we measure an upper critical field of the weak links rather than intrinsic upper critical field of the interior of the grains. These weak links are also a reason of extremely low critical current densities J_c which was found to be about 1/100 times smaller in bulk specimen compared to powder one [6]. This is because of the reduction of the effective cross-section for the superconducting transport current caused by the granular structure [1] of the bulk samples.The effective cross-section is limited by the links between the grains and thus it limits the upper critical field and critical current.

Acknowledgments

This work is financially supported by the Polish Committee of Scientific Research. We thank to Dr Z.Kąkol and to Dr J.Żukrowski for the help in computer elaboration of the drawings and of typesetting of the text.

References

1. G.Deutcher, in *Earlier and Recent Aspects of Superconductivity*, eds. J.G.Bednorz and K.A.Müller (Springer Series in Solid-State Sciences, 1990),**Vol.90**, p.174.
2. J.Chmist, *Thesis*, University of Mining and Metallurgy, Cracow, Poland, (1991).
3. T.P.Orlando, M.R.Beasley, *Phys. Rev.* **B19** (1979) 4545.
4. A.Kozłowski, Z.Tarnawski, A.Kołodziejczyk, J.Chmist, T.Ściężor and R.Zalecki,

Physica C - submitted (1991).

5. J.J.M.Franse, K.Kadowaki, in *High-T_c Thin Films and Single Crystals*, eds. W.Gorzkowski, M.Gutowski, A.Reich and H.Szymczak, *Progress in High Temperature Superconductivity, Vol.24* (1990) 72, World Scientific Publishers, Singapore
6. R.Wördenweber, H.C.Freyhardt, *Physica C* **153-155** (1988) 870.

THE EFFECT OF 211 PHASE ON Jc VALUE OF YBCO SUPERCONDUCTOR BY POWDER MELTING PROCESS

Zhang Pingxiang, Zhou Lian, Ji Ping, Wang Jingrong, Wang Keguang, Du Shejun and Wu Xiaozu

Northwest Institute for Nonferrous Metal Research,
P.O. Box 71, Baoji, Shaanxi 721014, P.R. China

ABSTRACT

The size and distribution of 211 particles can be controlled by "powder melting process" (PMP). SEM and TEM observations indicate that the microcracks which result in "weak links" were eliminated in PMP samples due to the mechanical pinning of 211 phase, and some microstructural defects were induced around the 211 particles, which can be act as effective flux pinning centers and responsible for high Jc values.

Introduction

Since S. Jin et al invented melt-textured growth (MTG)[1], the melting process have been the major approach to fabricate high Jc YBCO superconductor. The prospect of application for high Tc superconductors depend on the introduction of flux pinning centers. It was found that there are a lot of 211 phase distributed in the melted YBCO bulks[1,2,3,4]. These 211 phase in some samples are smaller than 1 μm and dispersively distributed within the 123 matrix[3,4], and were considered have some advantages for improving the critical current density. In this paper we report some Jc-B results and the investigation of microstructure on the samples with different 211 phase distribution.

Experimental Procedure

The samples were prepared by "powder melting process" (PMP) with different processing parameters. The details of which have been reported elsewhere[3,5]. the microstructure of the samples were investigated by SEM and TEM. The thin specimens for TEM observation were prepared by dimple grinder and dual argon ion mill. The microstructural parameter measurement was performed on the micrographies by LEITZ texture analyze system. Critical current was measured by standard four-probe method, both continuous dc and pulse current were used for measurement. Applied magnetic fields up to 5 T were generated by an NbTi magnet.

Results and Discussion

The 211 and BaCuO powders were used in PMP as starting materials to prepare the YBCO bulk. During the continuous zone melting process, the 123 superconducting phase were formed around the 211 particles through the peritectic reaction of 211 and liquid phase. The unreacted extra 211 phase were remained as inclusions within the 123 matrix, as shown in Fig. 1. Since the 211 phase were in solid state during the whole melting process, the dispersion of 211 particles in precursor pellet was almost unchanged the size and distribution of inclusions in the PMP bulk can be controlled by adjusting the composition and powder size of precursor materials.

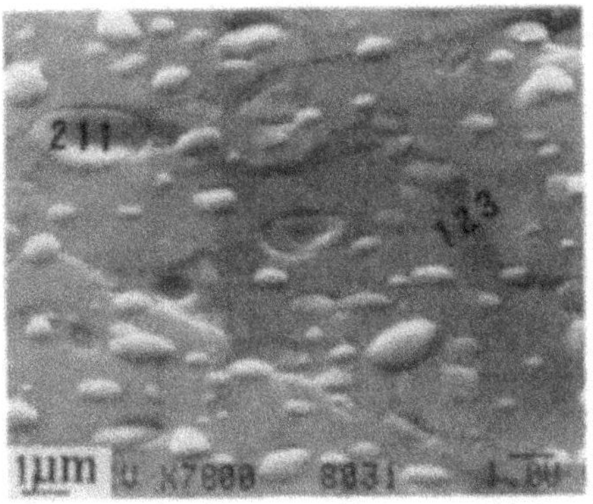

Fig. 1. 211 inclusions within 123 matrix.

It was found that the properties of PMP YBCO bulk are sensitive to the melting temperature (Tm). High Tm (more than 1040 °C in air) results in formation of large 211 particles, as shown in Fig. 2a. The morphology is similar to that of MTG samples[6,7] and critical current density is only about 1000 A/cm^2 (77 K, 1 T). In other hand, the specimens melted at lower temperature (less than 1010 °C in air) are not homogeneous, the crystals do not well aligned, which are due to the insufficient peritectic reaction. Fig. 2c, 2d are the fracture surface of a sample melted at proper temperature (about 1025 °C in air). There are a lot of fine 211 particles dispersively distributed in almost continuous 123 matrix. The thin 123 sheet are well-oriented along the lengthwise of the sample. In good samples, the 211 inclusions exhibit elliptic ball in shape, have average size of 0.4 - 0.8 μm, and no microcracks were found.

The transport Jc values of four samples prepared in different processing parameters are shown in Fig. 3. Since heat generation on the contacts, the pulse current was used for measurement. The high Jc character in magnetic field indicate that the PMP samples have strong flux pinning effects. The different Jc-B dependence of sample M2 perhaps due the insufficient content of oxygen, since M2 are only annealed at 500 °C for 20 hours.

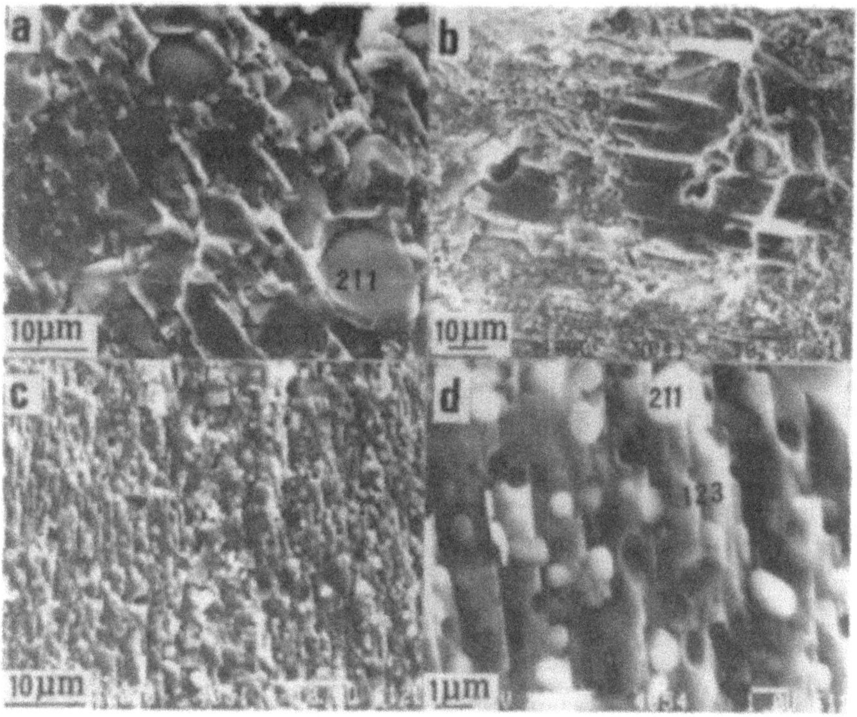

Fig. 2. Fracture surface of PMP samples. (a). melted at higher Tm, (b). at lower Tm, and (c), (d). at proper Tm.

In order to investigate the anisotropy of PMP bulk, the transport Jc of specimen (as shown in Fig. 2c, 2d) were measured with different applied field direction.The Jc value in the case of magnetic field parallel to the lengthwise of sample is about 2.1 times larger than that of perpendicular situation as shown in Fig. 4. This indicate that the PMP bulk have strong anisotropic properties.

TEM investigation found that there are dense crystal defects around the 211 particles (Fig. 5). The defects produced image contrast were similar to that of stacking faults. The thickness of the defects can be compared with the coherence length of YBCO superconductor. Other defects such as dislocation and dislocation loops are also found around the 211 phase[4,7].

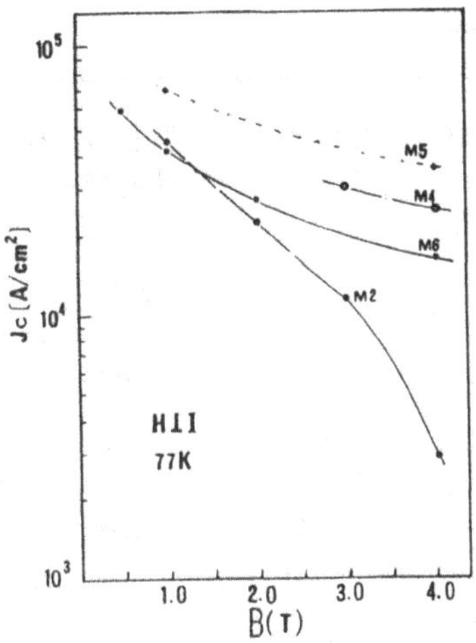

Fig. 3. The dependence of Jc on B of PMP samples.

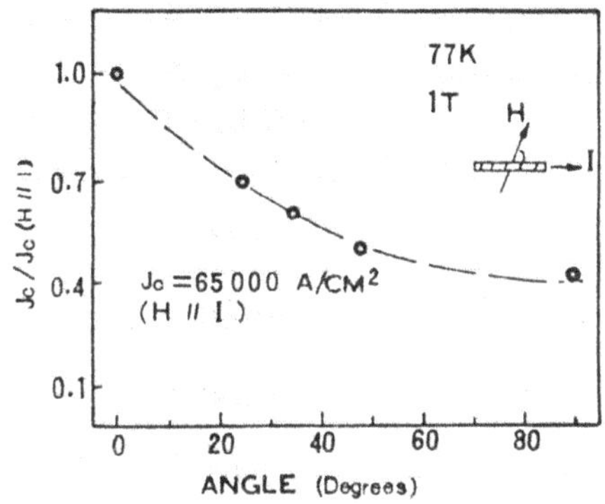

Fig. 4. Jc values in different applied field direction.

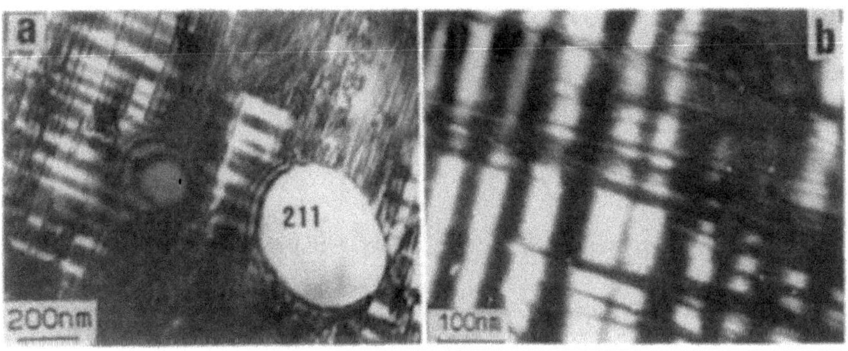

Fig.5. TEM photograph of crystal defects of PMP samples
(a) dense planar defects around 211 particles
(b) image contrast fringes of planar defects.

As the size of 211 particles much larger than the coherence length, they can not pin magnetic flux themselves. From the data in table 1, we can see that the Jc values are increased with the decrease of 211 particles spacing and average size. These results indicate that the dense planar defects around 211 particles are responsible for the high Jc values in PMP samples.

Table 1. Jc values and microstructural parameters of PMP samples.

No. samples	Jc, A/cm^2 (77 K, 1 T)	211 Vol%	211 size (1 μm)	211 spacing (1 μm)
M5	71000	21	0.4	0.77
M3	65000	32	0.66	1.02
MG4	45000	29	0.70	1.15
MG6	1000	44	2.44	3.26

Conclusion

The size and distribution of 211 particles in PMP sample can be well controlled by change the properties of precursor materials and melting parameters.The fine dispersive 211 inclusions in 123 matrix were found have two advantages: (a). as mechanical pinning centers to eliminate the microcracks; (b). induce dense planar defects as flux pinning centers to improve critical current.

Acknowledgements

This work was supported by the China National Center for R & D on Superconductivity.

References

1. S. Jin, T.H. Tiefel, R.B. Sherwood, R.B. Van Dover, M.E.Davis, G.W. Kammlott, and R.A. Fastnacht, Phys. Rev. B **37**, 7850 (1988)
2. M. Murakami, M. Morita, K. Doi, and K. Miyamota, Jpn. J. Appl. Phys. **28**, 1189 (1989)
3. Zhou Lian, Zhang Pingxiang, Ji Ping, Wang Keguang, Wang Jingrong and Wu Xiaoxu, Supercon. Sci. Technol. **3**, 490 (1990)
4. Zhou Lian, Zhang Pingxiang, Wang Keguang, Wang Jingrong, Du Zehua and Ji Ping,

Presented in ICMC'91 Conference, June 11-14, Huntsville, Alabama, USA
5. Zhang Pingxiang, Zhou Lian, Ji Ping, Du SheJun, Wang Jingrong, Wang Keguang and Wu Xiao Zu, Presented in ICMC'91, June 11-14, Huntsville, Alabama, USA
6. F.S. Liu, H.L. Mou, T.C. Wang, L. Zhou and X.Z. Wu, Proc. BHTSC'89, Progress in High Temperature Superconductivity, Vol.22, p. 14
7. Wang Keguaang, Zhou Lian, Zhang Pingxiang, Wang Shuqian, Wang Jingrong, Ji Ping, presented in ICMC'91.

CRITICAL CURRENT DENSITY AND PREPARATION OF $YBa_2Cu_3O_{7-x}$ TEXTURED SAMPLES

A.M.Balbashov,
Moscow Power Institute,

S.G.Karabashev, V.A.Pitov,
Moscow Steel and Alloys Institute,

D.A.Komarkov, V.V.Moshchalkov, A.A.Zhukov,
Moscow State University.

1. Introduction

Preparation of HTSC samples with high J_c values at T=77K in magnetic field 1T is still an important problem to be solved for wide HTSC technical application. The most encouraging results are known to be gained on 123 samples obtained by gradient melt solidifying method GMSM, which was developed by Jin [1] and lately modified in some other works [2,3]. Usually, the process includes 3 steps: firing of a proper Y-Ba-Cu-O mixture in $T_{melting}$ - $T_{liquidus}$ temperature interval, slow cooling in temperature gradient to solidification and subsequent heat treatment for optimization of 123 phase superconducting parameters. This method does not exclude a crucible material contamination of growing 123 crystals and, besides, they are to be solidified from the melt in nonsuperconducting tetra modification.

Earlier we have reported on FZM method for 123 single crystals preparation [4], which naturally excludes contamination of crucible material and ensures the growth of low twins density in 123 crystals in superconducting ortho modification directly from the melt. In this work we performed technological experiments aimed at preparation of 123 textured samples by the same method.

2. Experiment

Floating zone processes were performed in FZM apparatus described previously [5]. Ceramic bars with stoichiometric composition, with diameter of 7-8 mm and 40-50mm long, were used as feed rods for each zone melting process. The bars were prepared by pressing and firing (≈820°C for 4 hours) of proper Y_2O_3, $BaCO_3$ and CuO powder mixture.

The phase content and chemical composition of as-grown bouls were determined by means of X-ray microanalysis (Link 860). For J_c measurements we have used diamagnetic shielding method [6]. The small Hall probe with sensitive area $0,45 \cdot 0,15$ mm^2 was placed on a center of the sample surface. The relations between J_c value and shielding magnetic field were found using the numerical calculations on the basis of know samples dimensions.

3. Results and discussion

Describing zone process itself, one can distinguish two stages. The first one starts from seeding and lasts about 3-4 hours. During it the melt temperature must be kept high enough to hamper formation of solid particles in liquid zone. On the other hand, melt overheating makes it difficult to prevent liquid zone from pouring down because of, judging it by a zone shape, decrease of surface energy density. As melting zone is moving, its liquidus temperature decreases and its shape comes to normal. Finally, zone parameters stabilize and it is a clear indication of second stage beginning, which can last up to the end of a feed rod. Examination of polished cleaves of as-grown bouls shows that those initial parts, corresponding to the first stage, contain 211, 011, 001 phases and are strongly enriched by Y. The remaining part of a boul has a quite different phase composition: 123 (70-80%) and also some amounts of 211, 011,

001 compounds (Fig.1). Its mean chemical composition was close to stoichiometric one. The 123 crystals were of a plate like shape, with its developed faces being basic plane of 123 crystal structure and parallel to growth axis, and nonstoichiometric phases crystallized as long inclusions (Fig.2).

Figure 1.

Figure 2.

The mean chemical composition of rapidly cooled floating zone was determined to be 2 mol%, 30 mol%, 68 mol% of $YO_{1,5}$, BaO and CuO consequently. It should be noted, that these values precisely correspond to 123 primary crystallization region on $YO_{1,5}$-BaO-CuO phase diagram.

Summarizing the obtained data, one can propose the following picture of the crystallization process. At the beginning of it a predominant 211 crystallization is taking place from the melt, having 123 composition. While moving, floating zone is changing its composition because of Y depletion up to the moment, when it reaches 123 primary crystallization region. Afterwards, predominant 123 crystallization is going on from the melt.

4. Properties and heat treatment of 123 crystals prepared by FZM

According to X-ray diffraction analysis of polished cleaves of as-grown bouls, 123 compound has orthorombic twin-free crystal structure with a=3,871A, b=3,876A and c=11,686A.

To perform measurements of T_c and J_c and heat treatment experiments, the bars $1\cdot1\cdot20$-30 mm^3 in size were cut from bouls, so their long dimension was parallel to growth direction (a- axis). The $T_{c\ onset}$ and ΔT_c (from susceptibility measurements) of as-grown samples were 70-75K and 15-20K consequently. After heat treatment, which includes slow heating up to 900^oC, firing for 20h and subsequent oxidation in the temperature interval of 450-600^oC , $T_{c\ onset}$ = 93K and ΔT_c = 3K was reached in magnetic field 0,5 T.

Measurements of J_c were performed on the samples $9,7\cdot1,1\cdot1,2$ in size. The trapped magnetic field (H_{tr}) dependence on the amplitude of external magnetic field (H_e), which was input and output before measurements, is given in

Fig.3. The measurements were made for both orientations of magnetic field - parallel and perpendicular to the growth direction (Z). In both cases the $H_{tr}(H_e)$ behavior is different from the known for the ceramic samples, where two steps are observed [7]. In our case the region of

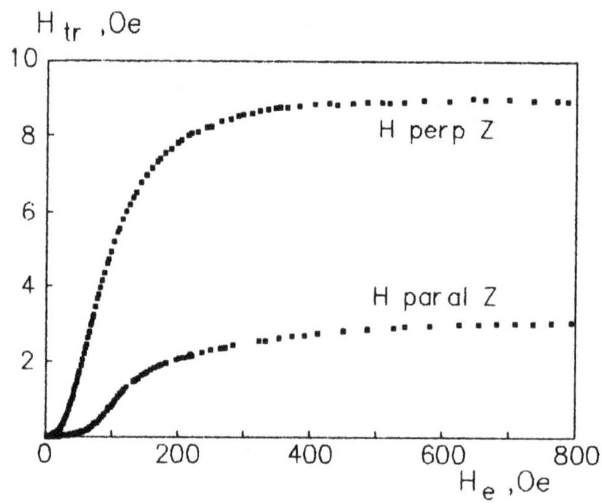

Figure 3.

inhomogeneous inter-granular penetration is absent. These facts point to the bulk nature of shielding currents in studied samples. The saturation value of trapped magnetic field gives high enough value of critical current density $(1,1\mp0,2)\cdot10^3 A/cm^2$, $(0,5\mp0,1)\cdot10^2 A/cm^2)$ parallel and perpendicular to growth direction respectively (T=77K, H_e=0). In correspondence with this observation the sample exhibits only one type of magnetization loop. For the studied samples the dependence of critical current or magnetic field (Fig.4) is much weaker than for ceramic samples [7] and is comparable with the results for GMSM samples [2]. The activation energy u_o=0,45 eV was determined from relaxation measurements.

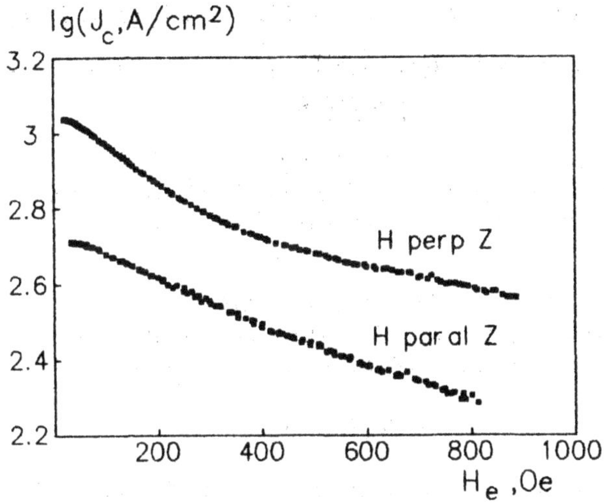

Figure 4.

5. References

1. S.Jin et al. Phys. Rev. B 37 (1988), p. 7850
2. T.Matsushita et al. Jap.Jour.Appl.Phys., vol.28, N9, 1989, L1545-L1548
3. P.I.McGinn et al. Physica C 161 (1989) 198-204
4. A.M.Balbashov, S.G.Karabashev, A.S.Nygmatulin, Proceedings of the European Conference on High-T_c Thin Films and Single Crystals, Sept. 30 - 4, 1989, Ustroń, Poland
5. A.M.Balbashov, S.K.Egorov, Journal of Crystal Growth, vol.52, (1981) p.498-504
6. A.A.Zhukov et al. Z.Phys. B 78 (1990) 195-195
7. A.A.Zhukov et al. Z.Phys. B 169 (1991) 661-662

CRITICAL CURRENTS IN *Ag*-SHEATHED *Y-Ba-Cu-O* WIRES AND *Bi-Pb-Sr-Ca-Cu-O* TAPES

G. FUCHS, A. GLADUN, K. FISCHER, C. RODIG
Central Institute of Solid State Physics and Materials Research,
Helmholtzstr. 20, O-8027 Dresden, Germany

and

R. EYEN, J. HÜDEPOHL
Department of Anorganic Chemistry, University Wuppertal,
Gausstr. 20, W-5600 Wuppertal 1, Germany

ABSTRACT

Transport critical currents of Ag-sheathed $YBCO$ wires and $BPSCCO$ tapes of the 2223 phase have been investigated.
The critical current density of the $YBCO$ wires is limited by weak-links between grains. An improvement of the weak-link parameter with decreasing sample diameter has been found, which is probably caused by a reduction of the junction length of the Josephson junctions and an enhancement by their number.
The critical current density of the textured $BPSCCO$ tapes is dominated by pinning mechanism. At high temperatures the field dependence of the critical current density shows a large anisotropy which is caused by an intrinsic pinning mechanism for the field orientation H || ab. The critical current density at high temperatures is limited by flux creep. The onset field of flux creep at T = 70 K vary between 0.5...1 T for H ⊥ ab and between 3...7 T for H || ab.

1. Introduction

The critical current density of Ag-sheathed $YBa_2Cu_3O_7(YBCO)$ wires is limited by weak-links between grains. The weak-link behaviour of polycristalline $YBCO$ can be overcome by melt texturing technique[1], but the development of melt textured $YBCO$ wires for magnet applications at T = 77 K is an extremly difficult technological problem.
In textured Ag-sheathed Bi-based tapes high critical current densities has been obtained at low temperatures up to high magnetic fields [2]. The critical current density in such tapes is dominated by pinning mechanism. However, the critical current density at temperatures above about 20 K is limited by dissipative flux motion. The

activation energy is for the Bi-based compounds lower than for $YBCO$ [3].
We investigated a series of Ag-sheathed $YBCO$ wires with different diameters and Ag-sheathed $Bi_{1.84}Pb_{0.34}Sr_{1.9}Ca_{2.1}Cu_{3.1}O_x(BPSCCO)$ tapes.
For the $YBCO$ wires the dependence of the weak-link properties on temperature and sample diameter was studied.
For the $BPSCCO$ tapes the critical current density limitations at T= 70 K were investigated by measuring the field dependence of the critical current density and the resistive transition in magnetic field.

2. Samples and measurements

The $YBCO$ and $BPSCCO$ samples were prepared by the 'powder in tube' technique.
Details of preparation of the $YBCO$ wires are described elsewhere[4]. The superconductor diameter of the wires varied between 0.25 and 0.5 mm, the ratio of the superconductor to silver was 42 %.
The $BPSCCO$ tapes were annealed at 870°C for 150 h. After pressing at room temperature, a final annealing was performed at 870°C for 200 h.
Resistive measurements yielded sharp transitions with $T_c(R = O) = 90\ K$ for the $YBCO$ wires and somewhat broader transitions with $T_c(R = O) = 100\ K$ for the $BPSCCO$ tapes. The critical current was determined at a criterion of 0.1 $\mu V/cm$ using a standard four - point resistive measurement technique.

3. Results and discussion

3.1. Weak-link properties of YBCO wires

The field dependence of the critical current density of the wire with d = 0.25 mm is shown in Fig. 1 for three temperatures together with the self field relation

$$H_s = \frac{1}{4} j_o\ d. \tag{1}$$

The self field H_s is produced by the zero field critical current density j_o on the surface of the wire with diameter d.
The zero field critical current density at $T = 4.2\ K$ in Fig. 1 is limited by the self field.
The self field can be neglected at high temperatures, where the self field becomes lower than the step field, and for higher applied magnetic fields.
The weak-link critical current density decreases as H^{-1}. That is the field dependence

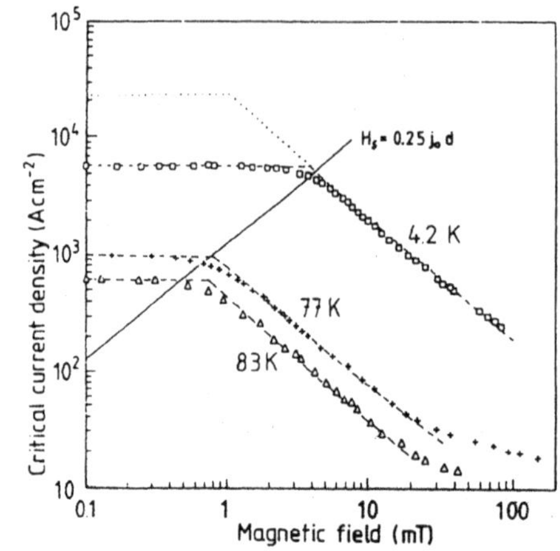

Fig. 1 Critical current density versus magnetic field at different temperatures for a *Y*-123 wire with d = 0.25 mm. — self field relation, ··· weak-link critical current density at T = 4.2 K

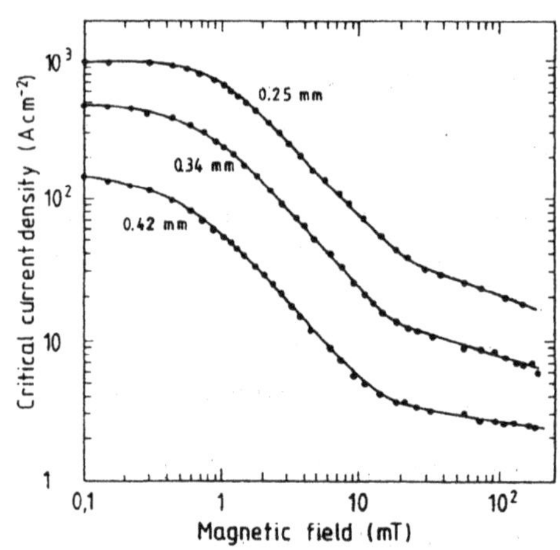

Fig. 2 Critical current density versus magnetic field at T = 77 K for *Y*-123 wires with different superconductor diameter.

of the critical current density for a randomly oriented distribution of rectangular Josphson junctions.
The corresponding step field H_w is given by

$$H_w(T) = \frac{\Phi_0}{\pi^2 \lambda(T) L} \tag{2}$$

with Φ_0 as the flux quantum, λ as the penetration length and L as the junction length. Because of the weak temperature dependence of the penetration length the step field H_w increases only by a factor 1.4, when the temperature is decreased from 77 to 4.2 K. The resulting zero field critical current density j_w $(4.2K) \sim 2 \cdot 10^4 Acm^{-2}$ is much higher than the experimental, self field limited critical current density.

In Fig. 2 the field dependence of the weak link critical current density at 77 K is shown for the investigated wires.
We found an improvement of the weak-link properties with decreasing diameter: the zero field critical current density j_w increases roughly as d^{-4} and the step field H_w increases as d^{-3}.
The increase of the step field can be explained by a reduction of the junction length L with decreasing diameter - from 3.1 μm for $d = 0.46\ mm$ to 0.9 μm for $d = 0.25\ mm$. Describing the temperature dependence of the weak link critical current density by a SNS model[5] for the Josephson junctions we found that the critical current density of the single Josephson junction remains constant, although the overall critical current density j_w rises with decreasing wire diameter. Therefore, we assume that the number of junctions increases with decreasing diameter.

3.2. Critical current density of the BPSCCO tapes

In Fig. 3 the field dependence of the critical current density at 4.2 K and 70 K of one of our best sample is shown in a semilogarithmic plot.
The critical current density at 4.2 K and the zero field value of the critical current density at 70 K of this sample is relatively low in comparison with other results, but we found a remarkably large anisotropy of the critical current density at 70 K with high critical current density in magnetic fields for the field orientation H || ab.
For epitaxial films of $YBCO$ and Bi-2212 the field dependence of the critical current density is qualitatively similar, but the anisotropy of the critical current density at higher temperatures is more pronounced[6]. This anisotropy is caused by an intrinsic pinning mechanism for the field orientation H || ab, in which the flux line lattice is directed parallel to the crystal layers. In the case of the textured Bi-2223 tape the intrinsic pinning effect depends on the degree of texture achieved in the sample. Therefore, it should be possible to increase the critical current density in magnetic fields for the field orientation H || ab by improving the texture.

By the lack of effective pinning centres for the field orientation H$\perp$ ab the critical current density drops to zero at relatively low magnetic fields.

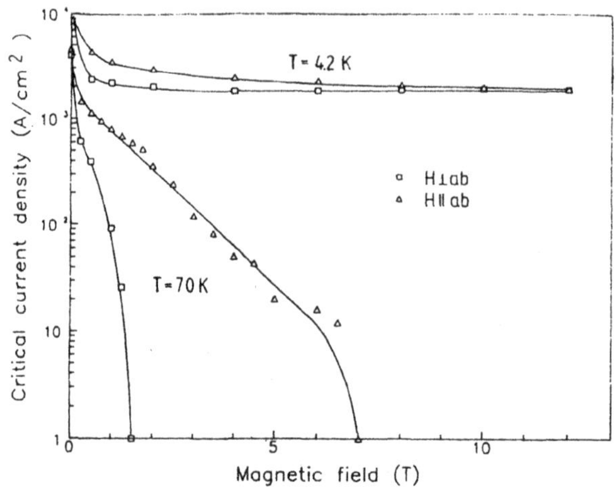

Fig.3 Critical current density versus magnetic field at T = 4.2 K and T = 70 K for a *Bi*-2223 tape. □ H ⊥ ab, △ H ⊥ ab

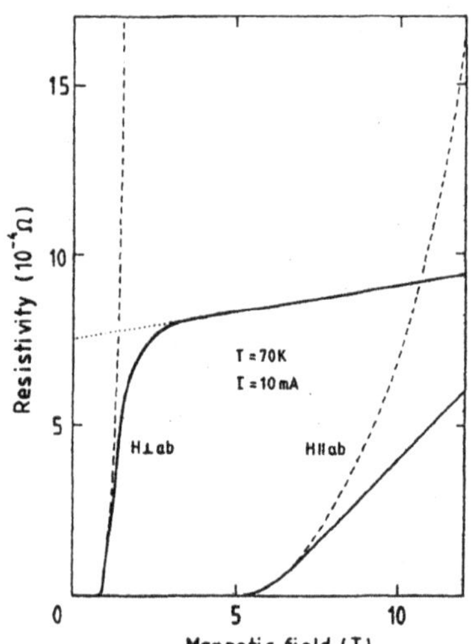

Fig. 4 Resistivity versus magnetic field at T = 70 K for a *Bi*-2223 tape. — measured curve, ··· Ag-sheath, - - - *Bi*-2223 core

We investigated the onset of flux creep by measuring of the resistive transition curves as function of magnetic field at T = 70 K (Fig. 4). The linear part of the measured curve for H $\perp$ ab corresponds to the *Ag*-sheath. Taking into account the resistivity of the *Ag*-sheath we obtained the lower part of the superconducting transition of the *Bi*-2223 core from the measured curves. The transition curves of *Bi*-2223 are dominated by flux creep and can be satisfactory described by

$$\rho(H) \sim exp[-U(H)/kT] \tag{3}$$

assuming for the activation energy

$$U(H) \sim H^{-1} \tag{4}$$

We found activation energies of U/k = 700 K at H = 1 T for H $\perp$ ab and U/k = 650 K at H = 5 T for H $\parallel$ ab. The lower onset field of flux creep for H $\perp$ ab in Fig. 4 reflects the lower activation energy for flux creep for this field orientation.
Systematic investigations of the activation energy for flux creep are known for *Y*-123, *Bi*-2212 and some other high-T_c materials[3], but not for *Bi*-2223.
Our values of the activation energy for *Bi*-2223 are quite similar to that found by Palstra et al.[3] for a *Bi*-2212 single crystal.
However, for the *Bi*-2212 single crystal a field dependence $U \sim H^{-n}$ with $n = 0.3 \ldots 0.5$ of the activation energy was observed which is weaker than our result $U \sim H^{-1}$ for *Bi*-2223.

4. References

1. M. Murakami, S. Gotoh, H. Fujimoto, K. Yamaguchi, N. Koshizuka and S. Tanaka, *Supercond. Sci. Technol.* **4**, (1991) S 43.
2. H. Krauth, K. Heine and J. Tenbrink, *Proc. ICMC'90 Topical Conf. on HTSC-Materials Aspects,* May 9 - 11, 1990, Garmisch-Partenkirchen, FRG
3. T.T.M. Palstra, B. Batlogg, R. B. van Dover, L. F. Schneemeyer and J. V. Waszczak, *Phys. Rev.* **B 41**, (1990)6621
4. P. Müller, M. Schubert, Ch. Rodig, G. Fuchs and K. Fischer,*Appl. Phys. Lett.* **55** (1989)917
5. J. K. Likharev, *Rev. Mod. Phys.* **51** (1979)101
6. P. Schmitt, P. Kummeth, B. Roas, L. Schultz und G. Saemann-Ischenko, *Proc. Statusseminar Supraleitung und Tieftemperaturtechnik,* VDI-Verlag Düsseldorf (1991) p. 234

DC and Microwave Properties of Epitaxial, *in-situ* YBaCuO Thin Films

W. Kula and P. Gierlowski
Institute of Physics, Polish Academy of Sciences,
02-668 Warszawa, Poland

ABSTRACT

$YBa_2Cu_3O_{7-X}$ superconducting thin films have been obtained *in-situ* by dc magnetron sputtering onto heated MgO substrates. As deposited films are c-axis oriented, have very smooth surfaces and exhibit T_c above 86 K. Their critical current density is $2x10^6$ A/cm^2 at liquid nitrogen temperature. Results of both microwave absorption and detection measurements in small magnetic fields are presented, and compared to the data obtained for non-epitaxial high-T_c thin films.

1. INTRODUCTION

In this contribution, we present details of thin superconducting YBCO film fabrication together with high-frequency properties of these films. Many different techniques of physical characterization of high-T_c material in the thin film form at microwave frequencies can be applied. Our films have been investigated by low magnetic field microwave absorption and detection measurements. Results of both types of measurements *vs.* temperature will be presented and discussed, and compared to the results obtained for granular high-T_c thin films.

2. FILM FABRICATION

The films were grown *in-situ* by high-pressure off-axis dc magnetron sputtering. The planar magnetron gun used in this work was described previously[1]. The magnetron was mounted in a conventional vacuum chamber with the base pressure of $5x10^{-6}$ Torr. Stoichiometric, 45 mm diameter superconducting targets were prepared by conventional solid state reaction from pure (at least 99.9 %) powders. During the film deposition, the substrates were positioned at 90° with respect to the target plane, and displaced 4 cm from the target axis. Additionally, a grounded copper reflection plate facing the target with a 45° angle was mounted above the target. This sputtering geometry canceled almost completely the resputtering effects caused by the bombardment of the substrate with negative oxygen ions emerging from the target. The substrates were placed on a heater constructed from alundum ceramic plates with a Kanthal-tape heating element. The substrate temperature was monitored by a chromel-alumel thermocouple located in close contact to the heater. Prior to the deposition process, the substrates were carefully cleaned in organic solvents and annealed for 1 hour at 1000^oC in flowing oxygen. The magnetron operating conditions during the film deposition are listed in Table I. Typically, the films were sputtered for 5 hours and the film thickness was about 250 nm. After deposition the vacuum chamber was filled by oxygen at 200 Torr and the films were cooled down to room temperature within 20 min. Most of our films were deposited on (100) MgO single-crystalline substrates, but other substrate materials such as $SrTiO_3$ and $CaNdAlO_4$ single crystals were also successfully tried. All the results presented in this paper were obtained for the films deposited on MgO.

X-ray diffraction measurements show that the films are purely *c*-axis oriented. The surface of our films, analyzed by scanning electron microscope, is very smooth down to submicrometer scale and does not show any precipitations, which are sometimes observed[2] in the *in-situ* deposited YBCO thin films.

TABLE I. DC magnetron operating conditions of a film deposition process.

Sputtering atmosphere	500 mTr of Ar:O_2=10:1 mixture
Substrate temperature	720 ^{o}C
Discharge voltage and current	115 V, 0.3 A
Sputtering rate	0.8 nm/min

3. FILM PROPERTIES

3.1. DC Superconducting Properties

Fig. 1. shows typical resistivity vs. temperature data obtained by the standard four-probe measurements of the YBCO film deposited on MgO. The superconducting transition temperature, T_{co} defined by the zero-resistance state was 86.5 K, and the transition width (10% to 90%) was 1.8 K. Transport and microwave detection measurements have been performed on films patterned photolithographically into 150 μm wide and 4 mm long strips, with evaporated silver contact pads. The temperature dependence of the critical current density, J_c is presented in Fig. 2. At liquid nitrogen temperature J_c was typically greater than $2x10^6$ A/cm^2.

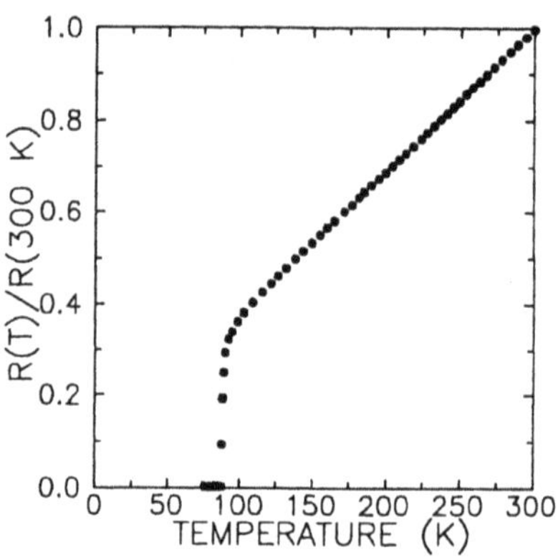

FIG. 1. Superconducting resistive transition of the YBCO-on-MgO film.

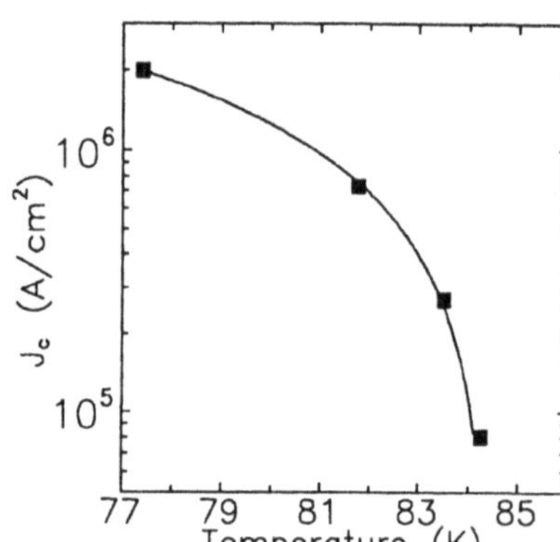

FIG. 2. Temperature dependence of the critical current density of the YBCO-on-MgO film.

3.2. MICROWAVE ABSORPTION MEASUREMENTS

The enhanced microwave absorption in low (below 100 Oe) magnetic fields is a well-known feature of high-T_c superconducting ceramics[3]. Generally, it comes from the interaction between the microwaves and various magnetic fluxons penetrating the weak-link system present in these materials[4].

Our experimental setup for low magnetic field microwave absorption measurements of the films is shown schematically in Fig. 3. The thin film sample, 2x2 mm^2 wide and 250 nm thick, was placed at the bottom of a 23 GHz TE_{116} cylindrical microwave cavity. The cavity was located in an external dc magnetic field perpendicular to the film surface. Parallel to the dc field was superimposed a modulation field at the frequency of 1 kHz. We measured the microwave power reflected from the cavity at the resonant frequency synchronously with the magnetic field modulation. This magnetically modulated microwave absorption signal reflects the changes in the film microwave losses induced by the changes of the external magnetic field.

The temperature dependence of the detected voltage for the film deposited on MgO is shown in Fig. 4. (rectangles). The signal was obtained with the dc magnetic field of 10 Oe and the amplitude of the modulation field of 7.5 Oe. For comparison, in Fig. 4. is also shown the modulated absorption signal for the granular YBCO-on-ZrO_2 film of similar geometry (triangles). The response of the *in-situ* film was nonzero only in the narrow temperature region just below the film superconducting transition temperature. The signal appeared at T_{co} and disappeared below 76 K. In contrast, the response of the granular film, which appeared also at the transition temperature (T_{co} of the granular film was 89 K) remained significant at lower temperatures. The response of the *in-situ* film shown in Fig. 4. did not depend on whether the signal was registered after the zero-field cooling or during the field cooling of the film. The modulated absorption signal observed in the *in-situ* films may be attributed to the microwave dissipation on Abrikosov vortices penetrating the films in the narrow temperature range close to the superconducting transition. The observed data do not exclude the presence of weak links in the films just below the super-

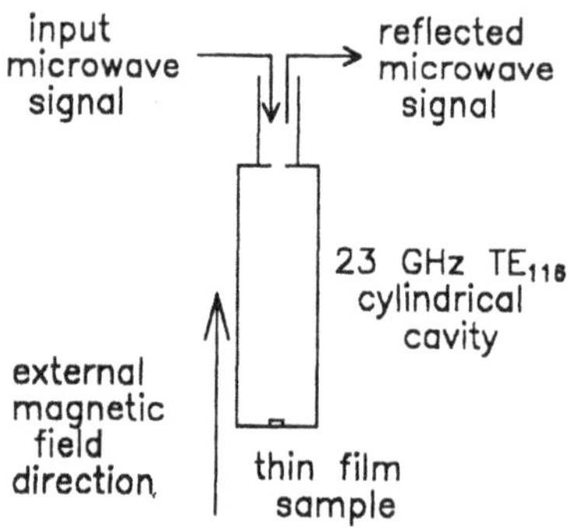

FIG. 3. Sketch of the modulated absorption measurements.

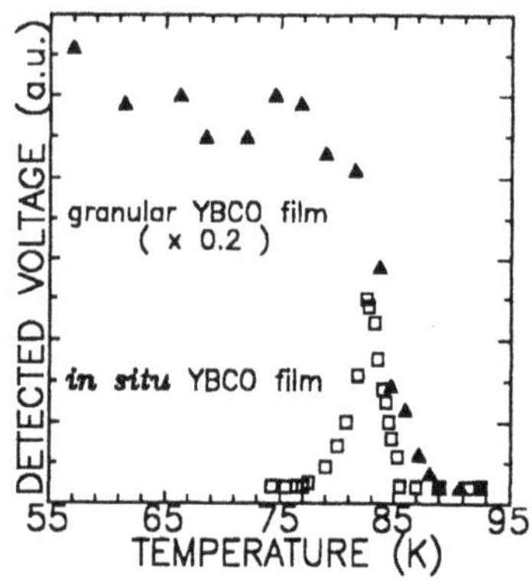

FIG. 4. Temperature dependence of the modulated absorption signal for the *in-situ* (rectangles) and granular (triangles) YBCO films.

conducting transition (and disappearing 3-4 K below T_{co})[5]. However, our investigation of the modulated absorption signal at the constant temperature of 82 K has not shown any features indicating the weak-link-like behaviour. The amplitude of the signal was one order of magnitude smaller then in the case of a granular film, and the signal did not change significantly its amplitude and was nonhysteretic within the external dc magnetic field sweep range from -40 to +40 Oe (with the same modulation field amplitude of 7.5 Oe). In the case of granular YBCO film, the modulated absorption signal showed a hysteretic behaviour during the dc magnetic field sweep, and the shape of hysteresis strongly depended on the film magnetic history (zero-field or field cooling, maximum applied dc magnetic field, etc.). Furthermore, the modulated absorption signal of the granular film revealed a very pronounced and reproducible "structure" during the dc magnetic field sweep, which could be attributed to magnetic flux quantization by intrinsic Josephson junction networks in the film and/or magnetic field induced changes in the critical currents of the network junctions.

3.2. Microwave Detection Measurements

A simplified drawing of the mounting structure used for microwave detection measurements in the 24 GHz frequency band is presented in Fig. 5.; more details can be found in Ref. 6. Simultaneously with detection measurements, dc I(V) characteristics have been recorded. In general, the response of high-J_c YBCO films to microwave irradiation is at least one order of magnitude lower than for granular high-T_c thin films investigated at the same temperature.

As can be clearly seen in Fig. 6., the response to a 1 kHz amplitude modulated 24 GHz signal recorded at 83.43 K is proportional to the incident microwave power, i.e. it is bolometric, while for a slightly higher temperature of 84.77 K, as presented in Fig. 7., a nonlinear behavior of the detected voltage *vs.* microwave power can be observed. This finding can be possibly explained by the presense of a weak-link system in the film at temperatures very close to T_c [5].

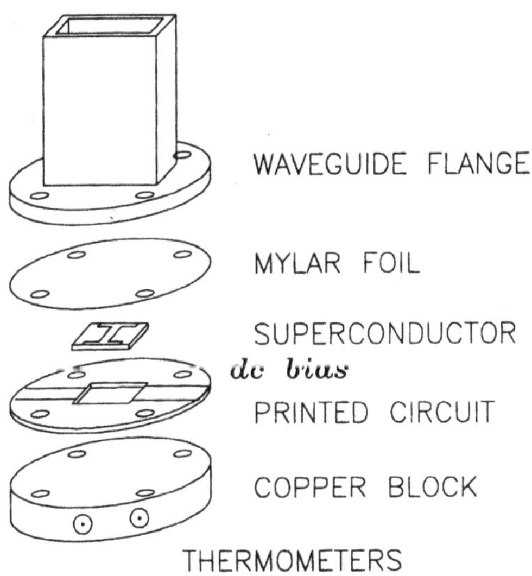

Fig. 5. Schematical drawing of the mounting structure used for microwave detection measurements.

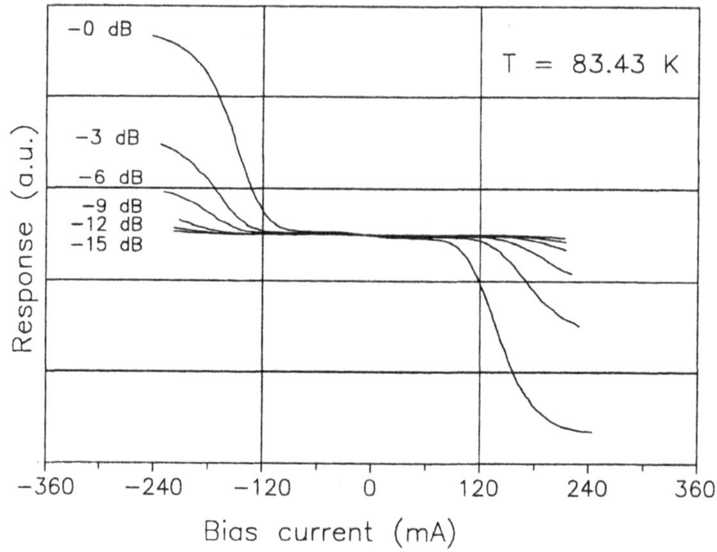

Fig. 6. Video response of the *in-situ* YBCO-on-MgO strip *vs.* bias current for different 24 GHz signal power levels recorded at 83.43 K.

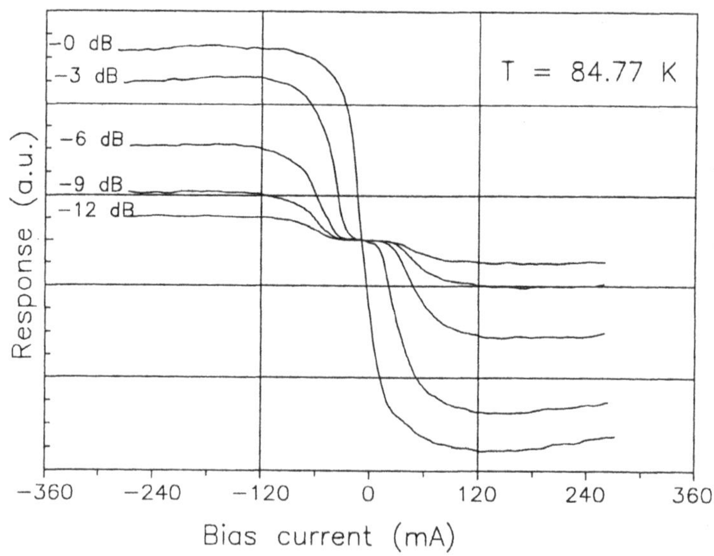

Fig. 7. Video response of the *in-situ* YBCO-on-MgO strip *vs.* bias current for different 24 GHz signal power levels recorded at 84.77 K.

4. CONCLUSIONS

High-J_c YBCO thin films have been prepared *in-situ* by high-pressure off-axis dc magnetron sputtering. Microwave absorption and detection measurements reveal no features attributed to the presence of weak links in the films up to temperatures very close to T_c.

5. ACKNOWLEDGEMENTS

The authors are grateful to J. Gorecka for X-ray diffraction measurements, Z. Kalinski for scanning electron microscope investigations, and to Dr. S. J. Lewandowski for a careful reading of the manuscript. This work was performed within the framework of the Government Research Project No. 2 0479 91 01.

6. REFERENCES

1. P. Gierlowski, S. Zarembinski, W. Kula, and R. Sobolewski, in *Weak Superconductivity, ed.* S. Benacka and M. Kedro (Nova Science Publishers, New York, 1990), pp. 7-12.
2. M. G. Norton, L. A. Tietz, S. R. Summerfelt, and C. B. Carter, *Appl.Phys. Lett.* **55** (1989) 2348.
3. J. Stankowski, P. K. Kahol, N. S. Dalal, J. S. Moodera, *Phys. Rev.* **B36** (1987) 7126; K. Khachaturyan, E. R. Weber, P. Tejedor, A. M. Stacy, and A. M. Portis, *ibid.* 8309.
4. A. Dulcic, B. Ravkin, and M. Pozek, *Europhys. Lett.* **10** (1989) 593 *and references therein.*
5. A. Dulcic, R. H. Crepeau, J. H. Freed, L. F. Schneemeyer, and J. V. Waszczak, *Phys. Rev.* **B42** (1990) 2155.
6. J. Konopka, R. Sobolewski, G. Jung, W. Kula, P. Gierlowski, A. Konopka, and S. J. Lewandowski, *IEEE Trans. on MTT*, **MTT-38**, (1990) 160-165.

THERMALLY ACTIVATED FLUX CREEP IN VIEW OF SOME RECENT EXPERIMENTS ON YBaCuO AND BiSrCaCuO SINGLE CRYSTALS

M.JIRSA, L.PŮST, S.DURČOK,

Institute of Physics, Czechosl. Acad. of Sciences, Na Slovance 2, CS-180 40 Praha 8, Czechoslovakia

ABSTRACT.

The Anderson - Kim's concept of thermally activated flux creep resulting in logarithmic decay of induced magnetic moment with time is compared with some recent experiments on different HTS single crystals. Relaxation experiments at constant magnetic field and hysteresis loop measurements are analyzed. The non-logarithmic decay of induced magnetic moment with time at sub-second region in YBaCuO and at wide range of times in BiSrCaCuO are examined. Fit of our experimental relaxation data by means of different recently proposed functions of time has not given a decision in favour of any particular model. Probably more relaxation mechanisms are active simultaneously.

1. Introduction

Extremely large time decay of the induced magnetic moment in HTS is usually investigated in the time range $10 - 10^5$ s allowed by technical possibilities of commonly used measuring equipment [1-5]. Many HTS samples, e.g. single crystals of YBaCuO, exhibit in this time interval logarithmic decrease of magnetic moment with time [2,3]. Anderson's concept of thermally activated flux creep [6,7] (TAFC) developed for conventional superconductors has been therefore accepted with slight modifications also for HTS [8]. Non-logarithmic magnetic moment decay has been observed in single crystals of BiSrCaCuO and other materials even at longer times of relaxation [9-11].

Analysis of relaxation processes at short (sub second) times is important for good understanding of the relaxation mechanism. Direct static recording of relaxation is not reliable in this time range due to experimental limits represented by finite time constants of magnet, power supply and magnetometer, and not defined origin $t = 0$ of time scale. Relaxation processes can be, however, analyzed indirectly by reconstruction from set of magnetic hysteresis loops (HL) measured at different rates of sweep $\dot{H} = \partial H_{ext}/\partial t$. This method eliminates above mentioned difficulties of direct measurement : (i) smooth sweep with constant $\dot{H}$ can be performed even on magnets with larger time constant, (ii) magnetic moment is not changing much on hysteresis loops and so the effect of finite magnetometer time constant is small and can be further simply minimized by proper data processing, and (iii) determination of corresponding t_{eff} from other parameters [4] is possible even for very short t_{eff}.

In this paper we compare experiments of magnetic moment time relaxation at constant magnetic field H_{ext} (conventional relaxation) with analysis of hysteresis loops measured at different sweep rates $\dot{H}$ (loop relaxation) [4,5].

2. Model

Size of HL in HTS depends on the value of sweep rate: the higher is $\dot{H}$ the larger is the induced magnetic moment m. In sweeping magnetic field relaxation decay $\partial m/\partial t$ is just compensated by magnetic induction due to $\dot{H}$ characterized by the differential susceptibility χ [12] as $\partial m(t)/\partial t = \chi \dot{H}$, where m(t) is a function describing the time relaxation of m at constant H_{ext}. Therefore HL measured at constant $\dot{H}$ can be interpreted as a frozen state of relaxation. For each HL we can evaluate some value of effective time t_{eff} interpreted as the time of conventional relaxation m(t) necessary to reach the same value of m. A set of HL recorded with different $\dot{H}$ enables indirect reconstruction of the process of relaxation down to very short times. In the case of pure logarithmic decay [6]

$$m(t) = m_1 + S \ln(t) \tag{1}$$

we can evaluate corresponding effective time simply as [4] $t_{eff} = S/(\chi \dot{H})$. Similarly we can evaluate for each time t of conventional relaxation corresponding sweep rate $\dot{H} = S/(\chi t)$.

Besides Anderson's model[6] also several alternative models has been reported with different analytical functions m(t) for the time relaxation. Power function for logarithmic conical potential well [10,13] yields

$$m(t) = a_1 + b_1 t^{-\beta} . \tag{2}$$

The 'power' logarithm resulting from the glass model [14,15] corresponds to

$$m(t) = a_2 / [1+ b_2 \ln(t)]^{\mu} . \tag{3}$$

Magnetic relaxation can be also described by the phenomenological function [11]

$$m(t) = a_3 + b_3 \ln[\ln(1+t/t_0)] . \tag{4}$$

3. Results and discussion

Induced magnetic moment was measured by vibrating sample magnetometer on single crystals of YBaCuO and BiSrCaCuO (T_C=82.5 K) in temperature range 7K - 44 K in fields between ±2 T. Experimental results for T = 16 K are shown in Figures 1 and 2 as a function of $\dot{H}$. Magnetic moment measured by both methods well overlaps in the common range of $\dot{H}$ or time t.

Decay of magnetic moment m at all analyzed temperatures in YBaCuO is well logarithmic at longer times (t> 100 s <-> $\dot{H}$ < 2 Oe/s), but relaxes more rapidly at short times (high sweep rates, Fig. 1). This effect was also recently observed in pulse and ac measurements [16,17].

Though any of the time functions Eq.2 - Eq.4 fits well experimental data at longer times. relaxation at short times can be fitted only by power function Eq.2 and dependence Eq.4.

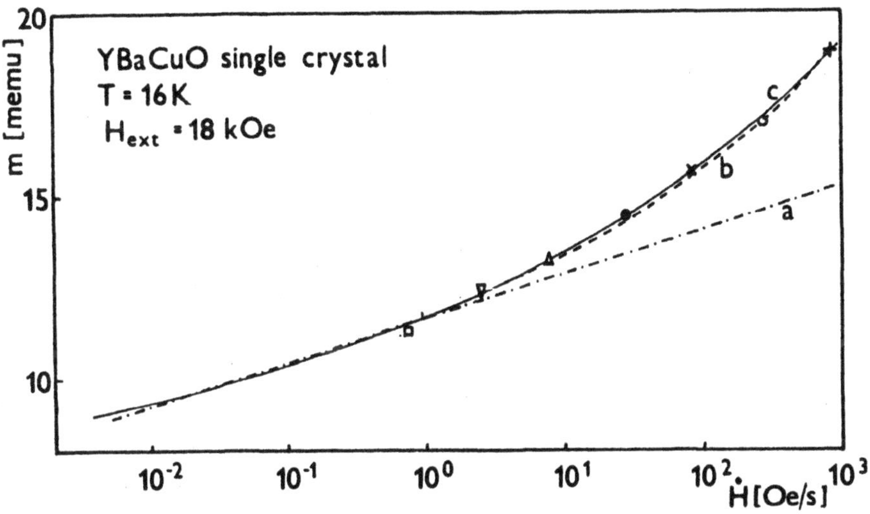

Figure 1: Plot of $m(\dot{H})$ representing magnetic moment relaxation measured: (i) at H_{ext}=const. (left side) and (ii) at $\dot{H}$=const (hysteresis loops) - marks +, o, x, •, Δ, ∇ and □ corresponding to $\dot{H}$= 816, 290, 89, 29, 8.7, 2.3 and 0.8 Oe/s, respectively. Sample area 2.12 mm^2, a-b plane, thickness 30 μm, χ= -0.036 memu/Oe, penetration field for high $\dot{H}$ is H^*= 3.2 kOe. Curves *a* to *c* are the best fits of Eqs. 1, 2, 4, respectively.

Our experiments on BiSrCaCuO single crystals exhibited a pronounced nonlogarithmic time dependence, Fig. 2, even in the region of conventional relaxation and only Eqs. 2 and 4 fit the experiment well.

Fitting parameters of all used functions are in Table 1. Detailed study of the dependence of fitting parameters on temperature and magnetic field is underway.

Table 1. Parameters of fitting functions, Eqs. 1,2 and 4.

	m_1	S	a_1	b_1	β	a_3	b_3	t_o
123	13.04	-0.52	6.60	7.99	0.152	21.30	-4.99	0.02
2212	---	---	-21.6	77.90	0.130	129.8	-49.00	0.0105

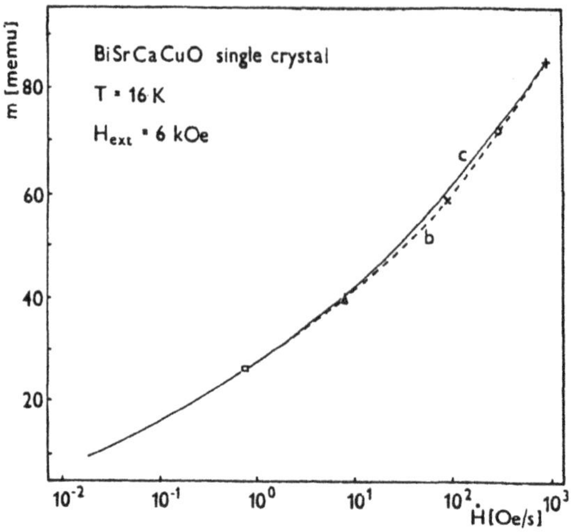

Figure 2: The same as in Fig.1 for BiSrCaCuO single crystal (T_c=82K), χ =-0.183 memu/Oe. Marks +, o, x, Δ and □ correspond to $\dot{H}$= 870, 290, 89, 8.7 and 0.08 Oe/s, respectively. Curves *b* and *c* correspond to fitting functions Eq.2 and 4.

The presented experiments indicate that the Anderson's model of TAFC describes the relaxation processes only under certain conditions and even then only at longer relaxation times. At the beginning of relaxation the decay rate of magnetic moment is faster than extrapolated logarithmic dependence from long time range. It may be due to a different mechanism of relaxation (collective flux pinning [18], flux motion under viscous forces[19], gradual activation of pinning sites with different pinning strengths with time[20]). Their identification is difficult as experimental data can be well fitted by different expressions.

Flux line dynamics at very short times corresponds more likely to flux flow than to flux creep.

In such samples as Bi-based single crystals the situation is further complicated by a strong dependence of irreversible magnetization on magnetic field. This leads in connection with low effective pinning energy to the rapid melting of flux lattice, namely at higher fields and temperatures.

5. References

1. K.A.Muller, M.Takashige, J.G.Bednorz, *Phys.Rev.Lett.* **58** (1987) 1143
2. A.C.Mota et al., *Phys Rev.***B36** (1987) 4011.
3. Y.Yeshurun, A.P.Malozemoff, F.Holtzberg, *Phys.Rev. Lett.***60** (1988) 2202 ;*J.Appl. Phys.***64** (1988) 5797
4. L.Půst, J.Kadlecová, M.Jirsa, S.Durčok, *J.Low Temp. Phys.***78** (1990) 179; *Cryogenics* **30** (1990) 886
5. M.Jirsa, L.Půst, J.Kadlecová, *Proc. EMMA'91*, Dresden, Germany (*J.Mag.Mag.Mat.1991*,accepted); M.Jirsa, L.Půst, J.Pačes, *Proc.ICAM'91*, Strasbourg, France
6. P.W.Anderson, Y.B.Kim, *Rev.Mod.Phys.***36** (1964) 39.
7. M.R.Beasley,R.Labush, W.W.Webb, *Phys.Rev.***181** (1969) 682
8. G.W.Hagen, R.Griessen, *Studies of high temperature superconductors*, Publ.Nova Science Publishers Inc., New York, 1989
9. P.Svedlindh et al., *Physica* **C176** (1991) 336
10. Xiao-Guang Li et al., *J.Appl.Phys.* **69** (1991) 7339
11. J.R.Thompson, Y.R.Sun, F.Holtzberg, *Phys.Rev.* **B44** (1991) 458
12. L.Půst, *Superc.Sci.Technol.***3** (1991) 598.
13. E.Zeldov et al., *Appl.Phys.Lett.***56** (1990) 680
14. A.P.Malozemoff, M.P.A.Fisher, *Phys.Rev.***B42** (1990) 6784
15. D.S.Fisher, M.P.A.Fisher, D.A.Huse, *Phys.Rev.***B43** (1991) 130.
16. L.Gao, Y.Y.Xue, P.H.Hor, C.W.Chu, *Physica* **C177** (1991) 438
17. B.Czyzak, J.Stankowski, *phys.stat.sol.*(*b*) **166** (1991) 219
18. M.V.Feigel'man et al., *Phys.Rev.Lett.***64** (1990) 2303
19. L.Půst, M.Jirsa, to be published
20. Griessen, *Physica* **C172** (1991) 441

SERIES JOSEPHSON JUNCTIONS AND TRANSPORT PROPERTIES OF HIGH-T_c SUPERCONDUCTORS

S. J. Lewandowski
Instytut Fizyki PAN, Al.Lotników 32, PL 02-668 Warszawa, Poland

ABSTRACT

It is generally recognized that high-T_c materials, both in bulk and thin film form, are prone to the creation of intrinsic Josephson weak links on grain boundaries, crystal lattice defects etc., and that transport properties of these materials can become dominated by the presence of Josephson junctions. Several theoretical models have been proposed to deal with this situation, including single junction and junction array models. We show that taking into account the possible presence of junctions connected in series introduces into these models an inherent source of instability and offers a possible explanation of some experimentally observed phenomena, e.g. excessive $1/f$ noise and spurious modulation of the critical current by the applied magnetic field.

1. Introduction

The existence of intrinsic Josephson junctions in high T_c materials is a well known fact. Its consequences for transport properties of these materials are also generally recognized. For instance, a single long, tunnel junction has been recently used as a model to explain the critical current dependence on applied magnetic field in a highly textured $YBa_2Cu_3O_7$ thin film[1]. However, many experimental facts indicate that junctions appear in the material rather as clusters, behaving in many respects like parallel arrays (*cf.*[2]). It can be reasonably supposed that some of the junctions in such parallel arrays can be connected in series. Theoretical analysis of such systems was carried out in Ref. [3,4]. In this paper we apply this theory to numerical analysis of the simplest possible arrangement involving series junctions, a (2+1)-junction superconducting interferometer.

2. General Theory

We consider stationary states of a superconducting interferometer composed of two parallel arrays of series connected Josephson junctions. The junctions are treated as lumped elements (point-like devices) and the effects of magnetic field are accounted for by assuming that the arrays include also series (lumped) inductances. The interferometer is supplied by *dc* current J from an external source and is linked by an externally applied magnetic flux Φ_e.

The system is described in terms of superconducting phase differences φ_{ni} across the junctions, where $n = 1, 2$ refers to the array and $i = 1, 2, \ldots N_n$ to the junction in this array. We need to consider also the critical currents I_{ni} of the junctions, the currents J_n through each array and the total magnetic flux Φ linking the circuit

$$\Phi = \Phi_e - \Phi_i,$$

where Φ_i denotes the induced flux.
Using the φ_{ni} variables, potential energy G of the system (current source and external magnetic field included) can be expressed as[4] (*cf.* also[5])

$$G = \sum_{ni} I_{ni}(1 - \varphi_{ni}) - \sum_{n} J_n \sigma_n + (\pi/L)\Phi^2, \tag{1}$$

where $L = L_1 + L_2$ and L_n is the sum of series inductances in each array. For fluxes we use the units of flux quantum Φ_o, the inductances are expressed in units of *inductance*$/\Phi_o$.

The first term on the right-hand side of Eq. (1) represents the potential energy of the Josephson junctions, the second is the total energy drawn from the current source in the process of increasing the phase across each junction from 0 to φ_{ni}, and the third is the magnetic energy stored in the system (and in the external field).

Eq. (1) must be supplemented with one other constitutive relation. It can be the fluxoid conservation relationship

$$\Phi = (1/2\pi)(\sigma_1 - \sigma_2 + 2\pi q), \tag{2a}$$

where σ_n denotes the sum of all phases in a given array and q is an integer, assumed further equal to zero. Instead of Eq. (2a) one can use simply the induced flux definition

$$\Phi = \Phi_e - \Phi_i = \Phi_e + J_1 L_1 - J_2 L_2. \tag{2b}$$

Let us now consider energy extrema relative to the condition of constant current J drawn from the source and constant externally applied flux Φ_e

$$dJ = d(J_1 + J_2) = 0, \quad d\Phi_e = 0. \tag{3}$$

Observe that this condition can be also interpreted as the stationary value problem for J and Φ_e. Another possible interpretation is that Eq. (3), with its implicit assumption of Kirchoff's law for currents, poses the problem of equilibrium distribution of a given current J into J_1 and J_2 with external flux as a parameter.

It can be easily shown[4] that the Josephson equations

$$J_n = I_{ni} \sin \varphi_{ni}, \quad n = 1, 2; \quad i = 1, 2, \ldots N_n, \tag{4}$$

constitute a necessary condition for the existence of an energy extremum. It suffices that these equations are satisfied locally at the extremum, but for reasons which will soon become apparent we will assume that they are global (differentiable). The second necessary (local) condition is always one of the equations (2), (2a) or (2b), depending on which of them was taken as global[4].

Let us assume that it is Eq. (2a) which has global character and let us proceed with effective solution of the considered problem. We need to conduct first some ordering operations. Equations. (4) are clearly redundant and the number of independent variables φ_{ni} can be reduced to two. Let the indices ni be ordered so that $I_{ni} \leq I_{nj}$ for $i < j$ and let $I_{11} \leq I_{21}$. The indices $n1$ will be further abbreviated to n in anticipation of the special role the weakest junction in each array is going to play. We introduce also the notation $a = I_1/I_2$, $a_{ni} = I_n/I_{ni}$. Then Eqs. (4) for $n = 1, 2$ and $i \geq 2$ are rewritten as

$$\varphi_{ni} \equiv \varphi_{ni}^{(m)} = (-1)^{m_{ni}} \varphi_{ni}^{(o)} + m_{ni}\pi, \quad i \geq 2, \tag{5}$$

where m_{ni} is an integer and $\varphi_{ni}^{(o)}$ denotes the principal branch of $\arcsin(a_{ni} \sin \varphi_n)$, i.e. $-\pi/2 \leq \varphi_{ni}^{(o)} \leq \pi/2$.

Observe that the ranges of $\varphi_{ni}^{(m)}$ and $\varphi_{ni}^{(m+1)}$ are separated by a phase gap of width[3]

$$2\Psi_{ni} = 2 \arccos a_{ni}, \quad 0 \leq \Psi_{ni} \leq \pi/2,$$

which cannot be entered without exceeding the critical current I_n.

The system is now completely described by the phase differences φ_1 and φ_2 across the weakest junction in each array and by the "state vector" $\langle \widehat{m} \rangle = \langle \widehat{m}_1, \widehat{m}_2 \rangle$ with $N_1 + N_2 - 2$ components $\langle \widehat{m} \rangle_{ni} = m_{ni}$. Moreover, the remaining two equations (4) and Eq. (2a) relate φ_1 and φ_2 to each other. Choosing φ_1, the phase of the weakest junction of the system, as the independent variable, and applying the method of implicit differentiation to Eqs. (3) we obtain after some manipulations[3] the necessary condition for a phase point (φ_1, φ_2) to be stationary

$$\frac{1}{\cos \varphi_1} + a \frac{1}{\cos \varphi_2} + \beta_1^{\langle \widehat{m} \rangle}(\varphi_1) + a \beta_2^{\langle \widehat{m} \rangle}(\varphi_2) = 0, \tag{6}$$

where

$$\beta_n^{\langle \widehat{m} \rangle}(\varphi_n) = -\beta_n + \sum_{i=2}^{N_n} \frac{(-1)^{m_{ni}} a_{ni}}{\sqrt{1 - a_{ni}^2 \sin^2 \varphi_n}},$$

and $\beta_n = 2\pi L_n I_n$. Eq. (6) must be solved numerically for $\varphi_2^{\langle\widehat{m}\rangle}(\varphi_1)$ [or $\varphi_1^{\langle\widehat{m}\rangle}(\varphi_2)$], the locus of the stationary points (φ_1, φ_2), dependent on the state vector $\langle\widehat{m}\rangle$ assigned to the system.

In general, the solution does not exist for an arbitrary value of φ_1 or φ_2. This can be easily shown in the case of a two-junction interferometer. Then $\beta_n^{\langle\widehat{m}\rangle} = \beta_n$ and Eq. (6) in $x = (\varphi_1)^{-1}$ and $y = (\varphi_2)^{-1}$ coordinates is a straight line

$$x + ay + \alpha = 0, \quad \alpha = \beta_1 + a\beta_2,$$

everywhere except inside the region bound by the lines $|x| = 1$ and $|y| = 1$. The importance of the parameter α in the analysis of a two-junction interferometer was first realized by Fulton *et al.*[6]. Intersection points of the solution line with the region boundaries define the forbidden ranges of φ_1 and φ_2, which are void only for $a = 1$ and $\beta_1 = \beta_2 = 0$. In the general case of nonlinear $\beta_n^{\langle\widehat{m}\rangle}$ the straight solution lines are replaced by curves, but the overall picture remains the same, as is demonstrated in Fig. 1 for the particular case of $N_1 = 2$, $N_2 = 1$, $I_1 = 0.8$, $I_{12} = 1.0$, $I_2 = 0.9$ and $\beta_1 = \beta_2 = 0.5$.

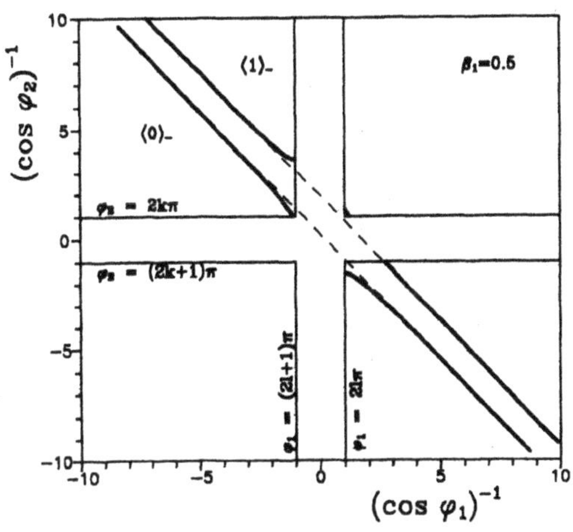

FIG. 1 Inverse cosine plots of stationary phases [solution lines of Eq. (6)] for a (2+1)-interferometer with $I_1 = 0.8, I_{12} = 1.0, \beta_1 = \beta_2, I_2 = 0.9$. The lower plot corresponds to the even $\langle 0\rangle_-$ state, the upper one - to the odd $\langle -1\rangle_-$ state. Dashed lines are to guide the eye only and to indicate that the asymptote of each plot is a straight line.

From the above observations it can be deduced that the plots $\varphi_2^{\langle\widehat{m}\rangle}(\varphi_1)$ form on the φ_1, φ_2-plane closed, periodically spaced loops (in direct analogy to fluxon vortices), separated by regions where stationary solutions cannot exist. In Fig. 2 we show the resulting phase patterns for the system of Fig. 1. The existence of such loops or phase vortices was recognized in the two-junction case by Tsang and Van Duzer[7].

The appearance of forbidden gaps in the ranges of *stationary* phases $\varphi_n^{\langle\widehat{m}\rangle}$ is not related to the gaps in φ_{ni}, introduced by Eq. (5). The latter cannot be crossed without expelling first the current from the relevant junction array[3] while the former can be bridged by any nonequilibrium process.

Assuming that the Josephson equations, Eq. (4) and fluxoid, Eq. (2a), are global, while the induced flux definition, Eq (2b), is satisfied only at the energy extremum we are essentially considering a situation in which the superconducting phases adjust immediately to the instantaneous values of J_1, J_2 and Φ but these values lag behind the changes in Φ_e. Under these assumptions it is also possible to calculate the second derivative $G''\langle\widehat{m}\rangle$ of $G^{\langle\widehat{m}\rangle}$ with respect to φ_1. The relevant expression is given in Ref.[4]. By calculating the sign of this expression it is possible to identify the stationary points as corresponding to energy minima, maxima and inflection points.

3. Numerical Analysis

Some insight into the system's behavior can be gained by numerical analysis of Eqs. (1) and (6). The simplest possible example is a (2+1)-junction interferometer, possessing a state vector with only one component. We assume the same circuit parameters as those used to calculate Fig. 1.

The key to understanding the stationary properties of the system is in the inverse cosine plots of Fig. 1 and the resulting phase plots, shown in Fig. 2.

As seen, the loops (vortices) of the stationary phase points $\varphi_2^{\langle\widehat{m}\rangle}(\varphi_1)$ are "paving" more or less tightly the phase plane and their centers form a square lattice with lattice constant of 2π. System properties are determined by the "unit cell" of this lattice with lattice points at the corners and in the center of a square with 2π sides. The gaps between phase vortices mark, as already observed, regions where stationary solutions do not exist and depend on the β_n's. Stationary phases can change continuously only on a vortex line. The existence of flat phase fronts, where the derivative of one phase with respect to the other tends to infinity, indicates that vortex-to-vortex transitions would be initiated at these fronts.

Each stationary phase vortex corresponds to a separate pattern of stationary current

$$J^{\langle\widehat{m}\rangle} = I_1 \sin\varphi_1^{\langle\widehat{m}\rangle} + I_2 \sin\varphi_2^{\langle\widehat{m}\rangle}$$

plotted in function of stationary flux $\Phi_e^{\langle\widehat{m}\rangle}$, calculated from Eqs. (2) and (5). The patterns of $G^{\langle\widehat{m}\rangle}(\Phi_e^{\langle\widehat{m}\rangle})$ can be constructed in the same manner. Both types of such patterns, obtained from the unit cell of phase vortices are shown in Fig. 3.

The current plot $J^{\langle\widehat{m}\rangle}(\Phi_e)$ corresponding to a separate phase vortex forms a closed loop, resembling in shape an ace of diamonds. In Fig. 3 we show only negative current values (normalized to $I_{max} = I_1 + I_2$) are drawn. Full current plots can be visualized

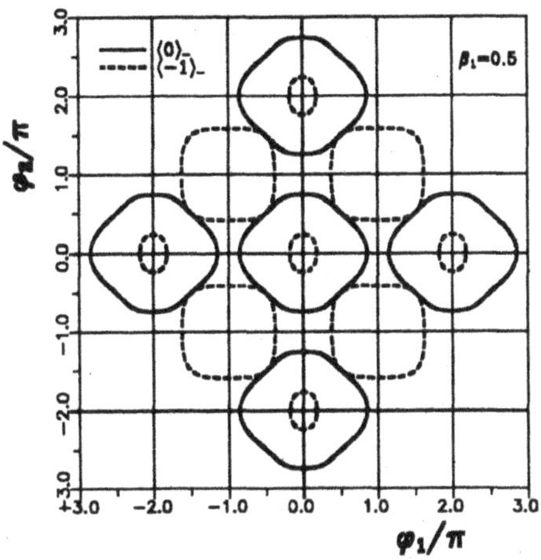

FIG. 2 Stationary phase patterns resulting from inverse cosine plots of Fig. 1.

by taking into account their point symmetry with respect to the coordinate origin: $J^{\langle\hat{m}\rangle}(\Phi_e) = -J^{\langle\hat{m}\rangle}(-\Phi_e)$. The sign of the second order derivative of energy with respect to φ_1 is indicated by the linewidth used for plotting: thick lines correspond to positive sign or local minimum of energy, thin lines - to negative sign or local maximum of energy. The branches drawn with thick lines are supposedly stable and will be referred to by this term.

The "aces of diamonds" corresponding to a single row or column of phase vortices form a periodical pattern along the Φ_e-axis. In general, vortices from two successive rows or columns are needed to reproduce the full functional dependence $J^{\langle\hat{m}\rangle}(\Phi_e)$, which must be Φ_o-periodic as required by the fluxoid conservation. If the corner sites in the elementary cell are numbered in a counterclockwise manner, starting from the upper left corner, the patterns in Fig. 3 correspond to the sequence 1 - 2 - 4 - 3, but the current patterns which correspond to 2 and 4 occupy the same flux range.

In contrast to the identical ace-of-diamond current patterns, the energy pattern or level produced by each phase vortex is unique. It is interesting to note that the energy levels corresponding to vortices which produce current patterns at identical flux ranges form closed loops. It can be also seen that at some flux ranges the energy levels correspoding to different phase states are nearly degenerate.

The transport properties of the system are determined by its *critical current.* Since $J^{\langle\hat{m}\rangle}$ are by definition stationary currents, we can define the critical current, $J_{max}(\Phi_e)$, as

$$|J_{max}(\Phi_e)| = max\ \left|J^{\langle\hat{m}\rangle}(\Phi_e)\right|,$$

i.e. the envelop of all $J^{\langle\widehat{m}\rangle}(\Phi_e)$. This definition assumes that transitions between

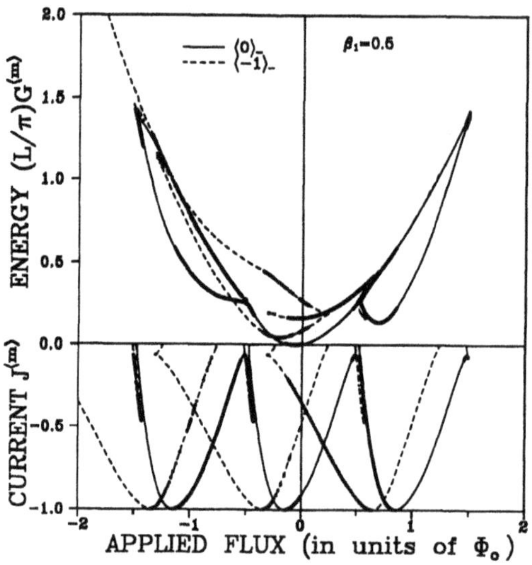

FIG. 3 Stationary energy function $(L/\pi)G^{\langle\widehat{m}\rangle}$ (upper part of the drawing) and stationary current $J^{\langle\widehat{m}\rangle}$ (lower part of the drawing) in function of externally applied magnetic flux Φ_e for $\langle\widehat{m}\rangle = \langle 0\rangle$ (solid lines) and $\langle\widehat{m}\rangle = \langle -1\rangle$ (dashed lines) for the system parameters as in Fig. 1. Both $G^{\langle\widehat{m}\rangle}$ and $J^{\langle\widehat{m}\rangle}$ are normalized to $I_{max} = I_1 + I_2$; only negative values of the current are shown. Thick lines are used to draw stable solution branches corresponding to local minima of energy.

different phase states are possible (*cf.*[3]. However, for the system to exhibit always this critical current, the transitions should occur at crossing points of relevant currents $J^{\langle\widehat{m}\rangle}$ (at least in the envelop region).

An enticement for a transition to occur can be either lowering of potential energy (spontaneous transitions) or a possibility to occupy a higher critical current (induced transitions under current bias conditions). The first type of transition is most easily attained when the stationary current of the initial state $J^{\langle\widehat{m}\rangle}$ descends to zero, at the phase vortex edge. Transitions caused by small phase fluctuations in regions where the vortex slope tends to infinity cannot also be excluded. In the presence of a constant current bias the transitions would occur whenever $J^{\langle\widehat{m}\rangle}(\Phi_e)$ descends below the bias value. If the bias current J is $\geq I_{max}$, the system can always leave superconducting state exhibiting the envelop value of the critical current. This observation might be helpful in the interpretation of experimental data on magnetoresistance of high-T_c materials[8].

The response of the system to external flux changes must be determined by its dynamic properties. The experiments of Fulton, Dunkleberger and Dynes[6] on highly inductive two-junction interferometers have shown that in dynamical situations the equilibrium currents below the envelop can take on the role of critical currents.

The envelop shown in Fig. 3 exhibits the effect of "spurious" or "secondary" modulation. The combination of this effect and that of hysteretic multiple critical currents was possibly observed in some high-T_c *dc* SQUID measurements. Let us also observe that the crossing energy levels provide a set of of two energy wells required by the two-level fluctuator model of random telegraph noise exhibited by high-T_c thin films[8].

4. Conclusions

In conclusion, we have shown from energy considerations that superconducting interferometers containing series junctions are generally unstable and should exhibit hysteretical behavior. Therefore, the presence of intrinsic series junctions in high-T_c superconductors can be the cause of some of the instabilities observed in these materials. However, the theory presented here is based on a very simplicistic model of Josephson junction and it stil needs a more definitive experimental verification carried out on classical tunnel junctions.

5. Acknowledgments

Many stimulating discussions with J. Zagrodziński are gratefully acknowledged.

6. References

1. P. Bernstein, S. Lamarti and J. Bok, *Solid State Commun.* **75**, 587 (1990).
2. G. Jung and J. Konopka, *Europhys.Lett.* **10,** 183 (1989).
3. S. J. Lewandowski, *Phys.Rev.* B **43**, 7776 (1991).
4. S. J. Lewandowski, *Phys.Rev.* B, to be published.
5. M. Klein and A. Mukherjee, *Appl.Phys.Lett.* **40**, 744 (1982).
6. T. A. Fulton, L. N. Dunkleberger and R. C. Dynes, *Phys.Rev. B* **6**, 858 (1972).
7. Won-Tien Tsang and T. Van Duzer, *J.Appl.Phys* **46,** 4573 (1975).
8. G. Jung, M. Bonaldi, S. Vitale and J. Konopka, *J.Appl.Phys*, to be published.

Thermally Activated Flux Flow and Anisotropy of the Pinning Force Density in Epitaxially Grown $YBa_2(Cu_{1-x}Zn_x)_3O_{7-\delta}$ Thin Films

M. Schmitt, A. Walkenhorst, C. Tomé-Rosa, M. Paulson, P. Wagner, Th. Kluge, G. Jakob, and H. Adrian
Institut für Festkörperphysik, Technische Hochschule Darmstadt, Hochschulstraße 8, 6100 Darmstadt, FRG

Abstract

The resistivity of dc-sputtered epitaxial $YBa_2(Cu_{1-x}Zn_x)_3O_{7-\delta}$ films on $SrTiO_3$-substrates was measured in magnetic fields up to 12 T. From $\rho(B,T)$ data the average activation energy for thermally activated flux creep $\langle U \rangle$ was determined. The pinning effectivity of the Zn-doping becomes evident by a distinct increase of $\langle U \rangle$ for small doping concentrations (x $\leq$ 0.005). For higher concentrations $\langle U \rangle$ decreases rapidly because of significant overlap of the local pinning potentials of the randomly distributed Zn-atoms. Further evidence for the pinning effectivity of Zn arises from the dependence of the critical current density on the angle between the applied magnetic field and the c-axis of the films. This dependence $j_c^{ab}(\theta)$ with $\vec{B} \perp \vec{j}$ shows strong deviations from the dependence $j_c^{ab}(\theta)$ of undoped $YBa_2Cu_3O_7$ films. The measurements can be explained quantitatively by assuming the existence of point-like pinning centers (created by the Zn-doping) and combining this pinning mechanism with the well known intrinsic pinning mechanism of the HTS materials and planar pinning centers due to twin boundaries. In these calculations it turns out that the component $\vec{B} \parallel \vec{c}$ determines the critical current density.

Introduction

For many applications of the high T_c materials high critical current densities j_c are necessary. This aim can be reached by increasing the pinning of the magnetic flux lines which penetrate the material. One of the possibilities is to dope the material with some pinning-effective impurities. In this paper we examine the pinning-effectivity of Zn in $YBa_2(Cu_{1-x}Zn_x)_3O_{7-\delta}$ films by measuring the specific resisistivity $\rho(B,T)$ and the dependence of the critical current density on the angle between the c-axis of the film and the applied magnetic field $j_c(B,T,\theta)$.

In the following we develop two models which are able to explain the obtained data for the two kinds of measurements. The first model describes the dependence of the activation energies on the Zn-concentration in the $YBa_2(Cu_{1-x}Zn_x)_3O_{7-\delta}$ films. It takes into account an overlap of the local pinning-potentials due to Zn-atoms together with the mean number of Zn-atoms that pin one flux line. The second model fits very well the $j_c(\theta)$-data. It is assumed that only the component of the field parallel to the

$\vec{c}$-axis of the lattice structure determines the dissipative behaviour. The pinning force on that component is produced by the following mechanisms: plane shaped pinning sites (due to twin boundaries) and point-like pinning sites (due to Zn-atoms).

Experimental

The films were prepared on $SrTiO_3$ by high pressure dc sputtering in pure oxygen from stoichiometric targets. The Zn-substitution does not influence the structure of the films. Details are described elsewhere[1]. $YBa_2(Cu_{1-x}Zn_x)_3O_{7-\delta}$ films with x between 0 and 10 % were examined. After evaporating silver contacts the films were patterned into an 8 probe geometry by wet chemical etching[1].

The resistance of the films was measured by standard 4-probe method in magnetic fields up to 12 Tesla for $\vec{B} \parallel \vec{c}$. In zero field the T_c-depression caused by the Zn-doping is -10.5 K/(at % Zn). The $j_c(\theta)$-data were obtained by measurements on a 10 μm wide and 100 μm long microbridge applying a 1 μV-criterion. The films were mounted on a rotatable copper block with an angular resolution of 0.04° driven by a computer controlled step motor as described in reference 2.

Results and discussion

The activation energy U

From the Arrhenius-plots of the resistive data the mean activation energies $\langle U \rangle$ for the examined Zn-concentations and magnetic fields were determined in a region of $\rho/\rho(100K) = 10^{-3} \ldots 10^{-2}$.

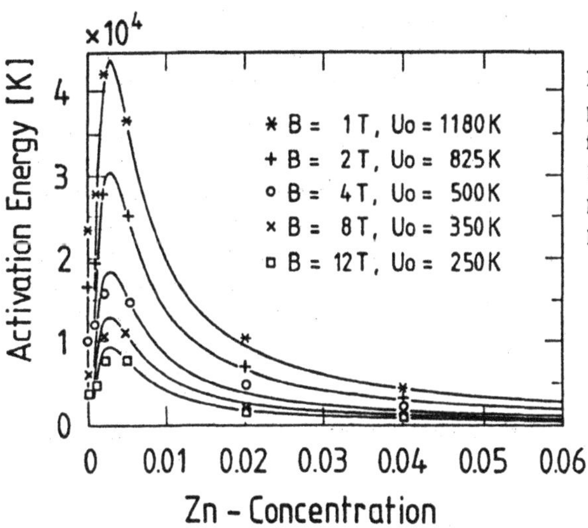

Figure 1: Activation energy in dependence of the Zn-concentration for different magnetic fields up to 12T. The solid lines are the fits proposed by the model of overlapping pinning-potentials

As can be seen from Fig. 1 the behavior of $\langle U(x)\rangle$ is qualitatively the same for all magnetic fields. There is a maximum in $\langle U(x)\rangle$ for $x \approx 0.3$ %. For $x > 0.5$ % the values for $\langle U(x)\rangle$ decrease rapidly. This behavior of $\langle U(x)\rangle$ can be explained by the following model: Assume a parabolic pinning-potential $U(\vec{r})$ of one Zn atom of the form

$$U(\vec{r}) = U_0 \cdot \min\left(\frac{2}{h^2}\cdot|\vec{r}-\vec{r_1}|^2 - 1, 0\right). \quad (1)$$

Here h is the FWHM, U_0 is the depth of the potential and $\vec{r_1}$ is the site of the pinning center. For the following calculations a dependence of U_0 on the temperature is not taken into account. Due to overlap of neighbouring Zn-atoms (as shown in Fig. 2) the effective potential barrier between the pinning sites is reduced to

$$U_{barrier}(d) = U_0 \cdot \min\left(\frac{2}{h^2}\cdot\left(\frac{d}{2}\right)^2, 1\right) \quad (2)$$

with d being the distance between nearest neighbour Zn-atoms. We can calculate the probability $P(d,x)$ that nearest neighbour Zn atoms are a distance d away from each other for a given Zn-concentration x assuming a random occupation. Considering a CuO_2-plane this is the product of the probability to find no Zn-atoms within a circle of radius r and the probability to find at least one Zn-atom in the ring of inner radius r and width of the lattice parameter a. Both probabilities are binomial-distributions. The probability to find a Zn-atom at any site we are looking at is given by the concentration x. Having calculated $P(d,x)$ we can calculate the mean activation energy for one flux line segment for any Zn-concentration. In addition we have to sum over all planes penetrated by the flux line within its correlation length parallel to the direction of the magnetic field. We therefore have to calculate the mean number of Zn-atoms which pin one flux line over its whole correlation length which is estimated to be 500 Å ($\approx$ 45 unit-cells in c-direction) due to the work of Fischer et al.[3].

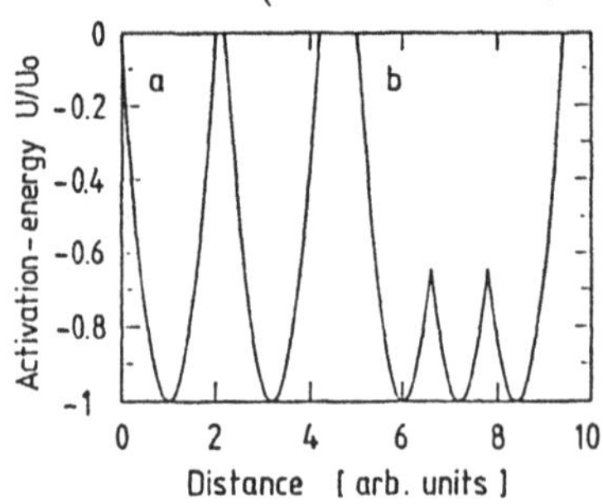

Figure 2: Pinning potentials for different concentrations. a: The concentration is small and the potentials do not overlap. The activation energy is the same as the height U_0 of the potentials. b: For increasing concentration the potentials begin to overlap and the activation energy is less than U_0.

For randomly distributed Zn-atoms the probability for pinning a flux line segment by at least one pinning center is

$$P_{pinning}(x) = 1 - \exp\left(-\frac{s(B)N_P(x)}{N_V(B)}\right) \quad (3)$$

$N_P(x) \propto x$ is the areal density of pinning sites and $N_V(B) = \frac{B}{\Phi_0}$ is the areal density of flux line segments. As the flux line can only slightly bend to reach a pinning center there is only a fraction $s < 1$ of the area of a flux line unit cell within which the flux line segment can find a pinning center. For $\vec{B} \parallel \vec{c}$ the vortex segments in adjacent layers are strongly coupled. In this regime the accesible area for a flux line segment is determined by this coupling and not by the area of the flux line unit cell, so the parameter $\frac{s(B)}{N_V(B)}$ turns out to be constant, i. e. independent of $\vec{B}$. The mean number of pinning active Zn-atoms in one correlation length of the vortex is $\langle n(x)\rangle = m \cdot P_{pinning}(x)$. $m = 90$ is the number of CuO_2-planes the flux-line has to cross over its whole correlation length. The mean activation energy for one correlation length of the flux line is then given by

$$\langle U(x)\rangle = \sum_{d_i \in n.n.} P(d_i, x) U_{barrier}(d_i) \langle n(x)\rangle \tag{4}$$

where d_i runs over all nearest neighbour-distances within a CuO_2-plane.

As can be seen from Fig. 1 the calculated formula fits very well the data obtained from the Arrhenius-plots. For the data shown in Fig. 1 the fit parameters U_0=250 K ($B = 12$ T) to 1180 K($B = 1$ T) were used whereas $h = 35$ Å and $\frac{s(B)}{N_V(B)} = c = 2000$ Å^2 were kept constant for all curves. It is noteworthy that the parameter h is of the order of the double correlation length ξ_{ab} which was to be expected as this is a measure of the region around the pinning centers with reduced order parameter. The dependence of the activation energy U_0 on the magnetic field could arise from the reduction of the order parameter in the material in higher fields. The disagreement of the model with our data for x=0% ($\langle U(x=0)\rangle = 0$) can be explained by additional pinning-centers in the film (e. g. defects, twin boundaries), which contribute to the effective pinning equivalently to a Zn-background of 0.05%. For the calculations made above we did not have to make any assumption of a correlation between the flux lines. It has been tested that the results depend only weakly on the detailed shape of the pinning potential.

The $j_c(\theta)$-model

Typical results of our measurements of the critical current density j_c in dependence of the angle θ for an undoped and for a Zn-doped film with x=0.5% are shown in Fig. 3. In comparison to the data of the doped film with that of the undoped one there are typical shoulders close to the maxima at 90° and 270°. These structures can be explained by a simple model which makes the following assumptions:

- Since for $B \perp c$ the flux lines are perfectly pinned by the intrinsic pinning of the layered structure of the material, the dissipative effects and thus the critical current density j_c is fully determined by the component of the magnetic field which is parallel to the c-axis of the film $B_{\vec{c}} = B_0 \cdot \cos(\theta)$ with θ being the angle between the c-axis and the magnetic field.

- Additionally to the intrinsic pinning of the high-T_c-materials and to the plane shaped pinning potentials of twin bounderies already discussed by Tachiki and Takahashi[4] there is a second point like pinning mechanism due to Zn-atoms.

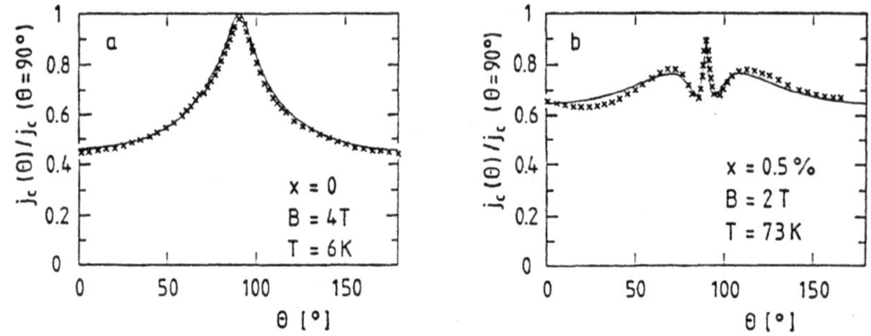

Figure 3: Critical current density j_c in dependence of the angle θ between the c-axis of the film and the magnetic field. a: For an undoped film. b: For a film with a Zn-concentration of x=0.5% with typical shoulders around the main peaks at 90° and 270°. The model discussed in the text is given by the solid lines.

The pinning probability for a flux line segment is given by eq. 3. In contrast to the model examined before the flux line segments are decoupled because of the kink structure of the flux line. In this limit $s(B)$ is assumed to be constant. Thus the pinning probability becomes dependent on the area of a flux line unit cell and on the magnetic field. For plane shaped pinning centers (i.e. twin boundaries) Tachiki and Takahashi[4] calculated

$$\frac{s(B)N_P(x)}{N_V(B)} \propto \frac{1}{\sqrt{\cos\theta}} \tag{5}$$

For point like pinning centers (i.e. Zn-atoms) an equivalent calculation[5] yields

$$\frac{s(B)N_P(x)}{N_V(B)} \propto \frac{1}{\cos\theta} \tag{6}$$

For both pinning mechanisms present at the same time we can calculate the total pinning force density

$$f_{tot} = f_1 p_1 + f_2 p_2 - f^* p_1 p_2 \tag{7}$$

with f_1 (f_2) the pinning force density of the first (second) pinning mechanism and p_1 (p_2) the probability for the flux line to be pinned by the first (second) pinning mechanism. If we looked seperately at each pinning mechanism, the total pinning force density could be described by the first two terms in equation 7, but as in some of the flux line lattice cells both kinds of pinning centers can be found, the first two terms of eq. 7 overestimate the total pinning force density. Thus we have to subtract the third term in the equation to reduce this overestimation. If we insert equation 3 for p_1 and p_2 in eq. 7 we can fit the obtained data very well as can be seen in Fig. 3.

The model suggested above does not need the assumption of dilutely spread pinning sites as is needed by the model of Tachiki and Takahashi[4,5].

Conclusions

We could explain the dependence of the activation energy of $YBa_2(Cu_{1-x}Zn_x)_3O_{7-\delta}$ films on the Zn-concentration by a model assuming overlapping pinning potentials due to the Zn-atoms. We conclude from this model that the pinning potentials have a FWHM of 35 Å and a depth between 250 K (B = 12 T) and 1200 K (B = 1 T). As for the high activation energy for $x = 0$ % we deduce to have pinning centers (like defects and twin boundaries) additional to the Zn-atoms which contribute to the pinning equivalently to a Zn-concentration of 0.05 %.

The $j_c(\theta)$-measurements and thus the effects due to Zn-pinning can be explained by a simple model assuming additional point-like pinning (by Zn-atoms) together with plane shaped pinning centers (by twin boundaries). In contrast to all previous calculations we take into account the reduction of the pinning potential at the locations where both plane-shaped and point-like pinning centers coincide. It turns out that the additional peaks around the main peaks at 90° and 270° can be explained in a natural manner. Furthermore the model does not assume dilutely spread pinning sites as the model of Tachiki and Takahashi does.

Acknowledgements

The authors acknowledge financial support by the Bundesminister für Forschung und Technologie der Bundesrepublik Deutschland and the Deutsche Forschungsgemeinschaft through SFB 252. One of us (C. T.-R.) was financially supported by the Conselho Nacional de Desenvolvimento Científico e Tecnológico do Brasil - CNPq.

References

1. C. Tomé-Rosa, G. Jakob, M. Maul, A. Walkenhorst, M. Schmitt, P. Wagner, P. Przysłupski, and H. Adrian, Physica C **171**, 231 (1990).

2. A. Walkenhorst, C. Tomé-Rosa, C. Stölzel, G. Jakob, M. Schmitt, and H. Adrian, Physica C **177** 165 (1991)

3. O. Brunner, L. Antognazza, J. M. Triscone, L. Miéville and Ø. Fischer, preprint

4. M. Tachiki and S. Takahashi, Solid State Commun. **72**, 1083 (1989).

5. A. Walkenhorst, C. Tomé-Rosa, P. Wagner, Th. Kluge, C. Stölzel, G. Adrian, G. Jakob, and H. Adrian, to be published.

TRANSPORT CRITICAL CURRENT LIMITATION IN THE ORIENTED AND NONORIENTED $Y_1Ba_2Cu_3O_{7-x}$ THIN FILMS

V.Štrbík, R.Adam, Š.Beňačka, Š.Chromik
Institute of Electrical Engineering, SAS, Dúbravská cesta 9, 842 39 Bratislava, Czecho-Slovakia

Abstract

The temperature dependences of transport critical current density of oriented and nonoriented $YBa_2Cu_3O_{7-x}$ thin films of thickness below 0.7 μm were studied. The nonoriented granular films exhibit the convex type of $j_c(t)$ dependences and limiting mechanism of critical current was considered in the frame of weak link network. The concave type of $j_c(t)$ dependences of oriented films were interpreted due to flux creep limiting mechanism and activation energy of flux creep $\approx$100 meV was estimated.

1. INTRODUCTION

The critical current density j_c of high-Tc superconductors strongly depends on the microstructure and composition of individual specimens. On the other side, it is known that bulk materials have smaller critical current densities than oriented thin films and single crystals. Extremely high j_c were published[1] on superconducting bridge connections of small dimensions realized from oriented films. The investigations of temperature and magnetic field dependences of critical current density as well as their absolute values may help to understand the mechanisms which limit the critical current in high-Tc superconductors.

In this paper we analyze the critical current density versus temperature dependences for samples with various critical current density. It will be shown that the main reason of critical current density limitation in granular thin films is the strength of intergranular connections and in epitaxial films flux creep effect.

2. SAMPLE PREPARATION

The superconducting high-Tc $YBa_2Cu_3O_{7-x}$ thin films were prepared by vacuum coevaporation, with substrate at relatively low temperature (600°C). During the simultaneous evaporation of Y, Ba and Cu from three separated sources the vacuum was degraded by oxygen to 10^{-2} Pa^2. Thin films (thickness $\approx$0.7 μm) prepared in situ on polycrystalline $SrTiO_3$ or Al_2O_3 substrates showed granular morphology with disordered grains of typical dimensions $\lessapprox$0.5 μm and relatively low critical current density j_c(4.2 K) $\approx 5\times10^4$ A/cm^2.

Another set of thin films (thickness smaller than 0.15 μm) was prepared at temperature 700-750°C on (200) MgO single crystal substrates. Here, we received highly oriented films with c-axis perpendicular to the substrate[3].

The higher crystallographic orientation influences electrical properties of thin films. The room temperature resistivity ρ_{300} decreases from 1 mΩcm to 0.5-0.3 mΩcm and ratio ρ_{300}/ρ_{100} increases from 2 to 3. The critical current density changed from 5×10^4 A/cm^2 to 6×10^6 A/cm^2 at T=4.2 K and B=0. The superconducting onset temperature was typically $T_{on} \approx 89$ K, the zero resistance critical temperature $T_{co} > 80$ K and the phase transition width ΔT_c smaller than 2 K for good films.

3. RESULTS AND DISCUSSION

In Fig.1 the typical dependences of critical current density on temperature for samples prepared by the both technological procedures are shown. The curves a,b are characteristic for granular disordered thin films, curves c and d characterize the $j_c(t)$ dependences of oriented thin film. The difference of $j_c(t)$ between these types of samples is evident. The critical current density is normalized to the value

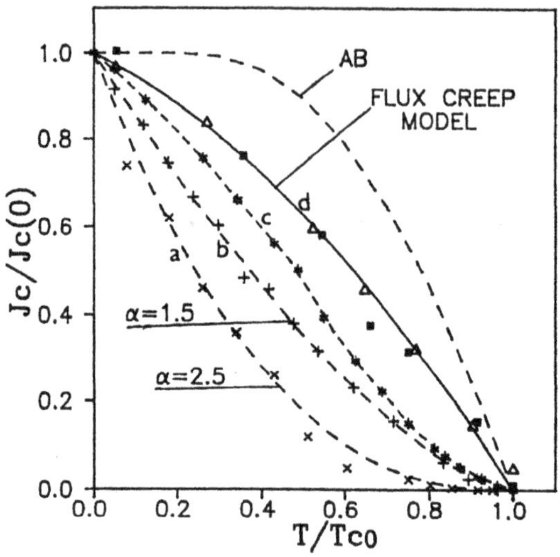

Fig.1 Dependences of critical current density on temperature for samples with various $j_c(o)$.
(a) $j_c(o) < 10^3 A/cm^2$, b) $j_c(o) \approx 10^4 A/cm^2$, c) $j_c(o) \approx 10^5 A/cm^2$, d) $j_c(o) > 10^6 A/cm^2$

of $j_c(o)$ received from the lowest temperature experimental data, the temperature T is normalized to $T_{co}(t=T/T_{co}$,

The $j_c(t)$ dependences of granular films (curves a,b) were analyzed in terms of weak link conections between grains where the power dependence $j_c(t) \sim j_o(o) x (1-t)^{\alpha}$ is expected and α characterizes the type of weak links and its value depends on dispersion of weak links critical current. The dependences with $\alpha \gtrsim 2.5$ are characteristic for samples with small T_{co} (typically $\lesssim 50$ K), broad phase transition ($\Delta T_c > 30$ K) and low critical current density $j_c < 10^3 A/cm^2$. From analogy with $j_c(t)$ dependence of low temperature nonhomogeneous superconductors e.g. Nb_3Sn [4] we suppose a large dispersion of individual connection in weak link network of our granular films, too.

In the case of samples where $\alpha \approx 1.5$-2 (as curve b), weak links of S-c-S (superconductor-constriction-Superconductor) or S-N-S (Superconductor-Normal metal-Superconductor) type dominate. The maximal critical current density of these samples was $j_c(o) \approx 10^4$ A/cm^2. On the basis of experimental observations we concluded that the transport properties of granular thin films are determined by weak superconducting links between the superconducting grains, thus reducing much higher critical current density of grains.

The temperature dependence of critical current density of oriented films is different with downward curvature of $j_c(t)$ (points on curve d). In this case a new mechanism of critical current limitation begins to dominate. We ascribe this current limitation to the flux creep[5], and full line (curve d) represent the fit of experimental data computed from relation[5]

$$j_c(t) = j_c(o) \quad (1-mt - nt)^2 \qquad (1)$$

where $t=T/T_{co}$, m=0.5, n=0.53 and $j_c(o)=2.6\times10^6$ A/cm^2. Because this type of $j_c(t)$ is characteristic for samples with $j_c(o) > 10^6$ A/cm^2 we suppose that the intergranular weak links are not responsible for limitation of critical current. The SEM pictures confirm this assumption because these oriented films show no granular morphology, and the substrate surface is imitated only. Due to these "substrate" defects the critical current density is about one order of magnitude smaller than that for ideal epitaxially grown thin films. The curve c represents a crossover from one type of $j_c(t)$ to another. In the low temperature region ($t < 0.6$) it is possible to recognize the flux creep limitation mechanism and for temperature $t > 0.6$ the power dependence of $j_c(t)$ is evident. The crossover curves are typical for samples with maximal critical current densities $\approx 10^5$ A/cm^2. It may be explained by weak link network in high temperature region, but at low temperatures the intergranular conections lost the weak link nature. The flux creep activation energy $W_{(B,0)}$ estimated from fitting parameter m[5] is about 0.1 eV.

The obtained $j_c(t)$ dependences showed in Fig.1 may be interpreted also in the frame of S-N-S weak link model. In reference 6 the $j_c(t)$ dependences with upward curvature (curves a,b) were interpreted by S-N-S weak links with various ratio $d/\xi_N(T_c)$, where d is a thickness of normal region between superconducting grains and $\xi_N(T_c)$ is a coherence length in this region. The downward $j_c(t)$ dependence (curve d) and crossover one (curve c) were explained in reference 7 as $j_c(t)$ of pure S-N-S and dirty S-N-S junction resp. But, if the flux creep mechanism is responsible for limitation of critical current, then the voltage induced by the flux creep has an exponential form[8]

$$U = U_c \exp \frac{I - I_c}{I_1} \qquad (2)$$

where I_c is critical current defined by the voltage criterion

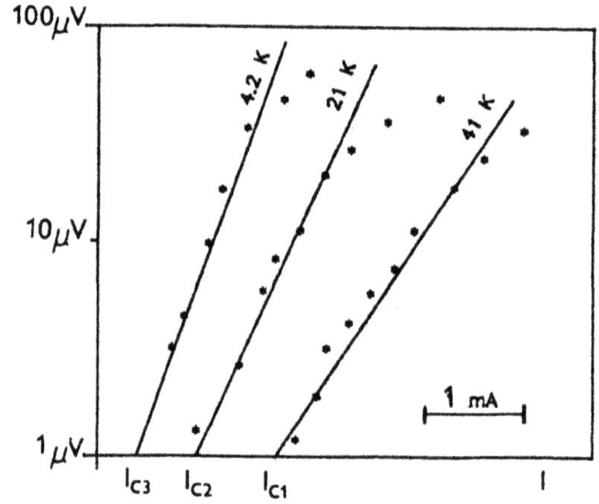

Fig.2 I-V characteristics of oriented film for voltages below 100 μV and different temperatures. The straight lines represent the relation (2).

U_c(=1 μV) and I_1 is a constant for given temperature ($I_1=k_BT\ \ I_{co}/W_{(B,T)}$). In Fig.2 the experimental dependences of low voltage versus the sample current are shown at different temperature. The theoretical prediction corresponding to Eq.2 is plotted as full line. From Fig.2 we see that flux creep mechanism accounts only in the region of low voltages (< 50 μV) but for voltages greater than 50 μV another dissipative mechanism takes place[8].

4. CONCLUSION

The set dependences of critical current density on temperature with various critical current density were investigated. The samples with low critical current density $j_c(o)<$ $<10^4$ A/cm^2 show the power dependence of $j_c(t)$ and the weak link network appears to be responsible for critical current limitation in this case. The samples with current density in the range or higher 10^6 A/cm^2 show flux creep limitation mechanism.

References

1. H.Jiang, Y.Huang, H.How, S.Zhang, C.Vittoria, A.Widom, D.B.Chrisey, J.S.Horwitz and R.Lee, Phys.Rev.Lett. **66** (1991) 1785
2. Š.Chromik, F.Hanic, R.Adam, M.Jergel, J.Liday and Š.Beňačka, Appl. Phys.Lett. **56** (1990) 2237
3. Š.Chromik, V.Štrbík, F.Hanic, V.Šmatko, R.Adam and Š.Beňačka, Physica B **165-166** (1990) 1493
4. R.H.Hammond, B.E.Jacobson, T.H.Geballe, J.Talvacchio, J.R.Salem, H.C.Pohl and A.J.Braginiski, IEEE Trans.Magn. **MAG-15** (1978) 619
5. J.Mannhart, P.Chaudhari, D.Dimos, C.C.Tsnei and T.R.McGuire, Phys. Rcv.Lctt. **61** (1900) 2476
6. N.Savvides, Physica C **165** (1990) 371
7. K.I.Konstantinian, G.A.Ovsyannikov, L.E.Amatami and Z.G.Ivanov, Zh.Eksp.Teor.Fiz. **99** (1991) 675 (in Russian)
8. M.Polák, P.Kottman, M.Majoroš, M.Kedrová and V.Plecháček, Supercond. Sci.Technol. **3** (1990) 67

THE EFFECT OF AN ELECTRIC CURRENT ON MICROWAVE ABSORPTION IN A YBaCuO CERAMIC SUPERCONDUCTOR

J. STANKOWSKI, S. WAPLAK, S. HUTTON,* and J. MARTINEK

Institute of Molecular Physics, Polish Academy of Sciences
Smoluchowskiego 17, 60-179 Poznan, POLAND

***Department of Physics, Montana State University**
Bozeman, USA

ABSTRACT

The process of microwave absorption in a high temperature superconductor was investigated in the presence of an external dc current. The data show that temperature of the Josephson junctions system changes proportionally to I^2.

INTRODUCTION

Microwave absorption in high temperature ceramic superconductors (JMA) is connected with the intra and inter grain defects. The inter grain absorption, observed by Stankowski et al.[1] and others[2-5] at small fields is attributed to weak Josephson links. We present a model in which the weak inter grain Josephson links can be decoupled by microwave field and transport current. This effect results in a large increase in the junction resistance and Joule heating appearance. We believe, that this mechanism is in fact the dominant one for the rapid increase in microwave absorption at small magnetic fields.[6-8] We cannot rule out, however, the possibility of fluxon motion, as suggested by Blazey et al.[3]

In this paper we preset the investigations of an influence of current flowing through the ceramic sample of YBaCuO (123) on Josephson Microwave Absorption (JMA).

RESULTS

Microwave absorption was investigated by a standard EPR spectrometer made by RADIOPAN with 100 kHz modulation of the magnetic field. We used a sample with dimensions 1x1x3 mm^3 with silver paint electrodes on the surface 1x3 mm^2. In our experiments the microwave field B_1 was perpendicular to constant external magnetic field B_0 equal to about 20 G. The external current was perpendicular to B_1 and B_0 as well. With this value of B_0 our microwave absorption shows a similar dependence to that previously observed by us[9] and other authors.[10]

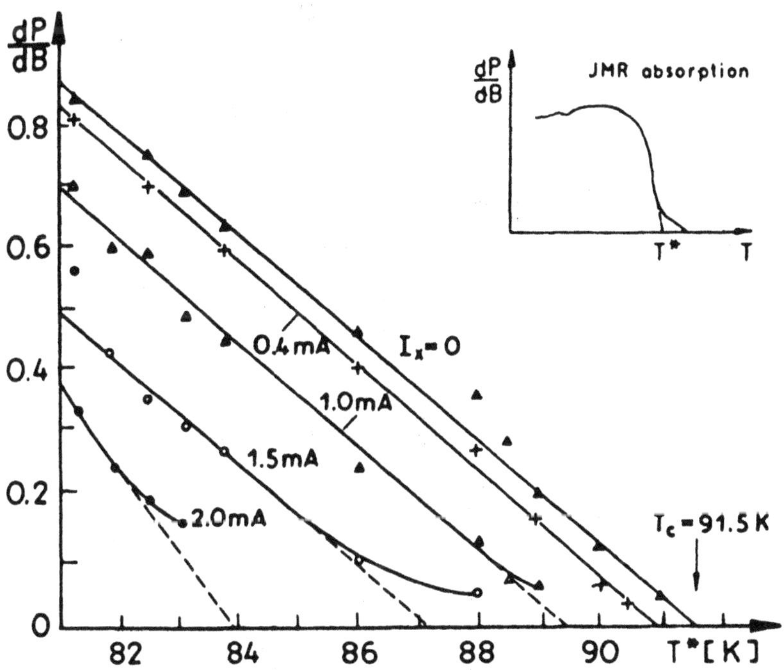

Fig. 1. Temperature dependence of the derivative of microwave absorption dP/dB in YBaCuO (123) sample for several currents

In Fig. 1 is shown the dP/dB dependence versus temperature for the several values of current flowing through the sample i.e. I=0 ,0.4 ,1 ,1.5, and 2 mA. In our experiment, the current was always decreased in a step wise manner from the maximum value to the value I=0. Each time the current was changed, it was also necessary to stabilize the temperature in order to provide isothermal conditions. Temperature readings were obtained from a thermocouple which was attached directly to the sample as is shown in Fig. 2a.

As we can see from Fig. 1, the microwave absorption level

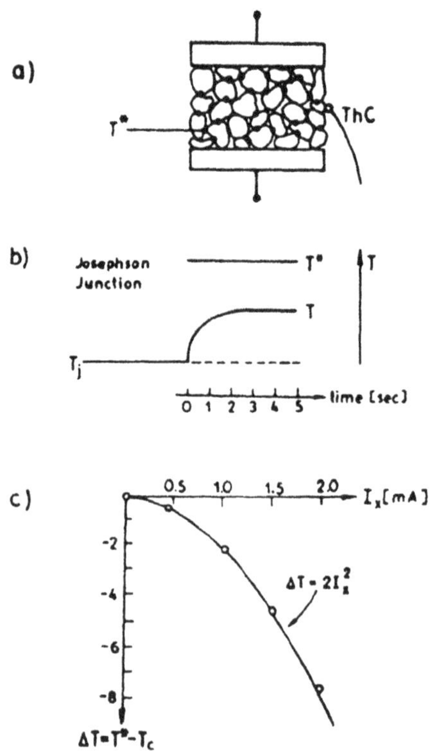

Fig. 2. Intergrain JJ contacts have the T^* temperature differ then bulk sample temperature T (a), equilibrium sample temperature T stabilized during the few seconds after current applying (b), difference of the T^*-T is proportional to the I^2 (c).

decreases as current increases. From the extrapolation of linear parts of the dP/dB curves versus temperature we obtain the following temperatures: T^*=91.5 K; 91.0 K; 89.5 K; 87.0 K and 85.0 K which reflects the effects of current increasing. A disappearance of the MA signal at T^* indicates that the temperature of junctions is equal to T_c=92 K, whereas the rest of the sample has the temperature $T^*<T_c$

It has been shown before[11,12] and it also seems reasonable that we have a reservoir of Josephson junction of the type S-I-S which favors a temperature difference T^*-T due to the differences in heat conductivity for dielectric and superconducting regions (Fig.2b).

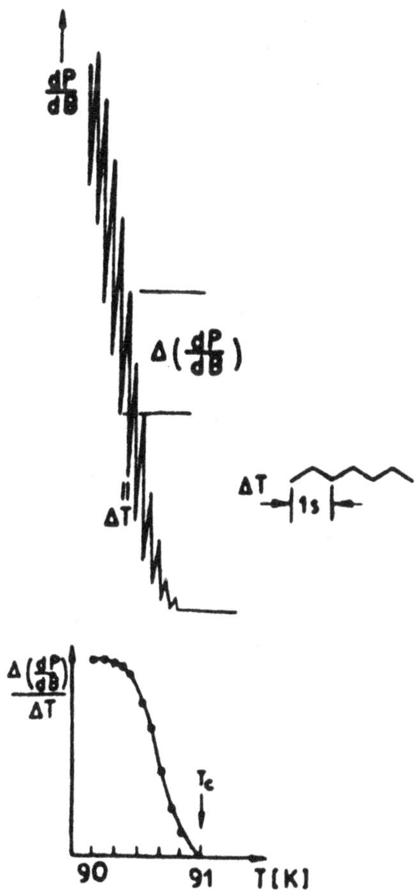

Fig.3. Current modulated JMR (top) and amplitude of oscillations (bottom) v.s. temperature.

CONCLUSION

The Josephson junctions system (JJS) can be characterized by a local temperature T^* and different from the whole sample temperature. This effect described in this paper is quite similar to a heating of JJS by the microwave irradiation on granular YBaCuO(123) superconductor[11].

AS a result of this work there is possibility to apply of an electric modulation of MA phenomena to obtain the curve d(dB/dP)/dT. In Fig.3a such a curve is presented for a modulation of frequency of 1Hz. The amplitude of signal reflects the value of the derivative d(dP/dB)/dT versus temperature as is shown in Fig.3b. This derivative was received by the phase detection method on 10 Hz frequency.

REFERENCES

1. J.Stankowski, P.K.Kahol, N.S.Dalal, and J.S.Moodera, Phys.Rev.B **36**, 7162 (1987)
2. K.W.Blazey, K.A.Mtller, J.G.Bednorz, W.Berlinger, G.Amoratti, E.Buluggiu, A.Vera, and F.C.Matacotta, Phys.Rev.B **36**, 7241 (1987).
3. K.Khachaturyan, E.R.Weber, P.Tejedor, A.W.Stacy, and A.W.Portis, Phys.Rev.B **36**, 8309 (1987).
4. S.V.Bhatt, T.V.Ramakrishnan, and C.N.Rao, J.Phys. C**20** L559 (1987)
5. C.Rettori, D.Davidov, I.Belaish, and I.Felner, Phys.Rev.B **36**, 4028 (1987).
6. T.Xia and D.Stround, Phys.Rev.B **39**, 4772 (1989).
7. C.Ebner and D.Stroud, Phys.Rev.B **31**, 165 (1985).
8. F.J.Owens, Physica C 171, 25 (1990).
9. J.Stankowski, J.Pichet, C.H.Poole, T.Datta, P.K. Kahol, N.S.Dadal J.S.Moodera, J.E.Drumheller, S.L.Hutton, and R.S.Rubins, Ferroelectrics **78**, 231 (1988).
10. R.Durny, J.Hautala, S.Ducharme, B.Lee, O.G.Symko, P.C.Taylor, D.J.Zheng, and J.A.Xu, Phys.Rev.B **36**, 2361 (1987).
11. J.Stankowski, B.Czyżak, and J.Martinek, Phys.Rev.B **42** 10245 (1990).
12. G.J.Nieuwenhuys, T.A.Friedmann, J.P.Rice, P.M.Gehring, M.B.Salamon, and D.M. Ginsburg, Sol. State Comm. **67** 253 (1988).

IRREVERSIBILITY OF THE MICROWAVE ABSORPTION IN HIGH T_c SUPERCONDUCTORS

J. STANKOWSKI, and B. CZYŻAK

Institute of Molecular Physics, Polish Academy of Sciences

Smoluchowskiego 17, 60-179 Poznan, POLAND

ABSTRACT

Hysteresis of the Microwave Absorption in magnetic field by means of standard EPR methodology in ceramic superconducting $YBa_2Cu_3O_{6.9}$ was investigated. The hysteresis of JMA strongly depends on 100 kHz modulation amplitude of the external magnetic field and the external magnetic field sweep rate. Hysteretic behavior disappears for the magnetic field sweep range lower than ± 25 G .

INTRODUCTION

Many authors have investigated strong microwave absorption (MA) below T_c temperature in high-T_c superconductors [1-7]. That absorption signal was not connected with paramagnetic ions. J. Stankowski et al. one of the first connected the low-field microwave absorption with Josephson current tunneling through internal Josephson junctions created between grains in ceramics and twin boundaries or defects in monocrystals [3]. MA was applied by us to develop a new method of T_c measurements [19] and to high-pressure studies [20] of high-T_c superconductors. In the external magnetic field the irreversibility (hysteresis) of the MA signal both in ceramic and monocrystalline samples were observed [7].

Irrevesible effects were observed also during magnetization [8-11], critical current and magnetoresistance [12] studies. Yeshurun and Malozemoff [15] have proposed model of thermally activated flux creep in II-type superconductors [16, 17], discussed by Zuo et al [18] for Bi-Sr-Ca-Cu-O single crystals. Bohandy et al [21] have presented comparison of this model with model of internal Josephson junctions for ceramic Y-Ba-Cu-O. The discussion confirmed efficiency of internal Josephson junctions model in granular superconductors. Recently, Dulcic et al presented new results connected with the irreversible behavior of the microwave absorption [22]. Erhardt et al presented results of studies of the microwave absorption decay [23]. In this paper we present the results of MA hysteresis investigations obtained using standard EPR technique.

RESULTS

The shape of hysteresis of the JMA signal depends on the field sweep rate dB/dt ($\dot{B}$), 100 kHz modulation amplitude (B_m) and on the sweep range (B_0).

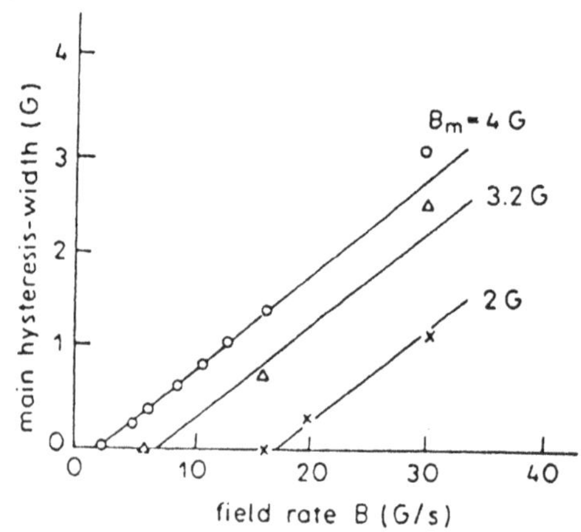

Fig.1. Microwave Absorption Hysteresis-width as a function of sweep rate.

For B_m < 1G reproducible "fine loops" on hysteresis loop appear, as it can be seen on Fig.2. Position of these "fine loops" is independent of the modulation amplitude and sweep rate, while their amplitude depends on modulation B_m.

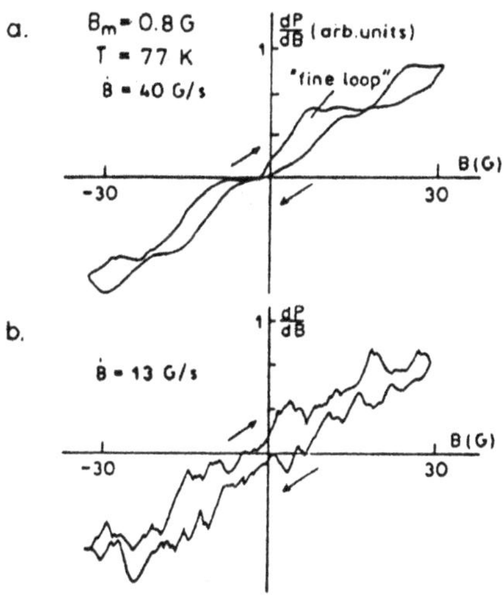

Fig.2. Hysteresis of the Microwave Absorption for modulation amplitude B_m=0.8G. Reproducible Josephson oscillations are clearly visible.

As it was stated before, the hysteresis of JMA depends also on sweep range B_0. For sweep range of 30 G (from -30 to 30 G), for $\dot{B}$ = 50 G/s and B_m = 3 G the hysteresis was observed while for B_0 = 25 G (from -25 to 25 G) it disappeared (Fig.3).

DISCUSSION

Microwave absorption provides an effective tool for studies of the nonequilibrium dynamics of quasi-particles in superconductors. Microwave irradiation induces nonequilibrium effects on the border between superconducting and normal state and increases resistivity in the presence of external magnetic field. This fast process is a consequence of the pair breaking caused by microwave irradiation.

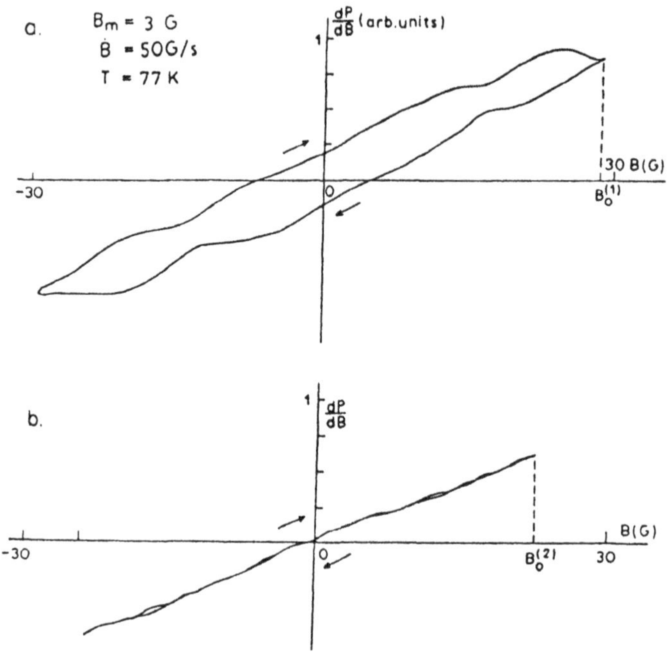

Fig.3. Disappearance of the Microwave Absorption irreversibility for low enough magnetic field sweep range.

The phonon- quasiparticle scattering has a short relaxation time t ≈ 10^{-10}- 10^{-13}s and quasiparticles adiabatically follows the driving field oscillating with it. This phenomenon can give so called fast component of the microwave absorption.

On the other hand in granular superconductors there exists the long time process caused by Josephson mechanism of the dissipation of energy in intergrain weak contacts. For the magnetically modulated microwave absorption it was called Josephson Microwave Absorption (JMA).

The irreversible phenomena for superconducting samples located in the external magnetic field have been observed by many authors [8-14]. Magnetization, magnetoresistance, critical current and microwave absorption were found to differ whether the external magnetic field increases or decreases. The dependence of the Microwave Absorption hysteresis width B_{hyst} on field sweep rate $\dot{B}$ can be described phenomenologically:

$$B_{hyst} = \tau \dot{B} - A \qquad (1)$$

where τ is the relaxation time for the process, and A is the factor dependent on 100 kHz modulation amplitude B_m . However, because of electronic limitations, it was not possible to investigate the hysteresis and examine the linear behavior of hysteresis-width for higher sweep rate. Usually, in classical magnetization measurements, after several seconds when homogenous fluxon distribution appears, logarithmic decay of magnetization is observed. The relaxation time for the process can be estimated from B_{hyst} $(\dot{B})$ depencednce (Fig.1), and it seems to be independent on modulation: τ = 0.8 s at 77 K. It is the time characterizing the establishment of homogenous fluxon distribution in the sample. In the other hand, when the field sweep does not reach the activation field (25 G) , the disappearance of JMA hysteresis is observed (see Fig.3).

Investigations of the irreversible phenomena gave us possibility to study the structure of the JMA line, especially reproducible "fine loops" visible on the line (Fig.2) Position of these loops does not depend on the modulation amplitude and sweep rate, what prove they are not the artefact. Amplitude of these "fine loops" is the highest for the 1.5s time of one sweep, what is comparable with the relaxation time τ derived from eq.(1). Amplitude of the "fine loops" also depends on modulation amplitude,because only for modulation optimal in comparison with the periodicity of line we obtain the high-fidelity signal. The JMA line, observed in ceramic materials, is a superposition (envelope) of many lines originated from many internal Josephson junctions.

For differant sizes of internal grains and superconducting loops, we obtain the current-field characteristics of Josephson junctions with different periodicities, what leads to broad multi-maxima JMA-line. However, observed "fine loops" must originate from the discrete distribution of the sizes of grains.

REFERENCES

1. A.Dulcic, B.Leontic, M.Peric, B.Rakvin, Europhys.Lett.4,1403 (1987)
2. R.Durny, I.Hautala, S.Ducharme, B.Le e , O.G.Symko, P.C.Taylor, D.I.Zheng, I.A.Xu, Phys.Rev.B 36, 2361 (1987)
3. J.Stankowski, P.K.Kahol, N.S.Dalal, J.S.Moodera, Phys.Rev.B36 7126 (1987)
4. K.W.Blazey, K.A.Mueller, J.G.Bednorz, W.Ber l inger, G.Amoretti, E.Buluggiu, A.Vera, F.C.Matacotta, Phys.Rev. B 36, 7241 (1987)
5. J.Stankowski, J.Pichet, C.P.Poole, T.Datta, P.Kahol, J.S.Moodera,
6. A.M.Portis, K.W.Blazey, K.A.Mueller, J.G.Bednorz, Europhys.Lett 5, 467 (1988),
7. A.Dulcic,R.Crepeau, I.H.Freed, Proc.XXIV Congr. AMPERE, L62, Elsevier, Poznan 1988
8. H.Kumakura, K.Toyano, M.Fukutomi, M.Uehara, K.Tachikawa, Jpn.J.Appl.Phys. 26, L655, (1987)
9. J.Marcus, C.Escribe-Filippini, C.Schlenker, R.Buder, J.Devonyi, P.1.Redet, Solid State Commun. 63, 129 (1987)
10. H.Takagi, S.I.Uchida, H.Obara, K.Kishio, K.Kitazawa, K.Fueki, S.Tanaka, Jpn.J.Appl.Phys.26, L434 (1987)
11. C.P.Poole, T.Datta, H.Farach, Copper Oxide Superconductors, John Wiley , 1988
12. K.Y.Chen, Y.J.Qian, Physica C 159 , 131 (1989)
13. S.Yan, S.Liang, H.Ma, Q.Feng, Y.Sun, Y.Gao, H.Zhang, Solid State Commun. 70, 553 (1989)
14. J.R.Fraser, T.R.Finlayson, T.F.Smith, Physica C 159 , 70 (1989)
15. Y.Yeshurun, A.P.Malozemoff, Phys.Rev.Lett. 60, 2202 (1988)
16. P.W.Anderson, Phys.Rev.Lett. 9, 309 (1962)
17. Y.B.Kim, C.F.Hempstead, J.Rice, D.Ginsberg, Phys.Rev.131, 2486 (1963)
18. F.Zuo, M.B.Salamon, E.D.Bukowski, J.P.Rice, D.M.Ginsberg, Phys.Rev.B 41, 6600 (1990)
19. B.Czyzak, T.Zuk, J.Stankowski, Acta Phys.Polon. A78 , 5 (1990)
20. J.Stankowski, B.Czyzak, M.Krupski, J.Baszynski, T.Datta, C.Almasan, Z.Z.Sheng, A.M.Hermann, Physica C 160 170 (1989)
21. J.Bohandy, B.F.Kim, F.Adrian, K.Moorjani, Phys.Rev.B 39, 2733 (1989)
22. .A.Dulcic, B.Rakvin , M.Pozek, Europhys.Lett. 10, 593 (1989)
23. P.Erhardt, J.E.Drumheller, B.Senning, S.Mini, L.Fransioli, E.Kaldis, S.Rusiecki, F.Waldner, Proc. of Congress AMPERE, 1990, eds.M.Mehring, J.von Schutz,H.Wolf, Springer, Berlin 1990, p.275.
24. B.Czyżak, J.Stankowski, Phys.St.Sol.(b) 166, 219 (1991)

SUPERCONDUCTIVITY OF POTASSIUM DOPED CARBON C_{60}

P.Byszewski[a,b], M.Baran[a], M.Kozłowski[b]

[a] *Institute of Physics, Polish Academy of Sciences Al.Lotnikow 32/46, PL-02-668 Warszawa, Poland*

[b] *Research and Development Center of Vacuum Electronics ul.Długa 44/50, PL-00-241 Warszawa, Poland*

ABSTRACT

The magnetic properties of fullerite superconducting samples doped with potassium, cesium and lithium were investigated using SQUID magnetometer. The critical temperature T_c was found to be weakly dependent on potassium concentration and on the admixture of lithium or cesium and it was estimated to be 16 - 18 K. The first critical field H_{c1}, for all investigated samples, was found to be very low (5 ÷ 9 Oe). The critical current density J_c was estimated to be $2 \cdot 10^4$ A/cm^2. It was observed that the shielding effect, for zero field cooled samples, strongly depends on the annealing procedure for a range of various types of dopants and its concentration.

1. The method of samples preparation.

To synthesize fullerite, the method described previously[1,2] was adopted with some minor changes. The reaction chamber where the soot was generated by the contact arc between graphite electrodes was filled to 0.7 atm. with helium. The fullerites were extracted from the soot with a mixture of hexane and benzene. When the solvents have been evaporated the obtained crystallites were checked by the X-ray diffraction and mass spectroscopy. The X-ray pattern (Fig.1) was best fitted by the f.c.c. structure with a=14.12A (as in Ref.1), but additional lines of low intensity indicated to a considerable admixture of grains with h.c.p. structure.

The mass spectroscopy measurements proved that up to 85% of C_{60} was

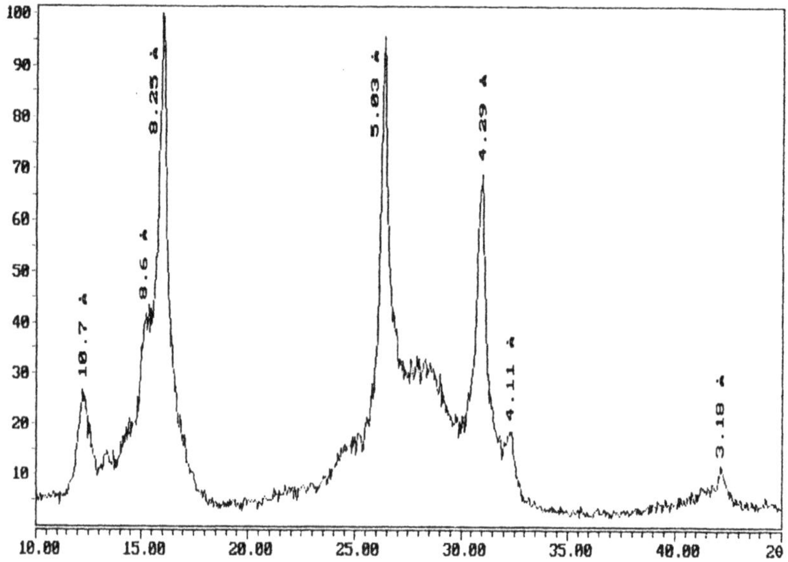

Figure 1.
The powder X-ray pattern of C_{60}. The strong lines have been identified as belonging to the f.c.c. structure. The background, relatively large width of lines and lack of some structurally allowed lines are resulted from a molecular structure of crystallites.

obtained in the total amount of dried fullerite crystallites. The powder was further enriched in C_{60} up to about 92% by consecutive recrystallization, selection of crystallites and their solution in dry toluene.

The superconducting samples were prepared by the method similar to those described previously[3-8] by other authors. The C_{60} portion (approx. 15 mg) was placed in quartz tube and next dried under vacuum at 150-200°C for an hour. Then the sample tube was filled with neutral gas in a glove box in order to add alkali metals-filled glass capillaries of a proper length. The tube was again vacuumed and next sealed. Then locally warming the ampoule the potassium (or another alkali metal) was evacuated from the capillary. The diffusion of dopants was carried out in two stages process. In the first stage the ampoules were annealed for 36 hours at 360°C. In the second one, repeating it few times, the powder was mixed inside the ampoules (by shaking them) and next heat treated at 210°C for 18 hours. After the first stage the ampoules were colored, what seems to be caused by potassium vapors condensation on its walls. Therefore, the average sample composition could be different from the starting one. It was found that repeated mixing and annealing appreciably improved superconducting properties of the samples, as checked by magnetic measurements.

Looking for methods testing the samples in the course of preparation, it was noticed that resonant microwave absorption is very sensitive on doping and homogeneity of the samples. Therefore, electron spin resonance (ESR) measured at room temperature, was used at various technological stages in order to determine the best thermal treatment (more detailed analysis of ESR signals, at various temperatures, to be presented elsewhere[9]). The room temperature ESR spectrum for pure C_{60} exhibits a very weak and narrow line ($\Delta H = 1.4$ Oe) at magnetic field corresponding to $g = 2.0022$. Doping with potassium significantly enhances the line, and annealing of the sample, even at low temperature, creates a new, relatively broad line. Further annealing stabilizes the ESR spectrum with a single broad line with $g = 2.0017$ and $\Delta H = 16$ Oe for low ($x < 3$) potassium concentration, and the same broad line with superimposed narrow line ($g = 2.0026$ and $\Delta H = 1$ Oe) for high ($x > 3.5$) potassium concentration. The intermediate spectrum, when three ESR lines of different width were observed, was probably connected with inhomogeneous dopant concentration.

The superconducting properties were also checked at different stages of thermal treatment by a squid magnetometer. The largest amount of superconducting phase was found in the samples which exhibited two ESR lines, but the narrow one was weak, almost disappearing.

2. Experiment and results.

The superconducting properties of the M_xC_{60} samples (where M = K, or combination of K with Li or Cs) were investigated by measurements of their magnetic properties with SQUID magnetometer (MPMS of Quantuum Design) at the temperature range from 4.5 K to 25 K and in various magnetic fields. The investigated samples were characterized by very unconvenient form for measurements in SQUID magnetometer (an ampoule of about 5 cm with the sample located at the end of it) and by relatively low magnetic moments in the superconducting state (the last point is discussed later). To avoid a superposition of a signal from the sample and a signal originated from the end of an ampoule an extra quartz tube was combined improving significantly the geometry. Applying this solution, for all the samples at 25 K and for nonsuperconducting samples at 4.5 K only very weak positive magnetic moments were observed from the region of the sample. To be sure, that the diamagnetism observed for majority of investigated samples is connected with its superconducting properties, the following measurement procedure was applied:

- first of all the sample was cooled down, in approximately zero magnetic field, to 4.5 K (zero field cooling regime - ZFC). After testing that the sample exhibits diamagnetic behaviour in magnetic fields up to 20 Oe, a relation M(T) was measured in 20 Oe up to 25 K;
- next M(T) was measured cooling the sample from 25 K, in the same magnetic field (20 Oe) to 4.5 K (FC - field cooling regime). In this regime the magnetic

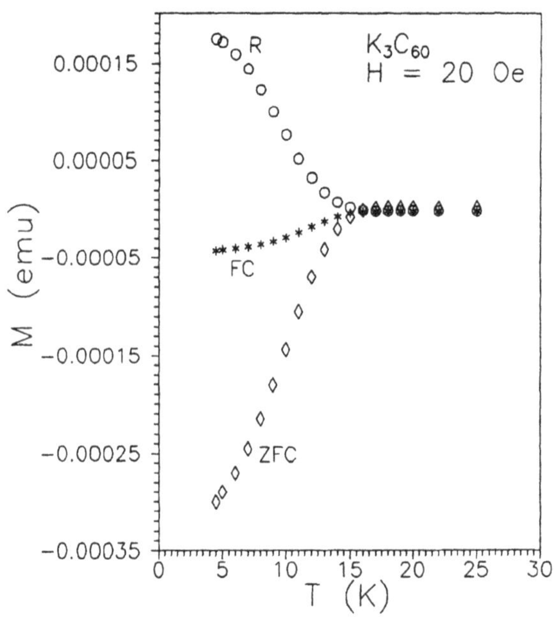

Figure 2.
The relation of magnetic moment versus temperature for zero field cooling (ZFC), field cooling (FC) and remanence (R) regimes, for sample K_3C_{60} after second annealing at 590K.

moment close to zero, as compared to that observed in previous regime, was observed. For some samples this moment was positive what means that a parasitic paramagnetic moment (originated either from uncompensated end of an ampoule or from uncontrolled paramagnetic admixture in the sample and the most probably from both sources);

- next at 4.5 K the magnetic field was lowered to zero and the remanent moment (originated from magnetic flux lines "frozen" in superconducting sample) could be observed at this temperature and next M(T) up to 25 K was measured. The absolute value of observed remanent moment was generally close to that observed for shielding effect in ZFC regime.

As an example of described above procedure the results of measurements of K_3C_{60}, after second annealing at 590 K, are presented in Fig.2. From such measurements the critical temperature T_c was estimated as the beginning of a plateau of M(T) relation both for shielding and remanent effect. The values of T_c, corresponding to the onset of superconducting properties, for various sample composition are given in Tab.I. For better illustration of obtained results, the shielding effect plots for various sample compositions are presented in Fig.3.

Table I. Characteristics (T_c, H_{c1}, and susceptibility χ related to that of ideal diamagnet) for various compositions of investigated samples

Sample composition	T_c [K]	H_{c1} [Oe]	χ [%]
K_2C_{60}	18	5	1.6
K_3C_{60}	18	7	2.5
K_4C_{60}	16	9	1.5
$K_{1.5}Cs_{1.5}C_{60}$	18	6	1.4
$K_2Li_2C_{60}$	16	5	0.33

The relation of magnetic moment versus magnetic field M(H) was measured in ZFC regime to characterize better the investigated samples. The special attention was paid to the beginning part of the virgin curve. From its linear part the first critical field H_{c1} was estimated, and from the slope of this part the rate of obtained susceptibility to the susceptibility of an ideal diamagnet was derived. The above parameters are presented in Tab.I for the choice of investigated compositions. The values of relative susceptibility are very low. At first, it can result from low contents of superconducting phase. But it should be also affected by small size of grains (a mean value of about 1 μm) in relation to the penetration depth.

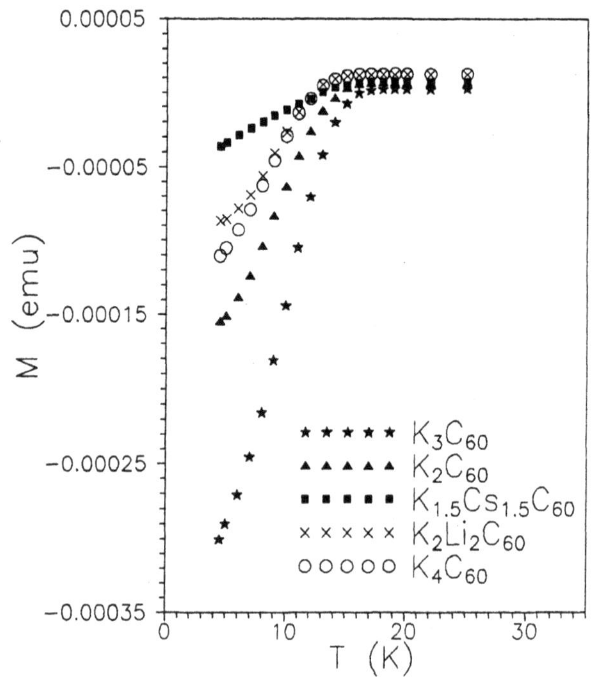

Figure 3.
The comparison of the shielding effect plots for various investigated samples, after the first 590 K annealing. Measurements done in H = 20 Oe.

In Fig.4 there is presented the hysteresis loop obtained at 5 K, for the sample of K_3C_{60} after first annealing. Using well known isotropic Bean formula, it was possible to estimate the critical current density in external magnetic field of 1 kOe, taking an average grain size of 1 μm (estimated from microscopic observations). The obtained value of the critical current density $J_c = 2 \cdot 10^4$ A/cm^2 seems to be hopeful as being related to the whole volume of the sample and taking to the account very low contents of superconducting phase (see relative susceptibility in Tab.I).

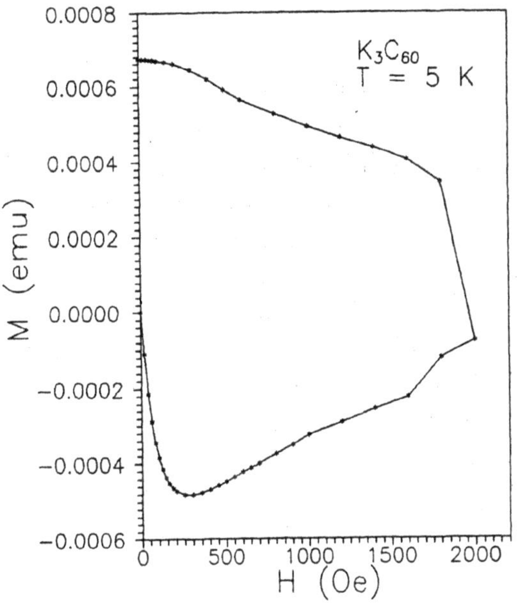

Figure 4.
The hysteresis loop for sample of K_3C_{60} measured at 5 K.

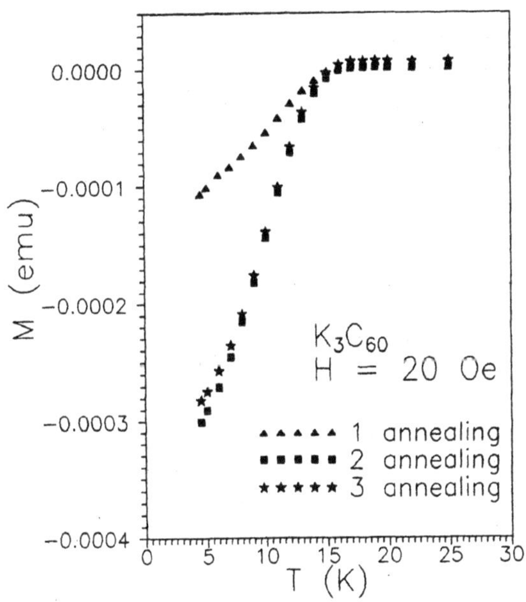

Figure 5.
The influence of multiple annealing of K_3C_{60} sample on the shielding effect measured in H = 20 Oe.

3. Conclusions.

Nevertheless, quite good result of critical current density was reached, it is necessary to stress that the critical temperature and the first critical field of our samples are lower then previously published[2-4]. We suppose that it may be caused by a small admixture of C_{70} and other fullerites in our samples and/or no properly chosen heat treatment. The shielding effect measurements performed for K_3C_{60} sample (see Fig.5) indicate that probably the equilibrium of potassium distribution was reached and further annealing would not improve the properties. This conclusion is supported by results of ESR experiments, performed both at room and low temperatures. It leads to the conclusion that the superconducting properties, as derived directly from magnetic measurements, are rather limited by the grains size rather, than by a potassium distribution in the C_{60} crystal lattice (a homogenous state seems to be reached). The rate of the mean grain size to the penetration depth ($\lambda = 0.24\ \mu m$ from Ref.10) suggests significant lowering of effectively observed susceptibility and it should also strongly affect the observed M(T) relations.

Because $C_{60}RbCs$ samples revealed the highest critical temperature between alkali metals doped fullerite superconductors[8], it seems plausible that the crystal lattice should be distorted by dopants for highest critical temperature.

4. References.

1. W. Kratschmer, L.D. Lamb, K. Fostiropoulos and D.R. Huffman, *Nature* **347** (1990) 354.
2. H.W. Kroto, J.R. Heath, S.C. O'Brien, R.F. Curl and R.E. Smalley, *Nature* **318** (1985) 162.
3. R.C. Haddon, A.F. Hebard, M.J. Rosseinsky, D.W. Murphy, S.J.Duclos, K.B. Lyons, B. Miller, J.M. Rosamilia, R.M. Fleming, A.R. Kortan, S.H. Glarum, A.V. Makhija, A.J. Muller, R.H. Eick, S.M. Zahurak, R. Tycko, G. Dabbagh and F.A. Thiel, *Nature* **350** (1991) 320.
4. A.F. Hebard, M.J. Rosseinsky, R.C. Haddon, D.W. Murphy, S.H. Glarum, T.T.M. Palestra, A.P. Ramirez and A.R. Kortan, *Nature* **350** (1991) 600.

5. R.M. Fleming, T. Siegrist, P.M. Marsh, B. Hessen, A.R. Kortan, D.W. Murphy, R.C. Haddon, R. Tycko, G. Dabbagh, A.M. Mujsce, M.L. Kaplan, and S.M. Zahurak, to be published in *Proceedings of the MRS Meeting 1991*.
6. K. Holczer, O. Klein, G. Gruner, S.-M. Huang, R.B. Kaner, K.-J.Fu, R.L. Wetten and F. Diederich, *Science* **252** (1991) 1154.
7. K. Holczer, O. Klein, G. Gruner, J.D. Thompson, F. Diederich and R.L. Wetten, *Phys. Rev. Lett.* **67** (1991) 271.
8. K. Tanigaki, T.W. Ebbesen, S. Saito, J. Mizuki, J.S. Tsai, Y. Kubo and S.Kuroshimo, *Nature* **352** (1991) 222.
9. P. Byszewski, J. Stankowski, Z. Trybula, W. Kempinski and T. Zuk, to be presented at the Symposium: Superconducting Materials, Wroclaw, Oct. 1991.
10. K. Holczer, O. Klein, G. Grüner, J.D. Thompson, F. Dederich, and R.L. Whetten, *Phys. Rev. Lett.* **67** (1991) 271.

Author Index

(The asterisk indicates that the author attended IW CCL)

Subject Index

(numbers are referred to the first page of the cited paper)

www.ingramcontent.com/pod-product-compliance
Lightning Source LLC
LaVergne TN
LVHW051925120926
847902LV00009B/74